AF334615

Microwave Induced Plasma Analytical Spectrometry

RSC Analytical Spectroscopy Monographs

Series Editors:
Neil W. Barnett, *Deakin University, Victoria, Australia*

Advisory Panel:
F. Adams, *Universitaire Instelling Antwerp, Wirijk, Belgium*, M. J. Adams, *RMIT University, Melbourne, Australia*, R. F. Browner, *Georgia Institute of Technology, Atlanta, Georgia, USA*, J. M. Chalmers, *VSConsulting, Stokesley, UK*, B. Chase, *DuPont Central Research, Wilmington, Delaware, USA*, M. S. Cresser, *University of York, UK*, J. Monaghan, *University of Edinburgh, UK*, A. Sanz Medel, *Universidad de Oviedo, Spain*, R. D. Snook, *UMIST, UK*

Titles in the Series:
 1: Flame Spectrometry in Environmental Chemical Analysis: A Practical Guide
 2: Chemometrics in Analytical Spectroscopy
 3: Inductively Coupled and Microwave Induced Plasma Sources for Mass Spectrometry
 4: Industrial Analysis with Vibrational Spectroscopy
 5: Ionization Methods in Organic Mass Spectrometry
 6: Quantitative Millimetre Wavelength Spectrometry
 7: Glow Discharge Optical Emission Spectroscopy: A Practical Guide
 8: Chemometrics in Analytical Spectroscopy, 2nd Edition
 9: Raman Spectroscopy in Archaeology and Art History
10: Basic Chemometric Techniques in Atomic Spectroscopy
11: Biomedical Applications of Synchrotron Infrared Microspectroscopy
12: Microwave Induced Plasma Analytical Spectrometry

How to obtain future titles on publication:
A standing order plan is available for this series. A standing order will bring delivery of each new volume immediately on publication.

For further information please contact:
Book Sales Department, Royal Society of Chemistry, Thomas Graham House, Science Park, Milton Road, Cambridge, CB4 0WF, UK
Telephone: +44 (0)1223 420066, Fax: +44 (0)1223 420247,
Email: books@ rsc.org
Visit our website at http://www.rsc.org/Shop/Books/

Microwave Induced Plasma Analytical Spectrometry

Krzysztof J. Jankowski
Department of Analytical Chemistry, Warsaw University of Technology, Warsaw, Poland

Edward Reszke
Ertec-Poland, Wroclaw, Poland

RSC Analytical Spectroscopy Monographs No. 12

ISBN: 978-1-84973-052-5
ISSN: 2041-9732

A catalogue record for this book is available from the British Library

Published by The Royal Society of Chemistry,
Thomas Graham House, Science Park, Milton Road,
Cambridge CB4 0WF, UK

Registered Charity Number 207890

For further information see our web site at www.rsc.org

Preface

Microwave frequencies have been used for plasma generation for many years. Advancements in the understanding of microwave plasmas (MWPs) have led to improvements in the design of the components that comprise the instruments used for MWP sources. As a result, MWPs have been successfully used as vaporization/excitation sources for optical emission spectrometry or as vaporization/ionization sources for mass spectrometry. Over the past 20 years, MWP analytical spectrometry has grown from being a niche technique to becoming a promising tool in the analytical laboratory. In particular, microwave induced plasma optical emission spectrometry (MIP-OES) has proved to be an essential technique for the direct analysis of chromatographic eluates. The aim of this monograph is to inform the reader of the present status of MWP analytical spectrometry by providing an overview of the technique. In more recent years, several commercial MWP-based analytical instruments have appeared on the market, and an increasing interest in the benefits which can be gained from the use of MWPs is observed. During this time, a number of key papers were published, instrumentation was significantly improved, and some technical limitations of MWP operation were solved. Today, MWP spectrometry, particularly atomic emission, is a highly developed measurement technique. Obviously, MWP analytical spectrometry is now entering its mature phase of development, with a large number of applications for a wide variety of sample types. This book provides a comprehensive discussion of the major aspects involved in MWP-based analytical spectrometry, with emphasis on practical issues, and documents the latest achievements in MWP spectrometry to stimulate their use on a wider scale in analytical and research laboratories. However, it is surprising that despite this remarkable growth, most of the research and practice of analytical spectrometry has been centred almost exclusively around inductively coupled plasmas (ICPs), even although it is clear that MWPs can offer significant advantages for certain applications. There are several sources of specific,

RSC Analytical Spectroscopy Monographs No. 12
Microwave Induced Plasma Analytical Spectrometry
By Krzysztof J. Jankowski and Edward Reszke
© The Royal Society of Chemistry 2011
Published by the Royal Society of Chemistry, www.rsc.org

detailed information about the MWP-OES technique available. However, up to now the MWP method was generally presented as an alternative plasma source and there was lack of a monograph concerning MWP analytical spectrometry. This book is designed for use by academics and postgraduates embarking on work in the field of MWP source spectrometry, ICP/MWP users, analysts and research groups who want to configure their own plasma spectrometry setup, and manufacturers of plasma spectrometers and MWP devices. The monograph will also be a useful source of information for those seeking to interface various sample introduction techniques with plasmas and for all those who would like to know more about the technique. The physical and chemical characteristics of the various MWP sources and sample introduction techniques available are all discussed and how these characteristics affect the design of the parts of the MWP setup, with inclusion of some very recent work with MWP sources carried out in our research group and that described by others.

A general introduction given in Chapter 1 concentrates on the fundamentals of MWP formation under different experimental conditions, presents basic physical characteristics of the MWP discharge, and discusses various spectroscopic techniques pertinent to the operation of a MWP. In Chapter 2, a general description of the instrumentation required for MWP analytical spectrometry is presented, including the benefits and limitations of MWP discharges. Chapter 3 addresses the art and science of MWP design and highlights very recent advances in the field, presenting a new advanced classification of microwave plasma sources. Chapter 3 also considers various MWP sources, including very recent approaches for generating an annular-shaped plasma, a rotating field three-phase plasma and a microwave frequency glow discharge. The implementation of these approaches, however, requires an understanding of the fundamental theory behind microwave plasma generation and maintenance. There then follows in Chapter 4 a review of microwave safety precautions. Chapter 5 deals with some simple, yet fundamental, concepts regarding optical emission spectrometry and spectra obtained by MIPs, including the most used emission lines in MIP-OES. Chapters 6 to 8 deal with the various techniques used for transporting samples into the plasma, probably the most challenging area for MWP-OES operators, including the introduction of gaseous, liquid and solid samples. Each sampling technique is described with respect to principles, instrumentation, operating parameters and limitations in the context of MWP-OES. Then the specific applications of MIP-OES are presented. In Chapters 9 and 10, more detail regarding the MWP-OES technique is given, including the optimization of operating parameters and analytical method development. Then follows additional information about line selection, instrument maintenance and performance verification, and trouble shooting. Analytical characteristics of MWP-OES are included. The chapter also compares analytical performance indices for various plasma sources. Chapter 11 is devoted to the applications of MWP-OES for a wide variety of sample types. Finally, Chapter 12 briefly describes non-emission MWP spectroscopic techniques, including some specific analytical applications of the MWP-MS method.

We are indebted to all our MIP-OES and MS teachers and colleagues who have provided significant contributions to the field. Their efforts behind the background to this book have brought us to a better understanding of the basic science of MIP-based techniques and has stimulated further studies. As a consequence, we came to the conclusion and after discussions we had been advised to understand that there was a new need for the publication of a comprehensive reference book on the theory and practice of the field. We are grateful to Dr Andrzej Ramsza for his expert assistance in setting up the scientific level of the work and his helpful comments on the draft manuscript. Special thanks are due to Professors Ramon Barnes and Gary Hieftje for some helpful suggestions and comments that encouraged us to write this book. Finally, we would also like to thank the publishers, who have, with great perseverance, enthusiasm and patience, brought this book to fruition.

Krzysztof J. Jankowski
Edward R. Reszke

Contents

RSC Analytical Spectroscopy Monographs No. 12
Microwave Induced Plasma Analytical Spectrometry
By Krzysztof J. Jankowski and Edward Reszke
© The Royal Society of Chemistry 2011
Published by the Royal Society of Chemistry, www.rsc.org

An Introduction to Microwave Plasma Spectrometries

1.1 Introduction

Plasmas generated by the interaction of electromagnetic fields with gases such as argon or helium were introduced in the middle of the 20th century as very promising media for atomic excitation. Microwave plasmas (MWPs) are included in this new generation of spectrochemical sources, which in the last 30 years have considerably widened the possibilities of trace analysis and speciation studies. MWPs operating at a GHz frequency have been especially used as emission sources for optical emission spectrometry (OES) and, later, also as ionization sources for mass spectrometry (MS). The microwave energy is coupled to the gas stream passing through the torch with an external cavity or antenna.

Generally, two groups of MWPs are mentioned according to the method of power transmission to the plasma gas, as well as the plasma shape and its position against the plasma torch. A flame-like plasma formed at the tip of the electrode, first developed by Cobine and Wilbur in 1951,[1] is commonly known as a capacitively coupled microwave plasma (CMP). A CMP torch is equipped with an inner conductor that forms a capacitance against the ground and transfers the microwave energy into the plasma gas through its tip. In the second type, the plasma is produced through the inductive transfer of energy from standing waves in a suitable resonator and sustained in a quartz or ceramic tube which is located within a resonant cavity. This electrodeless system is commonly referred to as a microwave induced plasma (MIP) and is the most successful and commonly used type of microwave discharge. The first application to spectrochemical analysis with the use of MIPs was published by Broida and co-workers in 1952.[2,3] The microwave energy is coupled either by an electrical field (E coupling) or a magnetic field (H coupling) to the working

RSC Analytical Spectroscopy Monographs No. 12
Microwave Induced Plasma Analytical Spectrometry
By Krzysztof J. Jankowski and Edward Reszke
© The Royal Society of Chemistry 2011
Published by the Royal Society of Chemistry, www.rsc.org

gas. However, such a classification has no strict scientific foundation. A more advanced one will be given in Chapter 3.

1.1.1 Historical Background

Historically, the development of CMP and MIP was in parallel, owing to some essential differences in technical design and operation of these two microwave approaches. One of these differences is related to the operating frequency. CMPs can operate over a wide range of frequencies and a great number of publications is devoted to radiofrequency plasmas. An individual approach can be relatively easily tuned to different frequencies. Resonant cavities for sustaining a MIP are dedicated to one operating frequency, mostly 2.45 GHz, and sometimes one kind of plasma gas. A more detailed comparison of CMP and MIP will be given below.

In the 1960s, analytical applications of the CMP were focused on the elemental analysis of solutions[4–8] and as a result a commercial CMP instrument was developed by Murayama *et al.*[6–8] at the Hitachi Central Research Centre in 1968. In the next decade, two commercial spectrometers were explored [Hitachi 300 UHF Plasma Scan and the Applied Research Laboratories (ARL) Model 31000]. However, these devices did not compare favourably with the inductively coupled plasma (ICP) method due to severe inter-element effects.[9,10]

In 1985, a renaissance of CMPs appeared thanks to Jin and others,[11,12] who developed the so-called the microwave plasma torch (MPT). The plasma operation was significantly improved and it offered a much better analytical performance for the introduction of aqueous aerosols. In the 1990s, various sample introduction methods as well as spectroscopic techniques based on the MPT were successfully introduced.[13,14] Finally, in 1999 a commercialized MPT-OES instrument (JXY-1010 MPT) was introduced in China. Other designs used for microwave-powered CMPs over time have included the "torch à injection axiale" (TIA).[15,16]

Until the mid-1970s, the MIP discharge was obtained almost exclusively in gases under reduced pressure. Because of this, this period of MIP technique development is often recognized as the low-pressure MIP era. A number of resonant cavities were constructed and examined at a wide pressure range (1–760 Torr), including foreshortened cavities and a tapered rectangular cavity.[17] When low-pressure discharges were used, analyte introduction was performed predominantly by gas chromatography, electrothermal vaporization or chemical vapour generation, owing to the difficulty in sustaining the plasma and its relatively low loading immunity. These difficulties caused part of the analytical spectrometry community to approach (and still approach) the MIP technique with reserve. On the other hand, the focusing of this early work on applications in which small amounts of sample were delivered to the plasma in the gas phase led to the fast development of the MIP technique with regard to its application to gas chromatographic detection. The first successful analytical application of the MIP method seems to have been the analysis of nitrogen isotopes by Broida and Chapman in 1958.[18] McCormack *et al.*[19] developed the first element-selective gas chromatography (GC) detector based on MIP emission spectrometry in 1965. The most successful commercial microwave plasma detector (MPD) used in GC systems was introduced by Quimby and Sullivan in 1990.[20]

A breakthrough in the development of the MIP technique was the designing by Beenakker in 1976 of the TM_{010} resonator,[21,22] in which the plasma discharge could be obtained under atmospheric pressure. This is often recognized as the end of the low-pressure MIP era and the beginning of the atmospheric pressure MIP era. However, the essential improvement introduced by Beenakker was not the sustaining of the plasma at atmospheric pressure (some successful studies were reported previously[23–25]), but the introduction of symmetrical coupling and improved transfer of electrical energy that allowed plasma stability. At the same time, in 1975, Moisan *et al.*[26] introduced a different microwave structure with symmetrical coupling based on surface wave propagation, called the "surfatron".

In next years, various types of resonators were devised, permitting the achievement of high discharge stability and excitation efficiency. Improved versions of Beenakker's cavity,[27,28] surfatrons[29,30] and strip-line sources[31,32] allowed a stable discharge to be obtained over a wide range of plasma gas flow rates and microwave power levels. Designs permitting the use of moderate and high power were worked out to increase the discharge energy.[33–36] Finally, the vertically positioned, aerosol-cooled MIP system based on the TE_{101} integrated cavity for OES was proposed by Jankowski *et al.*[37]

Microwave plasma analytical spectrometry (MWP-AS) has gained new importance for trace analysis since 1981, when Douglas and Frech[38] applied the plasma as an ionization source for MS. In 1990, Okamoto[39] developed a surface-wave-excited non-resonant cavity for MIP. This high-power source was applied firstly for OES[40] and secondly, and more importantly, for MS. In 1994, a nitrogen MIP-MS spectrometer (Hitachi P-6000) appeared on the market. An annular-shaped helium plasma was obtained by Okamoto[41] in 1999 for improving the analytical characteristics of non-metals. Recently, this type of plasma geometry, obtained in a transverse electromagnetic mode (TEM mode) resonant cavity at a helium flow rate below $3 \, L \, min^{-1}$ and at a low power level, has been reported by Jankowski *et al.*[42] Even more recently, a special three-phase microwave power source, applying non-stationary fields which are rotating around the plasma axis causing a flat planar plasma geometry with a triangular-shaped annular centre, was announced by the same group as a promising method for analytical spectrometry.[43]

In the last decade, a number of interesting fields for applying MWP-AS have appeared. In 1995 the particle-sizing instrument (Yokogawa, PT1000) based on a helium MIP[44] was manufactured (see Chapter 12). It offers the possibility of the determination of both the chemical composition as well as the size and basic physical structure of micro- and nanoparticles. The second hot topic seems to be the design and application of microplasma sources in miniaturized analytical systems. In 2000, Engel *et al.*[45] proposed an MIP-based microstrip device,

Table 1.1 Milestones of MWP sources for analytical spectrometry and analytical applications.

Date	Event	Ref.
1951	Development of a CMP excitation source	1
1958	Determination of nitrogen isotopes by MIP-OES	2,3
1965	Development of a microwave plasma detector for GC	19
1968	Commercial CMP-OES instrument (Hitachi 300 UHF)	7
1976	Development of TM_{010} cavity operating at atmospheric pressure	21
1981	Development of elemental analysis by MIP-MS	38
1985	Development of MPT	11
1989	Commercial MIP-OES instrument (Analab MIP750MV)	54
1990	Commercial MPD for GC (HP 5921A AED)	20
1990	Development of Okamoto cavity	39
1994	Commercial N_2 MIP-MS instrument (Hitachi P-6000)	40
1995	Commercial MWP instrument for particle-sizing (Yokogawa PT1000)	44
1999	Commercial MPT-OES instrument (JXY-1010 MPT)	–
2000	Development of MW microplasma system	45
2000	Development of MPT-TOFMS (time-of-flight MS)	55

taking advantage of the ability to form so-called cold plasmas using microwaves.

The development of MWP instrumentation and analytical applications has been regularly reviewed and some references can be recommended.[14,46–53] The milestones in MIP and CMP development are summarized in Table 1.1.

1.1.2 The Present Status of Microwave Plasma Spectrometry

MWPs have evolved considerably over the past 15 years as an excitation source for OES and as an ionization source for MS. At present, MWPs can be produced under a large variety of operating conditions, and the devices that are now available allow the production of stable and reproducible plasmas. Several common forms of this plasma source exist, including the low- and high-power MIP, the CMP, the surface-wave plasma and the MPT. On the other hand, this variety of designs causes a "hotchpotch" and breeds reservation to this technique among many analytical spectroscopists. There is a need to give a comprehensive theory and practice of the field for a better understanding of the basic science of MWP-based techniques.

Summarizing their present status in terms of spectroanalytical applications, MWPs have found a consistent need in plasma spectrometry. A specific excitation mechanism permitting the determination of many metals and non-metals with good sensitivity and low running costs for the apparatus are advantages of this technique. MWPs have found openings in various fields of chemical analysis. Gas chromatography microwave induced plasma optical emission spectrometry (GC-MIP-OES) holds a prominent position as the hyphenated technique for speciation and metallomics. Other spectacular applications, such as continuous emission monitoring, particle sizing, microanalytics and soft fragmentation of molecules, as well as tandem sources, continue to receive the attention of spectroscopists. There is no doubt that an increasing interest is observed in the benefits which can be gained from the use of the MWP method.

With regard to commercial acceptance of the technique, progress can be observed in the past 15 years. Five instruments have appeared on the market and some others have been announced as a preserial. However, there is still a lack of apparatus used worldwide, excluding MPDs. The appearance of such a spectrometer would undoubtedly cause rapid development of this technique. Unfortunately, despite many obvious assets of MWPs, until now no adequate interest has been shown by analysts or by apparatus manufacturers. The authors are aware of this, since they themselves participated in the design and commercialization of an MIP-OES (Analab, MIP 750 MV), produced in the 1980s.[54]

1.2 Energy Flow between Microwave Plasma and Analyte

1.2.1 Microwave Power Absorption by the Plasma

There are several ways by which microwave energy can be absorbed by a plasma. In general, the energy is transmitted to the plasma *via* the microwave electric field, which accelerates the electrons until they have sufficient energy to cause further ionization in a chain reaction. The electrons are the only ones to follow the oscillations of the electric field. On the other hand, an electron is able to gain energy from the field only if its ordered oscillatory motion is changed by collision with a plasma gas atom. The electric field may be applied within a resonant cavity or microwaves may be guided along the plasma column. In oscillating fields the electron gains energy from the field and, in addition, receives further energy from the field as a result of elastic collisions. However, the electron can also dissipate its energy through a great number of elastic and inelastic collisions with neutral and ionic species, causing excitation and ionization processes. At low discharge pressures (up to 50 Torr), a heavy particle will experience approximately 10^7 collisions before being excited. At atmospheric pressure the collision frequency is so high that an increase of microwave power input is required to ensure sufficient ionization of the plasma gas. The electron acceleration and collision energy exchange mechanisms in high-frequency fields have been discussed in detail by Brown.[56]

The mean power absorbed by electron from the field is given as follows:[57]

$$P = \left(\frac{e^2 E_{\max}^2}{2m\nu}\right)\left(\frac{\nu^2}{\nu^2 + \omega^2}\right) \qquad (1.1)$$

where e and m are the charge and mass of the electron, respectively, $E_{\max}$ is the maximum field amplitude (strength), ν is the collision frequency between the electron and the gaseous atoms and ω is the field frequency. As one can deduce from eqn (1.1), only very limited energy can be absorbed directly by the plasma gas ionized species, considering their considerable mass and the brief time period before the field reverses polarity.

A distinctive feature of a MWP is that free electrons gain kinetic energy from the microwave all over the discharge. The whole discharge remains in ionizing mode, *i.e.* free electrons are underpopulated almost everywhere. This means that plasma heating occurs more homogeneously, and no plasma decay will take place.[58] On the other hand, the presence of the microwave electric field induces plasma filamentation[59] and makes the power transfer to the discharge limited by the skin effect. This causes both plasma inhomogenity and a radial gradient of the gas temperature; as a consequence there is limited penetration of the sample into the plasma. The plasma inhomogenity varies inversely as the thermal conductivity of the gas and is more pronounced in argon then in helium.

1.2.2 Plasma–Sample Interaction

When the sample is introduced to the plasma in the form of a wet aerosol, an individual droplet at first undergoes desolvation at high temperature. The resulting microscopic salt particles (called "dry aerosol") explode and vaporize into a gas of individual molecules ("molecular vapour") that are then dissociated into atoms (atomization). These processes, which occur very fast in a peripheral plasma zone, are exactly the same as those taking place in flames and other plasma sources.

However, for conventional MIP sources the plasma is sustained inside the torch, forming a filament located at the axis surrounded by a low-temperature zone. In this case, the sample may partially travel around the outside of the discharge, where it experiences a lower temperature. As a result, the sample is not homogeneously mixed with the plasma. This may lead to more severe matrix effects and poorer plasma stability. The degree of lateral diffusion of the aerosol into the central plasma zone depends on the plasma operating parameters, including plasma gas properties, as well as sample composition. Moreover, the aerosol passing around the discharge interferes with the transfer of the energy and sustainment of the discharge. To alleviate these problems, in modern MIP sources some improvement has been made concerning microwave coupling, cavity and torch construction and application of a higher power level. The other possibility is to arrange the sample introduction through the central channel of the plasma, similarly to the ICP method. Previously, this was realized with the use of tangential torches[60] and, recently, by a more advanced way in the MPT,[14] Okamoto cavity[41] and TEM-based[43] MWP sources.

The other important factor that is largely responsible for the efficiency of the above-mentioned processes is the residence time of the analyte particles in the plasma. This depends on the plasma height and linear velocity of the gas stream and varies considerably for different MWPs. For annular-shaped microwave plasmas, mentioned above,[14,41–43] it is similar to that for ICP spectrometry (approximately 2–3 ms). However, for filament-type MIPs it could be comparatively long (about 10 ms) due to the much lower gas flow rates used. This is beneficial for the sample–plasma interaction and compensates in part the above discussed limitations of the lateral diffusion of the aerosol into the plasma.

1.2.3 Analyte Excitation and Ionization

After desolvation, vaporization and atomization of the sample aerosol, the subsequent processes are excitation and ionization. They occur predominantly in the hotter part of the plasma. Interestingly, as calculated by van der Mullen and Jonkers[61] for an argon TIA microwave plasma, 56% of the energy transfer from electrons to heavy particles is used for excitation and ionization processes.

Various mechanisms of analyte excitation and ionization have been proposed and identified for MWPs. They are briefly discussed in Chapter 5. Here, we only mention particles originating from the plasma that participate in excitation and ionization processes. The distinctive feature of MWPs is the extremely high electron temperature, especially for helium plasmas and low-pressure plasmas $(20\text{–}100 \times 10^3\,\text{K})$.[50] Free electrons can gain large amounts of energy in an extremely short period of time because of the very high frequency and amplitude of the microwave field. Then they collide inelastically with plasma gas atoms, causing ionization. In helium MWPs, free electrons can reach a higher kinetic energy than in argon MWPs owing to the higher ionization potential of helium.[58] Although the fraction of high-energy electrons in the plasma is small compared to the total number of free electrons, they play a specific role in analyte excitation and ionization, particularly with regard to hard-to-ionize elements.[62]

In MWPs, there exist a large number of energetic heavy particles (atoms, ions or molecules) capable of exciting or ionizing almost any elements present in the plasma. Particularly, metastable species that are specific excited states, because they can decay only by energy transfer through collisions, should not be neglected. They play a crucial role in the Penning ionization process, which is of great importance in MWPs, particularly for excitation of non-metals having high excitation energies. Also, for more readily excited metallic elements, excitation by electron impact may be expected to compete with excitation by metastable species, particularly in atmospheric pressure MWPs and helium MWPs.[61]

The introduction of a sample to a MWP may cause significant changes in the plasma characteristics, much more than in the ICP method. The sample matrix components, especially a solvent and easily ionized elements, disturb the energy exchange between the plasma particles, affect plasma temperature and electron density, and also the spatial distribution of particles in the plasma and the mechanism of excitation.[63] The overall effect depends on the sample composition and plasma parameters and is difficult to predict. Generally, the introduction of a water aerosol causes plasma contraction. However, decrease of the plasma volume usually results in a remarkable increase of the plasma temperature and electron density. By contrast, the introduction of molecular gases or powdered samples, as well as easily ionized elements, leads to a plasma suspension, resulting in a more homogeneous spatial distribution of the plasma temperature and electron density.[64]

1.2.4   Summary: Energy Flow Diagram

The scheme of energy transfer in a MWP is presented in Figure 1.1. The electric field transmits energy to the free electrons, which are very mobile charged species and thus participate in collisions and transitions phenomena. This electronic energy is then transmitted to the neutral species by inelastic exciting or even ionizing collisions. Successive collisions progressively ionize the plasma gas until the energy absorbed from the microwave field is balanced between the moving electrons and the generated ions. The energetic electrons can also transfer their energy to ions because of the relatively large collisional cross-section between them. Subsequently, ions can easily transfer energy to neutral atoms because of the similarity of their mass. Other possible inelastic collisions, *i.e.* collisional de-excitation and three-body recombination, may also participate in energy transfer between free electrons and heavy particles.

The energy transfer from electrons to heavy particles is inefficient, particularly in the helium plasma. Depending on the operating frequency and the amounts of energy transferred to the plasma, the properties of the plasma change, in terms of electron density or temperature. Fast plasma heating and low electron number densities in both argon and helium MWPs are responsible for their remarkable non-thermal features, manifested, in particular, by a non-maxwellian electron energy distribution.[58,61] The lack of efficient elastic collisions between electrons and heavy particles also leads to a low gas temperature in MIPs, which in turn causes problems in vaporization, dissociation and atomization of the analyte introduced into the plasma. Nevertheless, the existence of a large number of high-energy electrons and metastable species of the working noble gas makes MIPs the source of choice for analyzing many non-metallic elements.

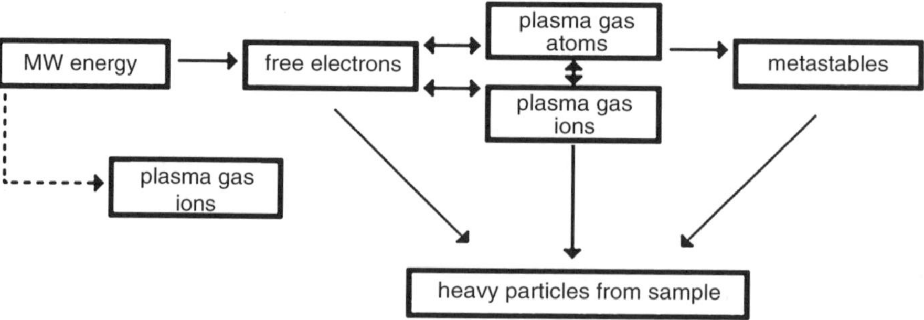

Figure 1.1   Energy flow diagram in a MWP.

1.3 Microwave Plasma Generation

The CMPs and MIPs do not operate at arbitrary frequencies. The International Telecommunication Union has defined certain narrow bands for industrial, scientific and medical uses. A MIP typically operates at 2450 MHz, although 915 MHz is used in certain approaches.

The MIP is ignited by the introduction of seed electrons in the plasma gas, usually by insertion of an electrically conductive and heat-resistive material while the power is applied. However, for non-resonant structures, *i.e.* surfatrons, a short burst from a Tesla coil is recommended. After introduction of these seed electrons, a properly tuned system promotes the ionization of the plasma gas, and a plasma forms. The initially tuned resonant cavity produces a standing electromagnetic wave with the maximum electric field at the centre of the cavity and directed axially to the plasma discharge tube. Following plasma ignition, the cavity is tuned to a minimum reflected power. Exceptionally, under specific experimental conditions, it is possible to initiate the plasma "spontaneously" due to heating of the discharge tube, gaining the seed electrons from the discharge tube material. A self-ignition of the discharge ensures flexible operating conditions that is beneficial, especially when low-pressure MIP sources are used for optical emission or mass spectrometry.

A CMP discharge can be initiated at the tip of the electrode by touching it with an isolated conductor, which releases electrons easily when heated by the microwave field. Subsequent minor retuning is usually required initially and after a short warm-up period.

In certain MWP sources, nitrogen or air plasmas are produced by igniting the plasma with argon under impedance matching conditions best suited to the nitrogen or the air discharge. The argon flow is gradually replaced by nitrogen or air immediately after the production of the argon plasma, and then a stable plasma in a molecular gas can be maintained.

MWPs can be generated in different gases, such as noble gases, nitrogen, air, oxygen, or carbon dioxide, under both normal as well as reduced pressure. The properties of the discharge depend on the pressure and the nature of the plasma gas. The MWP sources for analytical spectrometry typically use argon or helium, but in the past decade nitrogen plasmas have gained great importance. The MWP uses very little plasma gas, from $50 \, \text{mL} \, \text{min}^{-1}$ to a few litres per minute. By limiting this discussion to a short description of argon, helium and nitrogen plasmas, it is possible to present the main differences in plasma properties.

A low or moderate power helium MIP is a suspended plasma, spatially stable and of low continuum. Both the plasma position in the discharge tube and the stability depend upon the applied power level and the gas flow rate used.[60] Helium plasma fills nearly the entire discharge tube section, but does not expand outside the discharge tube. However, Zander and Hieftje[65] noted that for the Beenakker-type cavity at a relatively low microwave power level it uniformly filled the discharge tube, but as the power was increased the plasma moved to the wall of the tube. Pak and Koirtyohann[66] applied moderate power and a high helium flow rate of $8\,L\,min^{-1}$ for sustaining a helium plasma in a tangential flow torch outside the cavity wall. A high-power helium MIP[67] obtained with the use of an Okamoto cavity also expands outside the cavity wall. The flame-like helium plasmas produced by the MPT[68] or TIA[69] devices are located outside the torch; however, they are of lower plasma volume with regard to the respective argon plasmas. A higher thermal conductivity of helium compared to argon is beneficial for heat transfer from the plasma to the sample, improving the efficiency of the desolvation and vaporization processes. The helium plasma is very suitable for the determination of halogens and non-metals owing to its background characteristics and the high excitation energies available. This is explained by the participation in the excitation process of helium atoms in the metastable state. Because of low atomic mass and nearly monoisotopic abundance of helium, a greater mass spectral selectivity can be achieved. The limitations of a helium plasma include a relatively low gas temperature and electron density and relatively high running costs.

An argon MIP discharge is characterized by a reduced diameter. It fills only part of the space in the discharge tube, forming a thin bright filament stretching along the tube axis. It is spatially much less stable than a helium plasma and exhibits a tendency both to adhere to the tube walls and for filamentation. When not completely stabilized it resembles an electric arc in shape, contacting in more and more new places with the torch wall. An argon plasma generated in dry, pure argon forms a long filament and easily expands outside the discharge tube. Microwave leakage to the surroundings then takes place. Introduction of the sample aerosol causes narrowing of the discharge region to the space inside the tube, limited by the thickness of the resonant cavity. This indicates a relatively high skin effect limiting sample aerosol penetration to the plasma. In an argon MWP a majority of the elements can be detected with satisfactory sensitivity, despite a relatively high background. However, the excitation of halogens undergoes considerable worsening in comparison with a helium plasma. Both Ar-MPT and Ar-TIA discharges are formed outside the torch. It should be noted that argon is a much cheaper gas than helium.

Similar to a helium plasma, a nitrogen MIP fills the whole discharge tube diameter owing to the low electron density. The sustaining of a nitrogen plasma usually requires higher microwave power compared to the noble gases.[70] Also, its thermal conductivity is high enough to avoid non-uniform heating. A low skin effect allows efficient mixing of aerosol with the plasma gas. At higher gas flow rates and power levels it can expand outside the discharge tube.[71] Nitrogen discharges sustained with the use of a CMP[1] or an Okamoto cavity[72] are observed outside the torch.

Molecular gases such as nitrogen, oxygen or carbon dioxide have been successfully used for sustaining MWPs and the possibilities of their analytical applications have been demonstrated.[50–53] Interestingly, a wide variety of gases can often be used for plasma generation in essentially the same MWP system without any modifications.[1,70,73] The choice of plasma gas results first of all from striving for the best matching of the excitation conditions with the current analytical applications, such as chromatographic analysis of organic substances, detection of compounds separated with the use of supercritical fluid chromatography, or air monitoring. Gas mixtures are also used, including helium–hydrogen, helium–oxygen, argon–helium, nitrogen–helium, carbon dioxide–helium, and air plasmas.[50–53,74–78] The use of small amounts of oxygen, hydrogen or other gases as gas dopants for improving plasma parameters in specific analytical applications should be mentioned.[79–82] Depending on the gas used, various plasma properties (temperature, excitation mechanism, possibility of chemical reaction occurrence) can be achieved and it is vital to influence the conditions of determining the elements of interest. This variety provides a possibility to determine samples of different origin, both inorganic as well as organic ones.

The MWP can be operated over a wide range of pressures from 0.001 Torr to above atmospheric pressure. However, it is easier to sustain the plasma in different gases under reduced pressure. Regarding the efficiency of energy transfer in the plasma, the optimum pressure for MWP operation is about 4 Torr. Low pressure means a small number of particles, lesser probability of collisions and longer path segments between collisions for individual particles. As a result, free electrons gain extremely high energies. However, the level of operating pressure appears to have little effect on plasma temperature. For an atmospheric pressure MIP, the applied power level is in the approximate range 100–1000 W, whereas a low-pressure plasma can be sustained even at 10 W. The plasma parameters appear to be only moderately affected by the power settings. In general, the response seems to be less affected by power when operating at low pressures than at high pressures. Therefore, for certain analytical applications utilizing microsampling techniques, the use of high power levels is unfounded.

Busch and Vickers[83] reported only small changes in plasma temperature and electron density as the pressure was varied in the range between 1 and 25 Torr. For argon and helium MIPs operating at very low pressures (0.05–2.0 Torr), Brassem and Maessen[84] observed the increase of electron density with increasing microwave power. According to Goode *et al.*,[85] the electron density of an argon MIP is directly proportional to the plasma pressure in the range 10–760 Torr. In general, for argon and helium MWPs the electron density ranges from 10^{11} to 10^{12} cm^{-3} at pressures of about 1 Torr to approximately 10^{14}–10^{15} cm^{-3} at atmospheric pressure.[49]

Among MWP sources, the surfatron seems to be the most flexible approach that can operate over a wide range of pressures without requiring any change in matching conditions.[86] The effect of plasma pressure, in the range 2–100 Torr and at atmospheric pressure, upon the detection limit for mercury both in argon and helium MIPs was investigated by Costa-Fernandez *et al.*[87] A low-pressure argon plasma provided the lowest detection limit.

The major disadvantage of plasma sources operating at low pressures is the difficulty of coupling with certain sample introduction techniques. Usually, the use of an interface is required. On the other hand, a low-pressure MWP source is very compatible with an MS detector.[88]

1.3.1 Microwave Plasma Geometries (Configurations)

MWPs can form a variety of plasma geometries. A typical, single, axially directed plasma filament is located on the axis, not entirely filling the radial cross-section of the discharge tube, provided heat flows symmetrically from it towards the tube wall. The plasma diameter increases with the decreasing thermal conductivity of the gas. Hence, helium and nitrogen plasmas are less contracted than an argon one. This filament MIP located inside the discharge tube can be divided in two zones: a hotter central zone or plasma core and a plasma surrounding zone of lower temperature (Figure 1.2). In some MIP setup arrangements, at appropriate conditions it is possible to obtain an argon multi-filament or toroidal ("hollow", "doughnut") plasma, in which the high-temperature zone surrounds, in the form of a ring, the cooler channel through which the sample flows.[60,89,90]

The conventional MIP systems cannot produce an annular-shaped plasma like an ICP because they have the electric-field distribution yielding the maximum at the centre of the discharge tube. Formerly, tangential flow torches[60] were developed to form a sheath around the aerosol stream and effectively channel the aerosol into the plasma discharge. However, it was observed that in comparison with the toroidal plasma, the suspended plasma has the advantage of not touching, and hence not etching, the discharge tube wall. Further, it requires a lower plasma gas flow rate, and provides significantly greater emission intensity than can be obtained from the toroidal configuration. The plasma configuration proved to be gas flow dependent. In the tangential flow

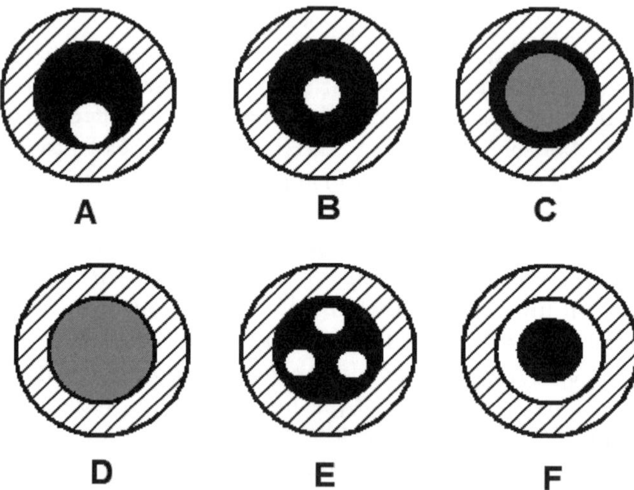

Figure 1.2 Possible geometries of a MIP discharge (axial viewing): A, unstable filament argon plasma; B, centred filament argon plasma; C, helium plasma; D, molecular gas or mixed gas plasma; E, three-filament argon plasma; F, toroidal plasma.

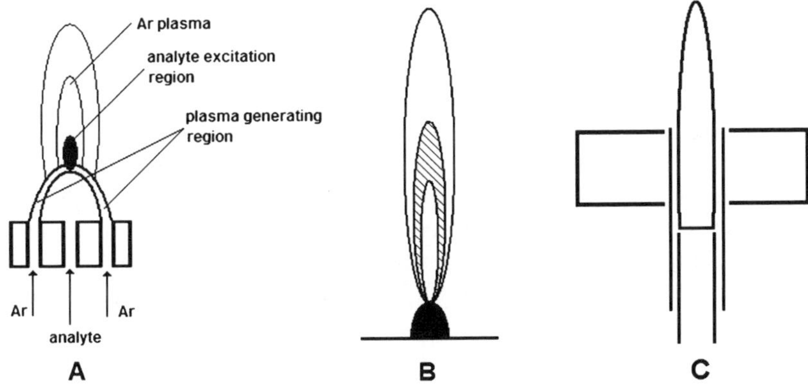

Figure 1.3   Possible geometries of radially viewed MWPs: A, MPT discharge; B, CMP discharge; C, annular MIP discharge.

torch, at a helium flow of approximately $3.0\,\mathrm{L\,min^{-1}}$ a toroidal shape is formed, while at approximately $4\,\mathrm{L\,min^{-1}}$ the plasma abruptly changes its configuration to become a suspended, filament-type plasma.

In the moderate power TEM-based MIP,[43] two distinctly different plasma configurations can be obtained by varying the plasma gas flow rate. An annular-shaped helium plasma can be generated at gas flow rate below $3\,\mathrm{L\,min^{-1}}$ without touching the wall of the outer tube, but a filament-type plasma is formed at about $1\,\mathrm{L\,min^{-1}}$. In the Okamoto cavity,[42] a high-power helium plasma can be obtained at a relatively high gas flow rate of $11\,\mathrm{L\,min^{-1}}$. For the same cavity the N_2-MIP[72] is a doughnut-shaped plasma, similar to an Ar-ICP at gas flow rates ranging from 6 to $15\,\mathrm{L\,min^{-1}}$.

Various CMPs and MPTs form flame-like discharges. Pless *et al.*[91] observed spherical and cylindrical configurations of a He–H_2 CMP, depending on the gas flow rate and microwave power selected. However, the MPT discharge expands uniformly at the tip of the electrode, forming a central channel surrounded by a hot plasma zone similar to annular-shaped ICPs.[92] Advantageously, the analyte excitation zone is separated from the plasma generation zone, thus causing little perturbation of plasma stability. Figure 1.3 shows a cross-sectional representation of the discharge along with the nomenclature for different regions of the plasma.

1.3.2   Power Density *versus* Plasma Stability

The power density in the generated plasma volume is one of external control parameters that are useful for the evaluation of plasma properties. The power density, or more precisely the rate of change of power density with applied power, is related to the plasma stability and its tolerance to sample loading. As a result, it influences the signal precision and the detection limit. Van der Mullen and Jonkers[61] used the power density for plasma modelling and characterization of plasma departure from the local thermal equilibrium (LTE). In general, a high power density induces LTE plasmas whereas non-LTE plasmas are favoured by either a low power density or a pulsed power mode.

In 1986 a comparison of three common plasma sources, *i.e.* ICP, direct current plasma (DCP) and MIP, was made by Zander.[93] On the basis of current developments of plasma spectrometric techniques he stated that each of those plasma sources has its own characteristic range of applied power as well as that the rate of change of power density with applied power is quite different for each plasma. For the MIP (and DCP), he observed a very steep rate of change of power density with applied power. In low-power MIP the vaporization process consumes a considerably large part of the energy, which results in a significant drop of the power density and discharge destabilization. He concluded that it is the major factor influencing both the poorer signal precision and the large fluctuation of background obtained from a MIP. By contrast, an ICP has a relatively constant power density and this can partially account for the excellent short- and long-term precision of the signal measured. However, the problem lies in the design of early MIP cavities. Recent MWP sources can operate up to about 1000 W and some of them, including the MPT, the Okamoto cavity and the TEM, can produce plasmas of a relatively large size. As a result, the power densities obtained for these plasmas are in the range 0.2–$0.8\,\text{kW cm}^{-3}$ and the rates of change of power density with applied power are similar to that for ICPs.

The fundamental problem of MWPs is the difficulty in forming a large-volume homogeneous discharge. This is due to the short microwave wavelength, the non-uniform gas heating through limited gas thermal conductivity, and the small penetration depth of the microwave into the plasma, called the skin effect, through a high-field frequency and a low electron density. However, the plasma volume depends on some details of the technical design that contains or supports the plasma, including the microwave coupling and applied power level used. As mentioned above, some MWP approaches can form relatively large plasmas of more than $1\,\text{cm}^3$. Moreover, a low-volume MIP discharge enlargement can be achieved by decreasing the operating pressure and preferably the use of a surface wave propagation mode.[94]

Even with MWPs operated at relatively high power, the limited size of the plasma makes sample introduction a challenging task. In the case of a relatively large-volume ICP discharge, sample penetration into the plasma is relatively easy. However, both the volume and shape of filament MWPs cause sample introduction to be more complex and less efficient. Additionally, the ability to vaporize the sample aerosol to molecular species and particles is limited. Bollo-Camara and Codding[60] have optimized the plasma height (volume) sustained in a TM_{010} cavity and stated that the best sensitivity of the optical emission measurement is obtained for a plasma height of about 2 cm. They concluded that it provides sufficient residence time of the sample in the plasma to excite atoms and ions to a significant extent. The CMPs operating at microwave power levels of several hundred watts are recognized as relatively large and more robust plasmas and therefore much more tolerant to sample introduction.[91] However, low-volume MIPs provide better emission signal stability (1–2%) than CMPs (5–10%).[51]

A result of the size of the physical plasma volume is the size of the viewing zone for spectroanalytical observation. Generally, the MWP discharge itself is used for analyte emission viewing.[91,95,96] In the case of MIPs, 1–4 mm i.d. discharge tubes are generally used, leading to a small total viewing zone. Unfortunately, the analyte emission intensity is often heterogeneously distributed across the viewed area, making the useful viewed area significantly smaller. Plasma mapping with the use of a spatial imaging system is reasonable to attain the maximum sensitivity and signal stability.[66,96]

1.4   Basic Physical Characteristics of a Microwave Plasma Discharge

The physical and thermodynamic description of the plasma state is beyond the scope of this book. In this section we only collect some MWP characteristics in order to compare various MWP sources themselves and with other plasma sources used in analytical spectrometry. The variety of conditions at which MWPs can be obtained means that basic plasma parameters can change over a wide range.

A MWP is definitely a non-thermal plasma, for which a considerable difference between the temperature of the electrons (T_e) and that of heavy particles (T_h) occurs. The greater the departure from LTE, the larger is the difference between T_e and T_h. On the other hand, a high T_e and considerable electron density (*ca.* 10^{15} cm^{-3}) favour the excitation of a larger number of atoms, and thus the better detectability of analytes. In general, each species in the plasma will have its own particular temperature if it obeys a maxwellian velocity distribution. The excitation temperature determines the distribution of the internal energy of atoms and ions of elements present in plasma, which results in the intensity of particular lines of the spectrum. Similarly, the rotational temperature, T_{rot}, excitation temperature, T_{exc}, and ionization temperature, T_{ion}, can be used to characterize the molecular behaviour, atomic excitation and ionization processes, respectively.

The literature data for temperatures in MWPs can be summarized as follows:

$$T_e > T_{ion} \approx T_{exc} > T_{rot}$$

The above inequality proves the lack of local thermodynamic equilibrium in MWPs. In Table 1.2 are shown, as examples, the MWP parameters measured by different authors for various MWP sources.

The electron density of an argon plasma is greater than that of a helium or nitrogen plasma. Also, the electron density, reaching at best 2×10^{15} cm^{-3}, is far away from the equilibrium value, which is estimated as 2×10^{16} cm^{-3}. The

Table 1.2 Plasma characteristics for selected MWP sources.

Mode	Plasma gas	Electron density (cm^{-3})	Electron temp. (K)	Excitation temp. $(K)^a$	Rotational temp. $(K)^a$	Ref.
TE_{013}	Ar	1.8×10^{15}	–	6280 (Ar)	1440–2440 (OH)	25
TM_{010}	Ar	3.8×10^{14}	–	4500 (Ar) 4000–5700 (Fe)	1150 (OH)	97–99
	He	1.3×10^{14}	–	3400 (He) 5700 (Fe)	1300 (OH) 1400 (N_2^+)	
TE_{101}	Ar	1.1×10^{15}	7900	4600–5900 (Fe) 4000–6400 (Ar)	2500–3600 (OH) 4900 (N_2^+)	38,100–103
	He	4×10^{14}	12 500	3000 (He) 5500 (Fe)	2200–2700 (OH)	
Surfatron	Ar	4×10^{14}	7800	1900 (Ar)	2250 (OH) 3600 (N_2^+)	104–106
	He	1×10^{14}	–	3000 (He)	2000 (OH)	
TEM	He	$5.5–7.5 \times 10^{14}$	–	3000–3300 (He)	3000 (OH)	43
MPT	Ar	7×10^{14}	13 000	5300–6000 (Fe)	1500–6000 (OH)	58,107
	He	1×10^{14}	21 500		2100 (OH)	
TIA	Ar	1×10^{14}	19 100	5500 (Ar)	3000 (OH)	16,69,108
	He	$1–5.7 \times 10^{14}$	26 000	3800 (He)	2400–2900 (N_2^+)	
Okamoto cavity	N_2	5×10^{13}	–	5400 (Fe)	5000 (N_2^+)	67,73,109,110
	He	2.3×10^{14}	–	5000 (Fe)	–	
CMP	N_2	$<1 \times 10^{14}$	–	4900–5500 (Fe)	4300 (N_2^+)	51
	He	4×10^{14}	–	3430 (He)	1620 (OH)	
Three-phase MIP	He	7.5×10^{14}	–	4000 (He)	3100 (OH)	44

^{a}In parentheses the "thermometric specimen" is included.

presence of water vapour appears to "thermalize" the plasma, which is manifested by an increase of the plasma temperature and electron density.

The difference in operating frequency between a MWP and an ICP seems to be the main reason for their differing behaviour as analytical sources. However, the operating frequency has a negligible effect on the rotational temperature, excitation temperature or electron density. On the other hand, the electron temperature in MWPs is much higher than that in ICPs (6000–8000 K), whereas the electron number density in MWPs is at least one order of magnitude lower than that in ICPs. Thus, one can conclude that plasma heating and energy transfer are considerably different for the two sources.[58]

The ICP and DCP have almost equivalent gas temperatures (4000–7000 K), with the MWP generally accepted to be thermally cooler (1000–4000 K). As a consequence of the relatively low plasma temperature, MWPs at low and moderate power have lower desolvation, volatilization and atomization capabilities, leading to more serious matrix interferences and similar or higher detection limits compared to ICPs. However, the helium MIP offers considerably better detectability for halogens and some other elements.

Comparing the degree of ionization of elements in high-power N_2-MIPs and an Ar-ICP, Ohata and Furuta[111] stated that elements which have an ionization potential (IP) lower than 7 eV are ionized almost completely in both plasma sources. However, for elements of IP higher than 7 eV the degree of ionization gradually decreases in both plasma sources but much faster in an N_2-MIP. Nevertheless, the N_2-MIP has proved to be a promising ionization source for mass spectrometric determination of such elements as K, Ca, Cr and As.

1.5 Spectroscopic Techniques Employing Microwave Induced Plasmas

Plasma-based analytical spectrometry can be performed by atomic emission, atomic absorption, atomic fluorescence or mass spectrometries. Presently, most of the work on MWPs is done by atomic emission, and an increasing number of papers have been reported on the use of MWP sources for MS, preferably coupled with solution nebulization and gas or liquid chromatography. The utilization of MWP as an atomization source for atomic absorption spectrometry (MWP-AAS) and atomic fluorescence spectrometry (MWP-AFS) is also approved. In particular, when employed as an atomizer for atomic fluorescence spectrometry the MPT offers large dynamic ranges over a concentration range of several orders of magnitude and relatively low background interference.[51,52] Recently, the use of MWPs for cavity ringdown spectroscopy has also appeared.

MWP source selection for an individual spectroscopic technique depends on the role that the source should play in it. In the case of AAS and AFS, a MWP is utilized only for the atomization, but in MS it is used for ionization of the sample. In OES, however, not only is thermal dissociation of the sample necessary, but also excitation of the atoms and ions formed. Here, the plasma source acts as an atomization reservoir with simultaneous excitation and ionization of neutral atoms, and excitation of ionized species. Thus, information of atomization/ionization and excitation efficiency of the MWP source to be chosen is valuable.

Among flames and plasmas used in analytical spectrometries, MWP is recommended for analysis of gaseous samples.[112] In addition to the introduction of gaseous samples and the combination with GC, various electrothermal vaporization devices, flow injection techniques and gas generation techniques have been developed. When using the MWP as an atom reservoir, tandem operation with direct solid sampling through laser ablation is very useful. Sealed samples are also excited for emission measurements at low microwave power to reach a high power of detection in a static system.[113,114] All the above-mentioned analytical spectrometric techniques will be discussed in Chapter 11 in more detail.

References

1. J. D. Cobine and D. A. Wilbur, *J. Appl. Phys.*, 1951, **22**, 835–841.
2. H. P. Broida and G. H. Morgan, *Anal. Chem.*, 1952, **24**, 799–804.
3. H. P. Broida and J. W. Moyer, *J. Opt. Soc. Am.*, 1952, **42**, 37–41.
4. R. Mavrodineanu and R. C. Hughes, *Spectrochim. Acta, Part B*, 1963, **19**, 1309–1317.
5. U. Jecht and W. Kessler, *Fresenius' Z. Anal. Chem.*, 1963, **198**, 27–35.
6. M. Yamamoto and S. Murayama, *Spectrochim. Acta, Part A*, 1967, **23**, 773–776.
7. S. Murayama, H. Matsuno and M. Yamamoto, *Spectrochim. Acta, Part B*, 1968, **23**, 513–520.
8. S. Murayama, *Spectrochim. Acta, Part B*, 1970, **25**, 191–200.
9. J. O. Burman and K. Bostrom, *Anal. Chem.*, 1979, **51**, 516–520.
10. G. F. Larson and V. A. Fassel, *Anal. Chem.*, 1976, **48**, 1161–1166.
11. Q. Jin, G. Yang, A. Yu, J. Liu, H. Zhang and Y. Ben, *J. Natl. Sci. Jilin Univ.*, 1985, **1**, 90.
    (in Chinese).
12. Q. Jin, C. Zhu, W. Borer and G. M. Hieftje, *Spectrochim. Acta, Part B*, 1991, **46**, 417–430.
13. Q. Jin, W. Yang, F. Liang, H. Zhang, A. Yu, Y. Cao, J. Zhou and B. Xu, *J. Anal. At. Spectrom.*, 1997, **13**, 377–384.
14. W. Yang, H. Zhang, A. Yu and Q. Jin, *Microchem. J.*, 2000, **66**, 147–170.
15. M. Moisan, G. Sauve, Z. Zakrzewski and J. Hubert, *Plasma Sources Sci. Technol.*, 1994, **3**, 584–592.
16. J. Jonkers, J. M. de Regt, J. J. A. M. Van der Mullen, H. P. C. Vos, V. P. J. de Groote and E. A. H. Timmermans, *Spectrochim. Acta, Part B*, 1996, **51**, 1385–1392.
17. F. C. Fehsenfeld, K. M. Evenson and H. P. Broida, *Rev. Sci. Instrum.*, 1965, **36**, 294–298.

18. H. P. Broida and M. W. Chapman, *Anal. Chem.*, 1958, **30**, 2049–2055.
19. A. J. McCormack, S. C. Tong and W. D. Cooke, *Anal. Chem.*, 1965, **37**, 1470–1476.
20. B. D. Quimby and J. J. Sullivan, *Anal. Chem.*, 1990, **62**, 1027–1034.
21. C. I. M. Beenakker, *Spectrochim. Acta, Part B*, 1976, **31**, 483–486.
22. C. I. M. Beenakker, B. Bosman and P. W. J. M. Boumans, *Spectrochim. Acta, Part B*, 1978, **33**, 373–381.
23. H. Kawaguchi, M. Hasegawa and A. Mizuike, *Spectrochim. Acta, Part B*, 1972, **27**, 205–210.
24. K. Fallgatter, V. Svoboda and J. D. Winefordner, *Appl. Spectrosc.*, 1971, **25**, 347–352.
25. F. E. Lichte and R. K. Skogerboe, *Anal. Chem.*, 1973, **45**, 399–401.
26. M. Moisan, C. Beaudry and P. Leprince, *IEEE Trans. Plasma Sci.*, 1975, **3**, 55–59.
27. G. L. Long and L. D. Perkins, *Appl. Spectrosc.*, 1987, **41**, 980–985.
28. K. J. Slatkavitz, P. C. Uden, L. D. Hoey and R. M. Barnes, *J. Chromatogr.*, 1984, **302**, 277–287.
29. J. Hubert, M. Moisan and A. Ricard, *Spectrochim. Acta, Part B*, 1979, **33**, 1–10.
30. M. Selby and G. M. Hieftje, *Spectrochim. Acta, Part B*, 1987, **42**, 285–298.
31. R. M. Barnes and E. Reszke, *Anal. Chem.*, 1990, **62**, 2650–2654.
32. K. A. Forbes, E. Reszke, P. C. Uden and R. M. Barnes, *J. Anal. At. Spectrom.*, 1991, **6**, 57–71.
33. P. G. Brown, D. L. Haas, J. M. Workman, J. A. Caruso and F. L. Fricke, *Anal. Chem.*, 1987, **59**, 1433–1436.
34. K. B. Cull and J. W. Carnahan, *Appl. Spectrosc.*, 1988, **42**, 1061–1065.
35. M. M. Mohamed, T. Uchida and S. Minami, *Appl. Spectrosc.*, 1989, **43**, 129–134.
36. H. Matusiewicz, *Spectrochim. Acta, Part B*, 1992, **47**, 1221–1227.
37. K. Jankowski, R. Parosa, A. Ramsza and E. Reszke, *Spectrochim. Acta, Part B*, 1999, **54**, 515–525.
38. D. J. Douglas and J. B. Frech, *Anal. Chem.*, 1981, **53**, 37–41.
39. Y. Okamoto, M. Yasuda and S. Murayama, *Jpn. J. Appl. Phys.*, 1990, **29**, L670–L672.
40. Y. Okamoto, *J. Anal. At. Spectrom.*, 1994, **9**, 745–749.
41. Y. Okamoto, *Jpn. J. Appl. Phys.*, 1999, **38**, L338–L341.
42. K. Jankowski, A. Jackowska, A. P. Ramsza and E. Reszke, *J. Anal. At. Spectrom.*, 2008, **23**, 1234–1238.
43. K. Jankowski, A. P. Ramsza, E. Reszke and M. Strzelec, *J. Anal. At. Spectrom.*, 2010, **25**, 44–47.
44. H. Takahara, M. Iwasaki and Y. Tanibata, *IEEE Trans. Instrum. Meas.*, 1995, **44**, 819–823.
45. U. Engel, A. M. Bilgic, O. Haase, E. Voges and J. A. C. Broekaert, *Anal. Chem.*, 2000, **72**, 193–197.
46. S. Greenfield, H. McD. McGeachin and P. B. Smith, *Talanta*, 1975, **22**, 553–562.

47. R. K. Skogerboe and G. N. Coleman, *Anal. Chem.*, 1976, **48**, 611A–622A.
48. A. T. Zander and G. M. Hieftje, *Appl. Spectrosc.*, 1981, **35**, 357–371.
49. J. P. Matousek, B. J. Orr and M. Selby, *Prog. Anal. At. Spectrosc.*, 1984, **7**, 275–314.
50. A. E. Croslyn, B. W. Smith and J. D. Winefordner, *CRC Crit. Rev. Anal. Chem.*, 1997, **27**, 199–255.
51. Q. Jin, Y. Duan and J. A. Olivares, *Spectrochim. Acta, Part B*, 1997, **52**, 131–161.
52. J. A. C. Broekaert and U. Engel, in *Encyclopedia of Analytical Chemistry*, ed. R. A. Meyers, Wiley, Chichester, 2000, pp. 9613–9667.
53. J. A. C. Broekaert and V. Siemens, *Spectrochim. Acta, Part B*, 2004, **59**, 1823–1839.
54. A. P. Ramsza and L. Starski, *Electronic Mater.*, 1985, **1**(49), 7–23. (in Polish).
55. Y. Su, Y. Duan and Z. Jin, *Anal. Chem.*, 2000, **72**, 2455–2462.
56. S. C. Brown, *Basic Data of Plasma Physics*, MIT Press, Cambridge, 1959.
57. G. Francis, *Ionization Phenomena in Gases*, Butterworth, London, 1960.
58. M. Huang, D. S. Hanselman, Q. Jin and G. M. Hieftje, *Spectrochim. Acta, Part B*, 1990, **45**, 1339–1352.
59. Y. Kabouzi, M. D. Calzada, M. Moisan, K. C. Tran and C. Trassy, *J. Appl. Phys.*, 2002, **91**, 1008–1019.
60. A. Bollo-Camara and E. G. Codding, *Spectrochim. Acta, Part B*, 1981, **36**, 973–982.
61. J. van der Mullen and J. Jonkers, *Spectrochim. Acta, Part B*, 1999, **54**, 1017–1044.
62. M. Huang, *Microchem. J.*, 1996, **53**, 79–87.
63. J. P. Matousek, B. J. Orr and M. Selby, *Spectrochim. Acta, Part B*, 1986, **41**, 415–429.
64. K. Jankowski and A. Jackowska, *Trends Appl. Spectrosc.*, 2007, **6**, 17–25.
65. A. Zander and G. M. Hieftje, *Anal. Chem.*, 1978, **50**, 1257–1260.
66. Y. N. Pak and S. R. Koirtyohann, *Appl. Spectrosc.*, 1991, **45**, 1132–1142.
67. H. Yamada and Y. Okamoto, *Appl. Spectrosc.*, 2001, **55**, 114–119.
68. K. H. Jo and Y. N. Pak, *J. Korean Chem. Soc.*, 2000, **44**, 573–580.
69. A. Rodero, M. C. Quintero, A. Sola and A. Gamero, *Spectrochim. Acta, Part B*, 1996, **51**, 467–479.
70. K. G. Michlewicz, J. J. Urh and J. W. Carnahan, *Spectrochim. Acta, Part B*, 1985, **40**, 493–499.
71. R. D. Deutsch and G. M. Hieftje, *Appl. Spectrosc.*, 1985, **39**, 214–222.
72. Y. Okamoto, *Anal. Sci.*, 1991, **7**, 283–288.
73. T. Maeda and K. Wagatsuma, *Microchem. J.*, 2004, **76**, 53–60.
74. B. Riviere, J. M. Mermet and D. Deruaz, *J. Anal. At. Spectrom.*, 1988, **3**, 551–555.
75. Z. Zhang and K. Wagatsuma, *J. Anal. At. Spectrom.*, 2002, **17**, 699–703.
76. P. Liang and A. Li, *Fresenius' J. Anal. Chem.*, 2000, **368**, 418–420.
77. Y. K. Zhang, S. Hanamura and J. D. Winefordner, *Appl. Spectrosc.*, 1985, **39**, 226–230.

78. S. A. Estes, P. C. Uden and R. M. Barnes, *J. Chromatogr.*, 1982, **239**, 181–189.
79. T. H. Risby and Y. Talmi, *CRC Crit. Rev. Anal. Chem.*, 1983, **14**, 231–265.
80. G. R. Ducatte and G. L. Long, *Appl. Spectrosc.*, 1994, **48**, 493–501.
81. M. McKenna, I. L. Marr, M. S. Cresser and E. Lam, *Spectrochim. Acta, Part B*, 1986, **41**, 669–676.
82. A. Besner and J. Hubert, *J. Anal. At. Spectrom.*, 1988, **3**, 381–385.
83. K. W. Busch and T. J. Vickers, *Spectrochim. Acta, Part B*, 1973, **28**, 85–104.
84. P. Brassem and F. J. M. J. Maessen, *Spectrochim. Acta, Part B*, 1974, **29**, 203–210.
85. S. R. Goode, N. P. Buddin, B. Chambers, K. W. Baughman and J. P. Deavor, *Spectrochim. Acta, Part B*, 1985, **40**, 317–328.
86. A. Garnier, E. Bloyet, P. Leprince and J. Marec, *Spectrochim. Acta, Part B*, 1988, **43**, 963–970.
87. J. M. Costa-Fernandez, R. Pereiro-Garcia, A. Sanz-Medel and N. Bordel-Garcia, *J. Anal. At. Spectrom.*, 1995, **10**, 649–653.
88. B. S. Sheppard and J. A. Caruso, *J. Anal. At. Spectrom.*, 1994, **9**, 145–149.
89. D. Kollotzek, P. Tschöpel and G. Tölg, *Spectrochim. Acta, Part B*, 1984, **39**, 625–636.
90. D. L. Haas and J. A. Caruso, *Anal. Chem.*, 1984, **56**, 2014–2019.
91. A. M. Pless, B. W. Smith, M. A. Bolshov and J. D. Winefordner, *Spectrochim. Acta, Part B*, 1996, **51**, 55–64.
92. A. M. Bilgic, C. Prokisch, J. A. C. Broekaert and E. Voges, *Spectrochim. Acta, Part B*, 1998, **53**, 773–777.
93. A. T. Zander, *Anal. Chem.*, 1986, **58**, 1139A–1149A.
94. E. Bluem, S. Bechu, C. Boisse-Laporte, P. Leprince and J. Marec, *J. Phys. D, Appl. Phys.*, 1995, **28**, 1529–1533.
95. H. Uchida, P. A. Johnson and J. D. Winefordner, *J. Anal. At. Spectrom.*, 1990, **5**, 81–85.
96. M. Selby, R. Rezaaiyaan and G. M. Hieftje, *Appl. Spectrosc.*, 1987, **41**, 749–761.
97. H. Schlüter, in *Microwave Discharges: Fundamentals and Applications*, ed. C. M. Ferreira and M. Moisan, Plenum Press, New York, 1993, pp. 225–245.
98. H. Tanabe, H. Haraguchi and K. Fuwa, *Spectrochim. Acta, Part B*, 1983, **38**, 49–60.
99. L. D. Perkins and G. L. Long, *Appl. Spectrosc.*, 1989, **43**, 499–504.
100. W. Zyrnicki and W. Waszkiewicz, *Chem. Anal. (Warsaw)*, 1996, **41**, 1075.
101. J. Mierzwa, R. Brandt, J. A. C. Broekaert, P. Tschöpel and G. Tölg, *Spectrochim. Acta, Part B*, 1996, **51**, 117–126.
102. A. Geiger, S. Kirschner, B. Ramacher and U. Telgheder, *J. Anal. At. Spectrom.*, 1997, **12**, 1087–1090.
103. K. Jankowski and M. Dreger, *J. Anal. At. Spectrom.*, 2000, **15**, 269–276.

104. W. Włodarczyk and W. Zyrnicki, *Spectrochim. Acta, Part B*, 2003, **58**, 511–522.

105. J. Cotrino, M. Saez, M. C. Quintero, A. Menendez, E. Sanchez Uria and A. Sanz Medel, *Spectrochim. Acta, Part B*, 1992, **47**, 425–435.

106. A. Besner, M. Moisan and J. Hubert, *J. Anal. At. Spectrom.*, 1988, **3**, 863–866.

107. M. H. Abdallah and J. M. Mermet, *Spectrochim. Acta, Part B*, 1982, **37**, 391–398.

108. Q. Jin, H. Zhang, A. Yu, Y. Duan, X. Liu and F. Wang, *Anal. Sci.*, 1991, **7**(suppl), 559–562.

109. E. A. H. Timmermans, J. Jonkers, I. A. J. Thomas, A. Rodero, M. C. Quintero, A. Sola, A. Gamero and J. A. M. van der Mullen, *Spectrochim. Acta, Part B*, 1998, **53**, 1553–1566.

110. K. Wagatsuma, *Appl. Spectrosc. Rev.*, 2005, **40**, 229–243.

111. M. Ohata and N. Furuta, *J. Anal. At. Spectrom.*, 1997, **12**, 341–347.

112. D. A. McGregor, K. B. Cull, J. M. Gehlhausen, A. S. Viscomi, M. Wu, L. Zhang and J. W. Carnahan, *Anal. Chem.*, 1988, **60**, 1089A–1098A.

113. A. Van Sandwijk, P. F. E. van Montfort and J. Agterdenbos, *Talanta*, 1973, **20**, 495.

114. A. Van Sandwijk, P. F. E. van Montfort and J. Agterdenbos, *Talanta*, 1974, **21**(360), 660.

Instrumentation for Microwave Induced Plasma Optical Emission Spectrometry

2.1 The Components of a Microwave Induced Plasma Optical Emission Spectrometry System

A typical microwave induced plasma optical emission spectrometry (MIP-OES) system consists of three main components: an excitation source coupled with a sample introduction system and a system of plasma gas feeding, a spectrometer and electronic system for signal processing, and data acquisition and instrument control (Figure 2.1).

The MIP excitation source consists of a microwave power generator, a coupling device for transferring the power from the generator to the load, a power regulator and the microwave cavity. Generally, three sources can be used for the generation of microwaves. The most popular are magnetrons, which contain all the necessary high-frequency components to produce an electromagnetic wave at a fixed frequency. At relatively low power levels, klystron tubes and generators based on solid-state technology can be also used. Depending on the microwave power used, three categories of microwave plasma (MWP) sources can be distinguished: low (below 150 W), moderate (around 500 W) and high power (about 1 kW). The low-power MIP is usually formed by transmitting the microwaves from the generator through a 50-ohm impedance coaxial cable to a resonant cavity. In many cases a tuning device, including a stub tuner, a slide tuner and a transformer, to match the impedance of the load to that of the generator, is also included. The other means of power transfer from the generator to the load are antennas, strip lines (or microstrips in the case of microplasma devices) and waveguides. The latter are useful for high-power transmission. Good gas flow controllers are essential since the

RSC Analytical Spectroscopy Monographs No. 12
Microwave Induced Plasma Analytical Spectrometry
By Krzysztof J. Jankowski and Edward Reszke
© The Royal Society of Chemistry 2011
Published by the Royal Society of Chemistry, www.rsc.org

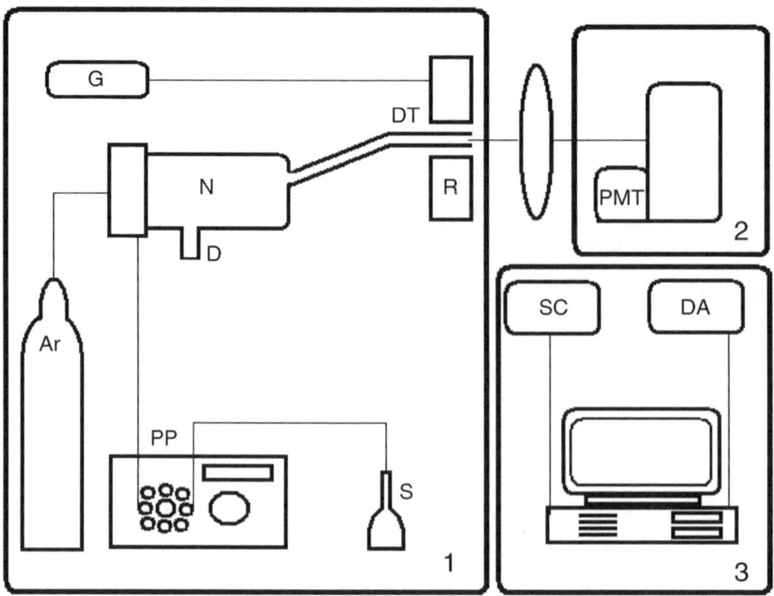

Figure 2.1 Schematic diagram of a MIP-OES system with solution nebulization: 1, excitation source; 2, optical system; 3, computer; G, microwave generator; DT, discharge tube; R, resonator; N, nebulizer; D, drain; PP, peristaltic pump; S, sample; PMT, photomultiplier tube; SC, spectrometer control; DA, data acquisition.

plasma is flow-rate sensitive. In the capacitively coupled microwave plasma (CMP) a magnetron coupled to a central electrode *via* a waveguide terminated by a tuning stub is usually used.

The resonant cavity is a structure for focusing the microwave energy inside the discharge tube as a standing wave. This assures the maximal utilization of energy and protects against emission to the surroundings of microwaves harmful to humans. The impedance matching is usually accomplished by two finely threaded screws located in a cylindrical wall opposite the coupling loop and in the bottom wall parallel to the discharge tube. The cavity is tuned to minimum reflected power. The other plasma containment devices, *i.e.* a surfatron or Okamoto cavity, are based on non-resonant microwave coupling. Some of the microwave structures that have been used in analytical spectrometry are listed in Table 1.2. A more detailed description of microwave system components is given in Chapter 3.

Sample introduction in plasma spectrometry aims at bringing a high fraction of the analytes into the plasma under such conditions that both the stability and the signal generation in the excitation source are optimal. Most emission spectrometers equipped with plasma sources are adapted for the analysis of liquid samples (solutions), but it is also possible to use sampling techniques for gases or solids. The solution introduction system consists of a nebulizer for converting the liquid stream, delivered by means of a peristaltic pump, into an aerosol. The fine droplets of the aerosol are separated in a spray chamber and introduced by means of a nebulizer gas directly into the plasma region or are earlier subjected to additional desolvation in a desolvation system. The other sample introduction techniques are based on the conversion of analytes into volatile species. The hydrogen generation technique and chemical vapour generation should be distinguished here, which are readily used in combination with MWPs. Another method consists in electrothermal vaporization of the solution microsample and the transport of the partially atomized sample in the form of a vapour. For solid samples, electrothermally heated filaments and furnaces are often used, as well as laser or spark ablation techniques. A more detailed description of sample introduction techniques coupled with MWPs is given in Chapters 6–8.

The radiation emitted from the MWP source is usually collected by focusing optics such as a convex lens or a concave mirror. This optic then focuses the image of the plasma onto the entrance slit of the wavelength dispersing device or spectrometer. The devices fall into two main categories: dispersive instruments, which are relatively complex and flexible in use optical systems with medium or high resolving power but relatively low light-gathering efficiency, and non-dispersive instruments, which are generally simple systems of large aperture and low resolving power. Owing to the way of emission measurement, sequential and simultaneous spectrometers are distinguished. Accurate analysis using scanning systems is only possible with a continuously atomizing stable source and the MWP fulfils this requirement. In the case of typical atomic spectrometry it is necessary to couple the MWP source with a spectrometer of at least average resolution, whereas when the MIP-OES system is applied as a detector coupled with a chromatograph or system for flow injection analysis (FIA), then a simpler optical system is sufficient for measuring a transient signal.

Photomultipliers or photodiodes are commonly used as detectors. In modern simultaneous spectrometers, semiconductor silicon detectors including a charge injection device (CID), semiconductor device (SCD) or other types are commonly used, permitting immediate recording of the sample spectrum in the 165–900 nm region while maintaining high resolution.

The electronic part of a MIP-OES system is similar to others typically used in analytical spectrometry and consists of hardware and software for signal processing, data acquisition and instrument control. Further details about monochromators, polychromators and detectors are included in Chapter 5.

It has been accepted in plasma-based analytical spectrometries that the measurement of emitted radiation is carried out radially (perpendicularly) to the torch axis, or axially, *i.e.* along the torch axis. For early MIP-OES systems the radial viewing was preferred, especially in the case of low-pressure plasmas.[1,2] A disadvantage of this viewing mode is the difficulty in maintaining high and stable light permeability through the torch's walls. At present an axial viewing through the open end of the torch dominates, which is not only technically simpler, but also allows a more selective measurement of atomic emission, and thus lowering of the detection limits of the elements determined. With axial viewing, the discharge tube can be made of nontransparent material, *e.g.* aluminum oxide, and its cleanup or replacement is not required frequently. Radial viewing is still used in those MWPs where the plasma discharge is partially located outside the torch, *i.e.* CMPs, microwave plasma torches (MPTs) and Okamoto cavities.[3–5]

For in-tube low-power microwave discharges, axial viewing has two additional advantages: a larger portion of the emitted radiation is collected along the plasma column and the usable spectral range is extended into the vacuum UV, assuming that the MIP system can be interfaced to a vacuum UV spectrometer. On the other hand, axial viewing may present major difficulties for operation at reduced pressures, self-adsorption may limit the linear dynamic range and this viewing mode is prone to spectroscopic matrix effects.

2.2 Microwave Induced Plasma Torches

A discharge tube placed in the centre of the microwave cavity is an essential part of the MIP-OES system. The properties of the material to construct the discharge tube are highly important because the plasma is located inside the tube and the tube material influences the microwave cavity tuning. The material should exhibit high strength, low absorption of microwaves, chemical inertness, low electrical conductivity, good heat resistance and thermal shock resistance.[6] Fused silica is commonly used because it has a very low thermal expansion and is highly resistant against thermal shock; however, it exhibits a relatively low melting point (1350–1500 °C). In contrast, alumina has a high temperature of use (2200 °C), but a relatively high thermal expansion and very low thermal shock resistance. Boron nitride is quite a popular material as it can be used up to 1800 °C; however, pure boron nitride shows the worst resistance against oxidation. It reacts with trace oxygen present in the plasma gas or with water vapour from the sample, forming volatile boron oxide. This leads to degradation of the discharge tube. Silicon nitride is more resistant against corrosion than boron nitride.

The choice of material for the discharge tube is particularly important when designing a microwave resonator for low-power MIP. The resonance frequency depends on the internal diameter of the cavity; however, when a dielectric material, such as quartz or alumina, is inserted into the cavity, the resonance frequency is shifted to a lower value. A somewhat greater frequency shift occurs for alumina due its greater dielectric constant. The magnitude of this shift also depends on the volume of dielectric inserted. A further frequency shift occurs when the plasma is ignited and when an aerosol is introduced. Thus, it is necessary to initially construct the cavity with a slightly smaller diameter to attain higher resonance frequency than desired, so the introduction of dielectric, plasma or aerosol will shift the frequency downward to near 2.45 GHz.

2.2.1 Torch Designs

The plasma torch should allow the formation of temporally and spatially stable plasma. For low-power MIPs a quartz tube is typically used as a discharge tube with an i.d. ranging from 1.0 mm to about 10 mm. This single-flow arrangement exhibits several limitations. The plasma, especially argon plasma, is often unstable. When helium is used as a plasma gas, the tube often devitrifies or may soften and the use of higher power levels significantly reduces the lifetime of the tube. A good design should assure the maintaining of a stable plasma placed in the discharge tube axis, resulting in a low level of background noise and measured signal, greater sensitivity of measurement and smaller erosion of the tube walls, and thus lower emission intensity from tube material components, and also prolongation of the tube lifetime. The most popular MIP torches are schematically drawn in Figure 2.2.

The inner diameter of the discharge tube significantly influences the plasma stability, its tolerance to sample loading as well as the sensitivity and reproducibility of the emission intensity measurement. Interestingly, when the MIP is sustained at reduced pressure, larger diameters of 2–10 mm are required to avoid discharge instability and overheating of the discharge tube. In contrast, at atmospheric pressure the plasma is more stable and more intense for smaller tubes and diameters in the range 0.5–2.0 mm are preferred.[7]

When designing a MIP torch it is also important to know what type of emission signal will be measured: transient or steady state. A discharge tube (DT) of 1 mm i.d. has been used for measuring transient signals from chromatographic eluates. This configuration provides low dead volume and assures that a significant fraction of the analyte passes through the plasma. However,

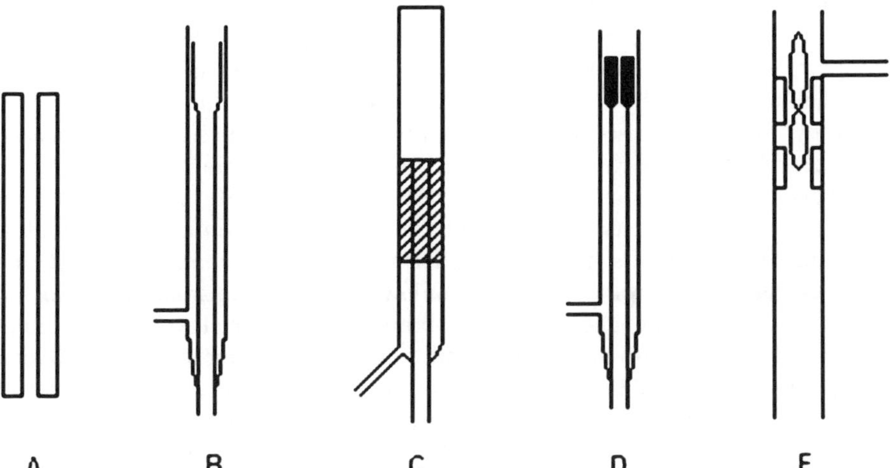

Figure 2.2   MIP plasma torch configurations: A, discharge tube; B, laminar flow torch; C, tangential flow torch; D, tulip-shaped tangential flow torch; E, side-arm torch.

several drawbacks are observed, including the possibility of the plasma extinguishing by even small amounts of organic solvent, the formation of carbon deposits on the inner wall of the plasma tube by organic samples and the loss of plasma energy to the tube wall. Moreover, the signal-to-noise ratio and the relative standard deviation of the signal are rather poor due to the plasma instability, resulting in increased flicker noise.[8] Müller and Cammann[9] used a ceramic discharge tube instead of a quartz one, which allows for injection of several microlitres of solvent.

Discharge tubes with a larger i.d. could also be used. However, plasma centring within the tube is necessary, especially for argon plasma, by using a tangential or laminar flow of plasma gas in a dual-flow torch.[10–15] The tangential flow torch (TFT) consists of a two concentric tubes arranged with a threaded insert to generate the tangential pattern of the outer gas flow. The sample is introduced to the plasma through the central tube with the inner gas flow. The spacing between the outer tube and the inner tube is kept narrow so that the gas introduced between them emerges at high velocity. A fluid dynamics approach has been used for torch design. By selecting the flow rates and the microwave power appropriately, the resulting plasma is stable and maintained far away from the torch walls without any external cooling. The annular flow with relatively high linear velocity acts as an insulating layer between the torch wall and the plasma.

A TFT designed by Bollo-Kamara and Codding[12] allows good stability and sensitivity of the measurements and greatly reduced matrix effects and problems of torch erosion. Unfortunately, flow rates higher than 1 L min^{-1} are required to keep the plasma centred. Consequently, the analyte residence time is short and the TFTs response is one to two orders of magnitude poorer than the DT.

Laminar flow torches (LFTs) have also been used with MIPs. The idea of LFTs is to centre the plasma with the use of two parallel, relatively low (laminar) gas flows of two different linear velocities. Fielden[16] studied a LFT for a helium MIP-OES system which was used as a detector for gas chromatography. Signal-to-noise ratios obtained with this torch design are influenced by a shearing gas flow rate, insertion depth and forward power.

An overall comparison of the three MIP torches, *i.e.* DT, TFT and LFT, in terms of design simplicity, cost, stability and lifetime, minimum operable flow rate, and detection limits (DLs) was undertaken by Bruce *et al.*[11] The LFT and the TFT lasted 2–4 weeks or longer, while the DT was normally good for no longer than a few days. The short-term stability of the three torches was similar (2–5%). The DT had no long-term stability unless operating in the limited low-power range. The LFT long-term precision was typically 6% and the TFT was 8%. Both DTs and LFTs can operate down to several mL min^{-1} of the plasma gas while TFTs require a minimum of 400 mL min^{-1}. The DLs for both DTs and LFTs were nearly the same; the DLs for the TFTs ranged from 3 to 100 times poorer than the other two torches. TFTs used with MIP-GC provided self-centring capabilities, increased emission and improved stability when compared with a DT.

The tangential flow air-cooled torch used by Geiger *et al.*[17] to form an argon plasma in a TE_{101} cavity operates at typical gas flow rates of 0.2–0.4 L min^{-1} for the inner gas flow and 0.7–1.5 L min^{-1} for the outer gas flow. Deutsch and Hieftje[18] reported the development of a TFT for the use with a nitrogen MIP that did not use a threaded insert, but flow rates were as high as 2 L min^{-1}. A low-flow (0.2 L min^{-1}) mini-MIP torch for argon plasma, similar to the TFT by Bollo-Kamara and Codding but made from alumina, was used by Jovicevic *et al.*[19] A TFT to obtain a toroidal-shaped helium MIP for use as an ion source in mass spectrometry has been designed by Satzger and Brueggemeyer;[20] the tangential helium flow was 4 L min^{-1}. Interestingly, the TFT was equipped with a tantalum injector penetrating all of the cavity depth and improving the efficiency of the microwave power transfer to the cavity.

The TFT used for containing the nitrogen plasma in an Okamoto cavity consists of two concentric quartz tubes, an outer tube with 10 mm i.d. and a tulip-shaped inner tube with a large o.d. of 9 mm and a small o.d. of 1 mm.[21] Owing to the small diameter at the end of the injector, the inner gas velocity is such that even the 1 L min^{-1} of nitrogen used for nebulization can punch a hole through the plasma. The outer and inner gas flow rates are typically 10–14 and 0.6–1.4 L min^{-1}, respectively. A dual-flow torch with side-arm sampling was proposed by Zander *et al.*[22] and subsequently used for interfacing MWPs with supercritical fluid chromatography (see Chapter 6). With this method, the MWP is not extinguished during solvent evaporation.

A demountable slotted plasma torch to maintain high-power argon, helium, nitrogen or air MWPs has been designed by Michlewicz *et al.*[10] The outer flow is formed by six gas flow streams which are directed tangentially within the quartz tube. The torch contains three concentric tubes for gas flow and aerosol injection and air can be used for the coolant gas. The inner gas flow varies from 1 to 2.5 L min^{-1} and the outer gas flow is up to 16 L min^{-1}, depending on the kind of the plasma gas used. The main advantages of the demountable torch lie in the ability to modify tube shapes and lower torch replacement costs.

A MIP operating at higher power levels usually causes melting or etching of the plasma torch. Approaches to increase the lifetime of the plasma include air cooling,[10,23] water cooling,[24–27] water aerosol cooling[28] and organic liquid cooling,[29–31] as well as the use of a pulse-operated MIP.[32] The DT used by Story *et al.*[23] to maintain a reduced-pressure helium MIP was modified by the addition of an air cooling jacket. As a result, a 100-fold improvement in the detection limit by MS for phosphorus present in pesticides separated by gas chromatography was obtained. However, oxide formation was still a problem even with this improvement. For atmospheric pressure MIPs the air cooling is generally not effective. When helium is used as the plasma gas the cooling of the torch becomes an important factor. Subsequently, Story and Caruso[25] adapted a water-cooled torch to provide a 1.5 mm thick water layer. With this cooled torch, a plasma could be maintained at power levels up to 450 W in a 2 mm i.d. tube and sub-nanogram detection limits for some non-metals were obtained with MS detection.

Water cooling, although very efficient, has certain disadvantages. Firstly, water absorbs microwaves and not only distilled water should be used but also it must appear in a thin layer to reduce attenuation in the microwave cavity. Secondly, water and other liquids are simply too efficient cooling agents for this specific application. Such over-efficient cooling leads to an over-cooled torch and this results in excessive energy loss in MIP systems. A temperature of, for example, 400–700 °C would still be acceptable to prevent quartz tubes from melting and at the same time this "hot regime" can save an appreciable amount of plasma energy. However, the water aerosol or vapour is characterized by low absorption of the microwave energy and could be easily used as a cooling agent.[28]

Various dielectric liquids were reported as highly efficient cooling media, including dimethylpolysiloxane, transmission fluids, hydraulic fluids and silicon oils. However, the formation of carbonaceous deposits on the hot wall of the cooling jacket was observed, leading to worsening of the cooling efficiency and plasma operation.

2.2.2 The Importance of Vertical Positioning of a Microwave Induced Plasma Torch

In most low-power MIP systems a horizontal position of the discharge tube has been preferred, owing to the common use of axial viewing of the plasma and horizontal positioning of the monochromator. However, distortion in the plasma symmetry can be expected at this positioning of the tube. Owing to the high temperature of the exhaust gases, characteristic bending of the plasma tail occurs. This distortion causes instability of the measured signal, connected with the changeability of the self-absorption effect. The second unfavourable phenomenon occurring with the horizontal position of the discharge tube is connected with the transport of the sample in the form of a wet aerosol through the tube. At a relatively low temperature the aerosol covering the distance of several centimetres through a tube of 1–2 mm diameter may undergo gravitational settling, which will lead to a non-uniform spatial distribution of the density of the aerosol approaching the plasma region. The phenomena described were reported for inductively coupled plasmas (ICPs).[33] It can be expected that they will be more essential for the MIP system due to the use of lower plasma gas flows. However, this induces the need for using a more complicated external optical system. The concept of vertically positioned axially viewed MIPs has been examined by Jankowski *et al.*[28] for the system consisting of the TE_{101} cavity and a water aerosol cooled plasma torch. Also, MPT and torch injection axial (TIA) approaches employ the vertical position of the torch; however, radial viewing is mostly used.[3,4]

2.3 Pros and Cons of the Microwave Induced Plasma Technique

Compared with other plasma sources, MWPs offer certain attractive or unique features, but also suffer from some limitations. As already mentioned, MWPs can be sustained with various working gases, at either reduced pressure or atmospheric pressure, over a wide range of gas flow rates between 0.005 and 16 L min^{-1}. MWPs posses high excitation efficiency, especially when helium is used as the plasma gas owing to its high excitation potential. The efficient excitation of halogens and other hard-to-excite elements is possible, which are not readily accessible to ICP detection with adequate sensitivity. Thus, MWPs offer detectability to almost all the elements in the Periodic Table. The non-thermal nature of the MIP has advantages and disadvantages for atomic spectroscopy. The MIP is very useful for the determination of non-metallic elements that are difficult to excite in more thermal sources. On the other hand, the gas temperature of the plasma is quite low, which gives rise to difficulties with sample vaporization and atomization.

A great advantage of the MWP technique is the compatibility of the gas flow used with the flows applied in many sample introduction techniques, *e.g.* hydride generation, gas and liquid chromatography, electrothermal vaporization. The gas flows applied in ICPs, direct current plasmas (DCPs) and flames are many times higher than that in MWPs, leading to technical problems with interfacing a sampling system with a plasma source. Also the combination of MWPs with MS is technically simpler than that in the case of ICPs. These features of the MWP technique also contribute to its versatility. It can be applied in emission, absorption, fluorescence and mass spectrometries.

Most MWPs can work at power levels less than 200 W and total gas flow rates below 1 L min^{-1}. Therefore, both the initial and running cost of a MWP system are lower than for other conventional plasmas. Such operating conditions also produce a plasma with a favourably low continuous background emission, much weaker than that of either the DCP or ICP methods. Moreover, at an operating frequency of 2.45 GHz there is no need for bulky shielding to protect electronic recording equipment, as is usually the case with ICPs. In addition, the need for cooling of the microwave setup is minimal, owing to the low power level used.

It should be noted that the above-mentioned advantages of MWP sources are weighty to a different extent with respect to an individual MWP approach. Bearing in mind the commercial availability of MWP systems, a low-power, low-flow helium MIP coupled with gas chromatography, MPT-OES or a nitrogen high-power MIP-MS should be pointed out as the most successful. The first one receives a benefit from the use of helium as a plasma gas, technical simplicity and low initial and running costs. The MPT can be operated with various gases in a wide range of power (0.03–0.6 kW) and gas flow rates (0.2–3.0 L min^{-1}) that assure both the flexibility and economic acceptance of the source for atomic spectrometry. Moreover, it is easy to tune and operate and provides good plasma stability and high tolerance to sample loading. Taking advantage of the central channel, the MPT also provides a relatively wide linear dynamic range and reproducibility. Nitrogen MIP-MS takes advantage of the kind of the plasma gas used that results in good selectivity and sensitivity of the MS detection toward a group of elements of high environmental importance.

In the past decade, significant progress has been observed in the field of miniature plasma devices for analytical spectrometry.[34–37] This is not surprising because MWPs are ready to be miniaturized, taking into account that they are fundamentally low-power, low-flow, low-volume sources. Microwave micro-plasma sources are integrated on a chip, preferably by using microstrip technology.[38–44] A strip line enables much smaller sizes and thus lower power levels and gas flow rates, while still enhancing higher accuracy, stability and reproducibility. The microstrip MIP is a very powerful device for elemental detection, provided the analyte species can be presented as a dry vapour to the MWP source.[45–49]

This technology has the potential for further integration with complementary devices onto a single chip, including on-chip spectrometers with optical fibre coupling and various sample introduction techniques. Here, microwave-powered plasmas are investigated with the special aim to realize them with low-priced generators and with minimal operation costs. This technology allows the integration of all necessary steps for sample preparation on a chip. It is the so-called lab-on-the-chip approach. For example, samples can be digested, transported, evaporated, separated and measured on the chip. The emitted radiation can be taken up by an optical fibre positioned in the plasma-gas channel, thus enabling axial observation and coupling to a miniaturized spectrometer.[35,37,50]

Limitations of MWPs influencing the practicability and commercial availability of these plasma sources are mainly the instability of the discharge, its low tolerance to sample loading and relatively serious matrix effects and a short lifetime of the torch. These limitations are connected with the low power levels used and low plasma volumes, as well as no central channel in the filament or flame-like plasma for the introduction of a sample. However, at present these limitations seem to be satisfactorily solved in the most successful MWP approaches.[3–5,28,51]

The specific limitations of individual MWPs, such as a relative small useful analytical region for MIPs and CMPs, electrode consumption and contamination for CMPs, a short lifetime of the torch for MIPs and the formation of carbon deposits when organic compounds are being determined by the GC-MIP coupled technique, should also be mentioned. Additionally, since the plasma is not restricted to the discharge tube axis, its position along the tube diameter may change, requiring a complex optical system. Nevertheless, the presence of complete, turnkey MIP-based systems in the marketplace is still limited. However, all of these limitations have been significantly overcome in recent years.

The production of stable plasmas at microwave frequencies requires mutual impedance matching of all the MWP system components. The introduction of a substantial amount of sample often causes changes in the plasma impedance and thus in plasma distortion or even extinguishment. The plasma instability is also influenced by inefficiency and irreproducibility of the power transfer from the generator to the microwave cavity. However, in modern MWP systems these limitations have been significantly improved owing to the application of highly symmetrical and strong coupling between the plasma load and the generator, especially with the use of coaxial lines and waveguides.[4,21,28,51,52] As a result, a self-tuning operation of the plasma load or operation at virtually zero watts of reflected power is possible.

When it is unstable or maintained in a discharge tube of small i.d., a helium or argon MIP etches the wall of the tube, thus shortening the tube lifetime. As mentioned before, the technical improvements in the design of MWP approaches and plasma torches as well as the use of plasma cooling lead to obtaining stable discharges with negligible local overheating of the tube. Etching may also cause analyte memory effects due to the occlusion of salt particles at the etched surface or by interaction of certain analytes with the components of the discharge tube material.

It was found that, at high temperature, phosphorus reacts with a quartz tube to form phosphorus oxides on the tube walls, which thus results in decreased sensitivity. A water-cooled torch was developed to reduce the phosphorus interaction with the tube walls.[23,25] The introduction of samples with a matrix containing alkaline metals, halogens or sulfate severely limits the lifetime of the quartz tube. When fluorine is present, tube degradation is even more accelerated. Apparently, the fluorine is partially retained in the discharge tube because of interactions with the wall.[53]

The deposition of carbonaceous materials on the inner wall of the discharge tube has been usually observed when organic samples are introduced to the MIP, leading to memory effects, nonlinear responses and unstable plasmas. The addition of a scavenger gas such as oxygen, hydrogen or nitrogen[26,54–57] is recommended in this case to minimize these effects and prolong the lifetime of the tube. Occasionally, a more thorough clean-up of the tube is required.

It is often stressed that MWPs do not accommodate large amounts of liquid samples and sometimes are even extinguished. Therefore, microwave discharges, especially at low power, are less suitable to take up any analyte contained in wet aerosol droplets or larger solid-phase particles, which both require a high amount of heat to be evaporated completely. Accordingly, all techniques where dry analyte vapours are generated, electrothermal evaporation and gas chromatography will be the favoured techniques for sample introduction. Indeed, for early MIP designs it was found that it was difficult to maintain a plasma with them when a sample is introduced at rates of about 1 mg min^{-1}.[58,59] However, more recently developed MIPs can withstand much higher aerosol loadings, ranging from 40 to 120 mg min^{-1},[51,60] while the commercial ICP-OES instruments usually operate at rates from 30 to 50 mg min^{-1} of water aerosol. MWP tolerance to sample loading will be discussed in more detail in Chapters 6–8.

An important advantage of the MWP technique is that the system design is relatively simple, especially when the plasma is to be operated at low power and at atmospheric pressure. Consequently, the overall system cost is relatively low, around €10 000. Moreover, miniaturized MWPs can operate at power levels below 40 W and the generators for these systems are inexpensive. Low power and small gas flows are decisive to the running costs of plasma excitation sources. A second economic advantage of the MWP system results from the low gas-flow requirements, which often are less than $1\,L\,min^{-1}$ of total gas consumption. This results in savings of 90% or more of the running costs relative to commercial ICP systems that require 15–$22\,L\,min^{-1}$ of argon. For ICPs the estimated gas cost approaches \$10 000 per year. In conclusion, the running costs of MWPs are about 10 times lower than that in other plasma excitation sources.

References

1. H. Kawaguchi, M. Hasegawa and A. Mizuike, *Spectrochim. Acta, Part B*, 1972, **27**, 205–210.
2. J. P. Matousek, B. J. Orr and M. Selby, *Appl. Spectrosc.*, 1984, **38**, 231–239.
3. Q. Jin, C. Zhu, M. W. Borer and G. M. Hieftje, *Spectrochim. Acta, Part B*, 1991, **46**, 417–430.
4. A. Rodero, M. C. Quintero, A. Sola and A. Gamero, *Spectrochim. Acta, Part B*, 1996, **51**, 467–479.
5. M. Ohata and N. Furuta, *J. Anal. At. Spectrom.*, 1997, **12**, 341–347.
6. P. S. C. van der Plas and L. De Galan, *Spectrochim. Acta, Part B*, 1987, **42**, 1205–1216.
7. A. J. McCormack, S. C. Tong and W. D. Cooke, *Anal. Chem.*, 1965, **37**, 1470–1476.
8. S. A. Estes, P. C. Uden and R. M. Barnes, *Anal. Chem.*, 1981, **53**, 1829–1837.
9. H. Müller and K. Cammann, *J. Anal. At. Spectrom.*, 1988, **3**, 907–913.
10. K. G. Michlewicz, J. J. Urh and J. W. Carnahan, *Spectrochim. Acta, Part B*, 1985, **40**, 493–499.
11. M. L. Bruce, J. M. Workman, J. A. Caruso and D. J. Lahti, *Appl. Spectrosc.*, 1985, **39**, 935–942.
12. A. Bollo-Kamara and E. G. Codding, *Spectrochim. Acta, Part B*, 1981, **36**, 973–982.
13. G. S. Sobering, T. D. Bailey and T. C. Farrar, *Appl. Spectrosc.*, 1988, **42**, 1023–1025.
14. S. R. Goode, B. Chambers and N. P. Buddin, *Spectrochim. Acta, Part B*, 1985, **40**, 329–333.
15. K. A. McCleary, G. R. Ducatte, D. H. Renfro and G. L. Long, *Appl. Spectrosc.*, 1993, **47**, 994–998.
16. P. R. Fielden, M. Jiang and R. D. Snook, *Appl. Spectrosc.*, 1989, **43**, 1444–1449.
17. A. Geiger, S. Kirschner, B. Ramacher and U. Telgheder, *J. Anal. At. Spectrom.*, 1997, **12**, 1087–1090.
18. R. D. Deutsch and G. M. Hieftje, *Appl. Spectrosc.*, 1985, **39**, 214–222.
19. S. Jovićević, M. Ivković, Z. Pavlović and N. Konjević, *Spectrochim. Acta, Part B*, 2000, **55**, 1879–1893.

20. R. D. Satzger and T. W. Brueggermeyer, *Microchim. Acta*, 1989, **99**, 239–246.
21. Y. Okamoto, *J. Anal. At. Spectrom.*, 1994, **9**, 745–749.
22. A. T. Zander, R. K. Williams and G. M. Hieftje, *Anal. Chem.*, 1977, **49**, 2372–2374.
23. W. C. Story, L. K. Olson, W. L. Shen, J. T. Creed and J. A. Caruso, *J. Anal. At. Spectrom.*, 1990, **5**, 467–470.
24. R. M. A. Bolainez, M. P. Dziewatkoski and C. B. Boss, *Anal. Chem.*, 1992, **64**, 541–544.
25. W. C. Story and J. A. Caruso, *J. Anal. At. Spectrom.*, 1993, **8**, 571–575.
26. B. D. Quimby and J. J. Sullivan, *Anal. Chem.*, 1990, **62**, 1027–1034.
27. R. L. Sing, C. Lauzon, K. C. Tran and J. Hubert, *Appl. Spectrosc.*, 1992, **46**, 430–435.
28. K. Jankowski, R. Parosa, A. Ramsza and E. Reszke, *Spectrochim. Acta, Part B*, 1999, **54**, 515–525.
29. H. Matusiewicz and R. E. Sturgeon, *Spectrochim. Acta, Part B*, 1993, **48**, 515–519.
30. J. Mierzwa, R. Brandt, J. A. C. Broekaert, P. Tschöpel and G. Tölg, *Spectrochim. Acta, Part B*, 1996, **51**, 117–126.
31. L. A. Schlie, *Rev. Sci. Instrum.*, 1991, **62**, 542–543.
32. M. M. Mohamed, T. Uchida and S. Minami, *Appl. Spectrosc.*, 1989, **43**, 129–134.
33. M. T. C. de Loos-Vollebregt, J. J. Tiggelman and L. De Galan, *Appl. Spectrosc.*, 1989, **43**, 773–778.
34. J. Hopwood and F. Iza, *J. Anal. At. Spectrom.*, 2004, **19**, 1145–1150.
35. M. Miclea, K. Kunze, J. Franzke and K. Niemax, *Spectrochim. Acta, Part B*, 2002, **57**, 1585–1592.
36. R. Stonies, S. Schermer, E. Voges and J. A. C. Broekaert, *Plasma Sources Sci. Technol.*, 2004, **13**, 604–611.
37. G. Feng, Y. Huan, Y. Cao, S. Wang, X. Wang, J. Jiang, A. Yu, Q. Jin and H. Yu, *Microchem. J.*, 2004, **76**, 17–22.
38. J. Deng, in *Proceedings of the 2001 IEEE International Frequency Control Symposium and PDA Exhibition*, IEEE, New York, 2001, pp. 85–88.
39. A. M. Bilgic, U. Engel, E. Voges, M. Kückelheim and J. A. C. Broekaert, *Plasma Sources Sci. Technol.*, 2000, **9**, 1–4.
40. A. M. Bilgic, E. Voges, U. Engel and J. A. C. Broekaert, *J. Anal. At. Spectrom.*, 2000, **15**, 579–580.
41. J. A. C. Broekaert, V. Siemens and N. H. Bings, *IEEE Trans. Plasma Sci.*, 2005, **33**, 560–561.
42. P. Siebert, G. Petzold, A. Hellenbart and J. Müller, *Appl. Phys. A*, 1998, **67**, 155–160.
43. J. Gregorio, O. Leroy, P. Leprince, L. L. Alves and C. Boisse-Laporte, *IEEE Trans. Plasma Sci.*, 2009, **37**, 797–808.
44. F. Iza and J. A. Hopwood, *IEEE Trans. Plasma Sci.*, 2003, **31**, 782–787.
45. U. Engel, A. M. Bilgic, O. Haase, E. Voges and J. A. C. Broekaert, *Anal. Chem.*, 2000, **72**, 193–197.

46. S. Schermer, N. H. Bings, A. M. Bilgic, R. Stonies, E. Voges and J. A. C. Broekaert, *Spectrochim. Acta, Part B*, 2003, **58**, 1585–1596.
47. P. Pohl, I. J. Zapata, N. H. Bings, E. Voges and J. A. C. Broekaert, *Spectrochim. Acta, Part B*, 2007, **62**, 444–453.
48. I. J. Zapata, P. Pohl, N. H. Bings and J. A. C. Broekaert, *Anal. Bioanal. Chem.*, 2007, **388**, 1615–1623.
49. P. Pohl, I. J. Zapata and N. H. Bings, *Anal. Chim. Acta*, 2008, **606**, 9–18.
50. V. Karanassios, K. Johnson and A. T. Smith, *Anal. Bioanal. Chem.*, 2007, **388**, 1595–1604.
51. K. Jankowski, A. Jackowska, A. P. Ramsza and E. Reszke, *J. Anal. At. Spectrom.*, 2008, **23**, 1234–1238.
52. W. Yang, H. Zhang, A. Yu and Q. Jin, *Microchem. J.*, 2000, **66**, 147–170.
53. F. A. Huf and G. W. Jansen, *Spectrochim. Acta, Part B*, 1983, **38**, 1061–1064.
54. D. Kollotzek, D. Oechsle, G. Kaiser, P. Tschöpel and G. Tölg, *Fresenius' Z. Anal. Chem.*, 1984, **318**, 485–489.
55. L. Ebdon, S. Hill and R. W. Ward, *Analyst*, 1986, **111**, 1113–1138.
56. P. C. Uden, Y. Yoo, T. Wang and Z. Cheng, *J. Chromatogr.*, 1989, **468**, 319–328.
57. K. B. Olsen, D. S. Sklarew and J. C. Evans, *Spectrochim. Acta, Part B*, 1985, **40**, 357–365.
58. F. E. Lichte and R. K. Skogerboe, *Anal. Chem.*, 1972, **44**, 1321–1323.
59. A. T. Zander and G. M. Hieftje, *Anal. Chem.*, 1978, **50**, 1257–1260.
60. K. Jankowski, A. P. Ramsza, E. Reszke and M. Strzelec, *J. Anal. At. Spectrom.*, 2010, **25**, 44–47.

Principles of Operation and Construction of Microwave Plasma Cavities

3.1 E- and H-type Discharges at Different Gas Pressures and Frequencies

The terminology of E- and H-type discharges was introduced by Babat in 1942.[1] A common feature of H-type discharges is the induction of a current around the circumference of the plasma column, with a characteristic minimum of the induced E field at the plasma axis. Although, since then, an inductive plasma is commonly used in technology, to the analytical spectroscopists it is better known as an inductively coupled plasma (ICP). The ICP is usually generated using RF power, with frequencies in the range of 3–150 MHz in a torch placed at the axis of a few turns of inductor.[2] There have also been trials working at higher frequencies, *e.g.* 413 MHz, which is considered to lie in the microwave (MW) band. At extremely high powers of industrial applications the H-type supply drops down to the lowest frequency bands, even as low as the frequency of the power mains (50 or 60 Hz). It is worth stressing that on moving to very high frequencies the beginning of the transition between the RF and MW frequencies is a question of agreement. This is so because microwaves may be considered to be a technique that makes use of the circuits among which at least one element has a dimension which is intentionally chosen to be comparable with the working wavelength. By this definition, one has to agree that miniaturization of modern circuits may extend the RF range towards still higher frequencies. This fact has already been noticed in the practice of MW integrated circuits.[3] Several hundreds of watts are normally required to run

RSC Analytical Spectroscopy Monographs No. 12
Microwave Induced Plasma Analytical Spectrometry
By Krzysztof J. Jankowski and Edward Reszke
© The Royal Society of Chemistry 2011
Published by the Royal Society of Chemistry, www.rsc.org

argon plasmas and a few kilowatts would be necessary to maintain plasmas in molecular gases. A comprehensive review of different aspects of ICP and microwave induced plasma (MIP) spectrometry has been given.[4]

The ICP-like conditions with the magnetic field patterns similar to those existing inside the lumped inductor can also be realized at MW frequencies by excitation of a TE_{011}-mode oscillation inside the cylindrical resonator. However, at atmospheric pressure the power required to maintain the microwave discharge is found to be relatively high.[5] It is shown later in this chapter that an H-type discharge can be obtained in the MW range with less power using multi-helix or multi-loop cavities.[6] Similar inductive-type excitation can be achieved using the so-called loop-gap resonators, as well as using single rectangular waveguides or a set of rectangular waveguides. The microwave-driven H-type plasma has already been proven to deliver analytical results of the same or even better quality as the best ICPs. Hammer[7] used a rectangular waveguide by placing the plasma torch tube along the lines of the standing magnetic field across the narrow waveguide walls and demonstrated stable H-type plasmas and very promising analytical results.[8]

The E-type discharges are those excited by the *E* component of the field. The simplest construction of an E-type applicator used in the RF range consists of two metal electrodes connected to an RF power generator that is capable of delivering a high voltage.[2] The induced high electric field strength between the electrodes or between a single electrode and a shield can cause electrical breakdown, which results in the formation of the plasma, sometimes called an RF arc. In the shielded environment of a microwave cavity the single electrode-initiated plasma is called a capacitive microwave plasma (CMP) in which at least one electrode, usually the inner conductor in the coaxial line, has immediate contact with the plasma.[9] The E-type discharge at high frequencies can be electrodeless, *i.e.* it can be created through the dielectric barrier, eliminating the plasma-to-electrode contact. An example of an electrodeless E-type RF discharge would be a capacitively coupled plasma (CCP) with the gas flowing inside the dielectric tube equipped with a pair of metal rings clamped around it and connected to the RF power source. Knapp *et al.*[10] described such small RF-excited plasmas energized from a RF power source and applied as an excitation source for emission spectroscopy.

The E-type RF discharges usually generate plasmas with substantially lower electron density than those delivered by H-type discharges.[2] The higher the RF frequency, the greater the electron density that can be obtained in the discharge. At the same time, the higher the frequency, the more important is the necessity to prevent electromagnetic radiation leaking from the circuit. Therefore a microwave circuit for generating the plasma cannot be as simple as those two rings connected to the RF power generator, but still it is not too difficult to make a MW cavity with a double ring coupling similar to the CCP.[10,11] The microwave plasma (MWP) version of that two-ring cavity has been demonstrated by Piotrowski *et al.*[12] One may notice good shielding of that cavity, which is a common requirement for all high-power MW circuits.

3.1.1 Choice of Operating Frequency

The obvious difference between the RF and MW techniques is the wavelength, which for microwaves is a few times shorter than for RF.

The choice of operating frequency must conform to the law. Normally, RF heating is performed at 13.56 MHz, while at MW frequencies there are only two common choices: 2.45 GHz or 915 MHz. The latter is usually applied at higher power levels, exceeding 6 kW, or when adopting solid-state power oscillators and amplifiers.

It is worth mentioning that the less common selection of high-power microwave sources goes up to millimetre wavelengths, with power levels at tens of kilowatts used for materials processing experiments.[13,14] Nevertheless, for commercially acceptable solutions the electromagnetic compatibility (EMC) regulations may be a cost-elevating barrier.

A typical working frequency for an RF-operated ICP is 13.56 or 27.12 MHz, which corresponds to wavelengths of 22 and 13 metres, respectively, while a typical MWP works at 0.915 GHz or 2.45 G Hz, which corresponds to only 31 or 12 cm. It is obvious that the working frequency of MWPs is usually placed at 2.45 GHz, which is the same frequency as that used in domestic microwave ovens. This makes 2.45 GHz not only the most popular frequency but also offers the cheapest hardware, such as magnetrons and power supplies. MW frequencies such as 5.8 GHz and even higher have also become available for scientific experiments.[15]

When very good shielding is used the requirements of the EMC directive can also be fulfilled even in cases when a non-standard frequency is used. Using lower frequencies may be beneficial because at such frequencies the availability of cheap solid-state devices is more reliable. For instance, a power oscillator can be adapted from a 144 MHz band transmitter,[16] while power amplifiers of the L or S bands can be adopted from cellular telephony broadcasting systems.[17] Concerning solid-state power oscillators, one should first consider the cheap realization of such sources at 915 MHz. The station amplifiers of cellular telephony can be used as a building block of free-running feedback oscillators coupled with a MW plasma in a strip-line ring.[18] One can expect adaptation of existing technical solutions along with very tight sealing against RF interference. When using standard MW frequencies it is generally easier to cope with disturbances than in the lower RF bands. Moreover, the availability of cheap microwave power sources used in domestic microwave ovens can be an important price cutting factor and still a good reason to keep at a standard MW frequency of 2.45 GHz.

3.2 Some Basic Knowledge about Microwave Transmission Lines and Resonant Cavities

The techniques of transmission of the RF and microwave energies are, in principle, similar. Because of the short wavelength the MWP designs use microwave cavities rather than lumped inductors and capacitors. It should be mentioned that the use of modern computer-aided designs and precise machining cause the broadening of the range of frequencies for lumped or semi-lumped circuits. According to Pozar,[3] every 1/8th of the wavelength of the microwave-transmitting line can be modelled by a single lumped element, whether inductor or capacitor. Transmission lines are therefore important passive microwave components. Microwave cavities and therefore the MIPs and CMPs consist of different sections of transmission lines, such as coaxial lines, strip-lines, waveguides, *etc.*[3] Coaxial lines have an inner conductor and a shield (Figure 3.1). An example of coaxial line is the cable used in CB radio or cable TV.

In old TV systems the two-wire symmetrical feeders with two parallel conductors embedded in plastic were used to fit better to the characteristic impedance of symmetrical Yagi antennae (Figure 3.2).

Such parallel conductors embedded in a shield have already been used to construct a MWP cavity generating microplasmas at atmospheric pressure. Another microwave-feeder line has the form of a strip-line which in its non-symmetrical appearance is often applied in MW printed circuits.[3]

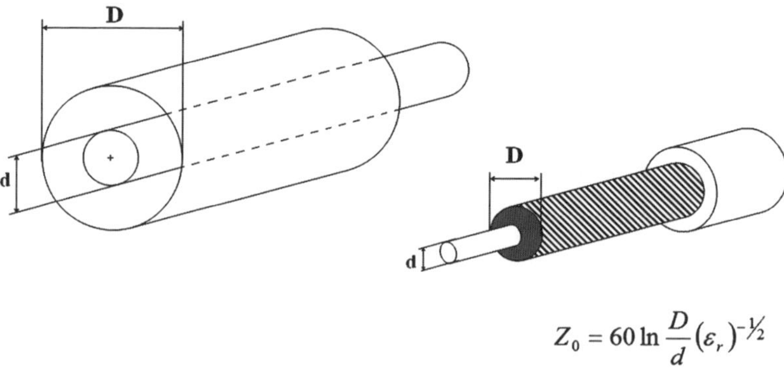

$$Z_0 = 60 \ln \frac{D}{d} (\varepsilon_r)^{-\frac{1}{2}}$$

Figure 3.1 Coaxial lines; Z_0 is the characteristic (wave) impedance of the line, ε_r is the relative dielectric constant.

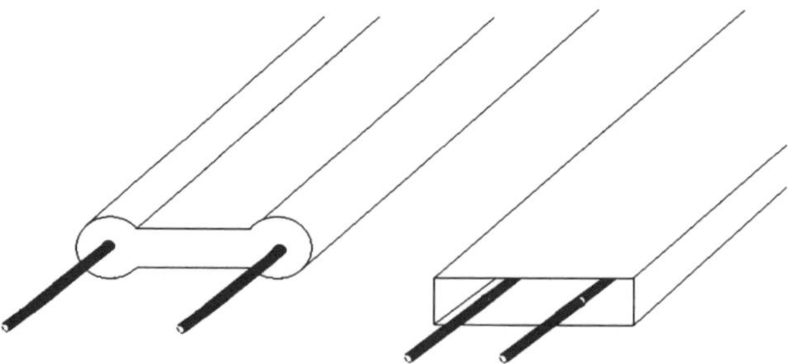

Figure 3.2 Symmetrical feeder lines.

The less popular but more power enabling shielded symmetrical strip-line is built using a bus-bar shaped central conductor placed within the hollow shield with a rectangular cross-section (Figure 3.3).

All coaxial, symmetrical and strip-line feeders belong to the group of so-called transverse electromagnetic (TEM) lines in which both E and H field vectors have only components transverse to the direction of the moving wavefront. TEM lines work in a conductive manner, *i.e.* both the inner conductor and the shield or both symmetrical conductor bars must conduct the RF currents. These currents flow only in the thin skin layer of metal, and the skin depth is thinner the higher the frequency. Therefore, generally the TEM lines exhibit high attenuation for microwaves and should not be used to send microwave power at long distances, but use of TEM lines is still a good technique to make small microwave plasma cavities where losses are not the most important issue. The use of waveguides is more reasonable from an attenuation point of view and they can transfer much more power than TEM lines. Although really high power is not always necessary, the use of waveguides in the construction of MWP devices is common (Figure 3.4). However, the smallest possible waveguide sizes must fit at least half of the wavelengths, while TEM lines enable the designs to be much smaller. Generally, the smallest designs will use TEM lines. This may meet the new trend in analytical applications towards using as small a power and gas consumption as possible, which in turn must be connected with miniaturization of plasma circuits.

Inside the hollow pipes of waveguides, different modes of oscillations can be propagated depending on the relation between the wavelength and waveguide dimensions. These modes are described as TE_{mn} or TM_{mn}, respectively, with a transverse E and a missing H component and a transverse H and a missing E component in the direction of propagation. The indexes denote eigenvalues that define the wave variation in a given direction. For a rectangular waveguide this would be a number of half-ways which can fit in an x–y cross section and thus the fundamental rectangular mode is TE_{10}. For a circular waveguide, mn denotes, correspondingly, the order of the Bessel function describing radial

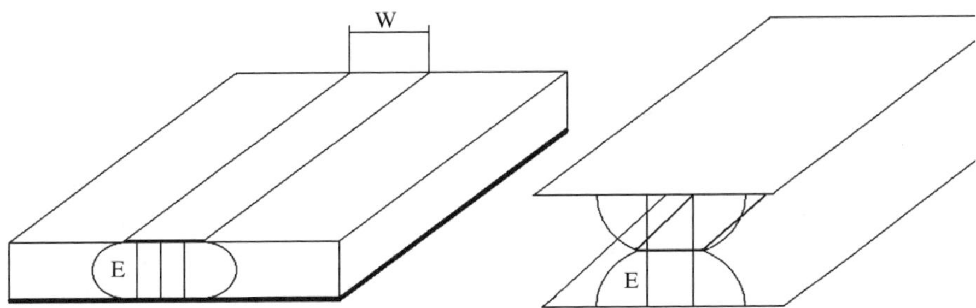

Figure 3.3 Asymmetrical (*left*) and symmetrical (*right*) strip-lines: W is the width of the strip, E shows electric field lines.

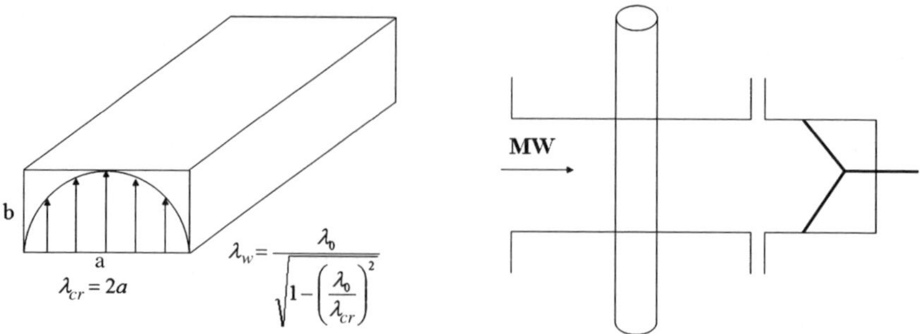

Figure 3.4 Rectangular waveguides with a fundamental mode (*left*) and the coupling of a discharge tube in a rectangular waveguide with a regulated plunger short (*right*): λ_o is the wavelength in free space, λ_w is the wavelength inside the waveguide, λ_{cr} is the critical wavelength.

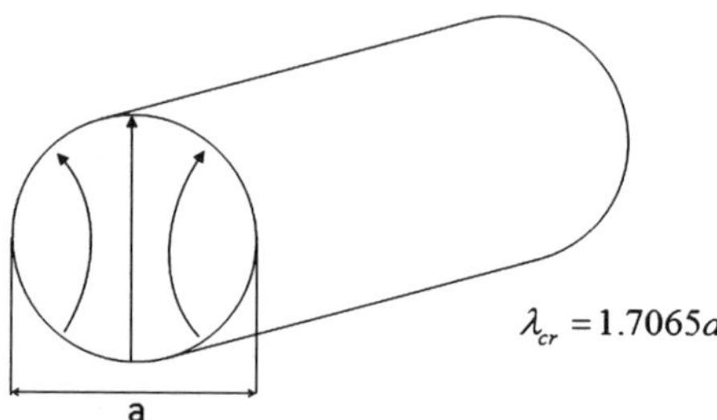

Figure 3.5 Circular TE_{11} mode of oscillation; above λ_{cr} the waveguide is cut off.

distribution of the field and the number of its root. The fundamental circular mode is TE_{11} (Figure 3.5).

If a cavity resonator is formed from a section of waveguide, the field forms a standing wave along the line and there will be another p index added to those indexes along the waveguide section. For example, a rectangular resonator with zero variation along the x axis, one-half variation along the y axis and p variations along the z direction would be described as TE_{103}. It should be mentioned that some authors introduce the notation "*nm*" instead of "*mn*". It is also worth knowing that the TE and TM modes are often named respectively as the H or E modes, according to which field vector has a transverse component.

The interesting mode in rectangular waveguides is practically only the fundamental mode TE_{10} or H_{10}. In circular waveguides, except the fundamental TE_{11} (H_{11}) mode, there are two more interesting modes, both having radial symmetry. One of those modes is TM_{01} (E_{01}) with the E field component directed along and having a single maximum on the axis, and the second one, TE_{01} (H_{01}), with the H field component directed along and having a single maximum on the axis. To describe the modes in resonators a third index is

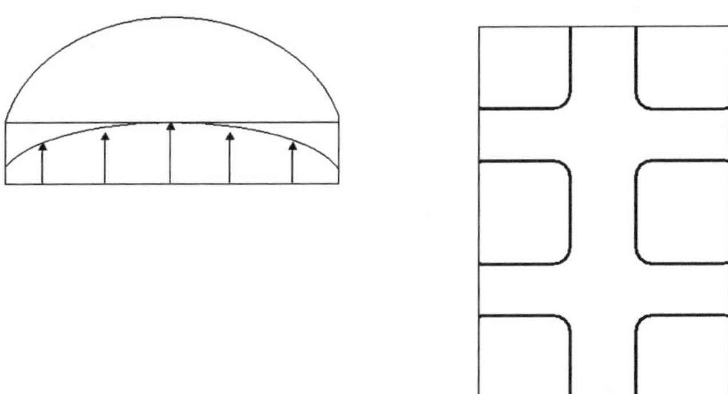

Figure 3.6 E field lines for circular TM_{010} (Beenakker) and TM_{012} resonator modes.

added which describes the number of half-ways along the z axis. The cavity introduced by Beenakker[19] was of the TM_{010} (E_{010}) type, while Asmussen *et al.*[20] and Kapica[21] used the TM_{012} (E_{012}) mode (Figure 3.6).

Note that the $p = 0$ index is allowed for the shortest slice of TE_{01} waveguide-based resonator, while for the resonator with a fundamental-mode rectangular waveguide the shortest length of waveguide section must fit to at least half of the propagated wavelength, which means $p_{min} = 1$. The field distribution in a circular TM_{010} cavity has both E and H field components transverse, so this cavity is identical to a radial TEM-mode cavity. In fact, this cavity can also be called a radial TE_{010} cavity and both names are correct, and a third name is also correct: the Beenakker cavity.

Concerning rectangular and square shaped cavities, it is interesting to know that a standing wave in a regular TE_{101} mode rectangular waveguide resonator can deliver the electric field pattern very similar to that of the Beenakker cavity. For instance, the rectangular design proposed by Plazmatronika[22] was found to deliver similar analytical results. This fact has been exploited in the cavities of Feuerbacher,[23] Matusiewicz,[24] Broekaert *et al.*[25] and Jankowski *et al.*[26] Rectangular cavities of this type can be tuned by moving a simple plunger and an antenna coupler, but in actual designs[24–26] the more efficient iris-type tuner and a longitudinal (side wall) plunger do the matching even better.

At the short end of the rectangular waveguide and at every half of the wavelength away from the short end there exist standing magnetic field lines crossing the waveguide parallel to the wider walls. That magnetic field across the rectangular waveguide has been used by Hammer[7,8] to excite the very stable H-type plasma and for further documenting the low detection limits obtained with this new cavity.

By choosing the width of the waveguide one can build a cavity having exactly a square shape. Its height can be small, which means that the cavity can be made very slim, and its square shape causes the field at the centre to be almost identical to the circular one in the Beenakker resonator. The latter TM_{010} circular mode cavity can also be very flat. This can be so because here the third

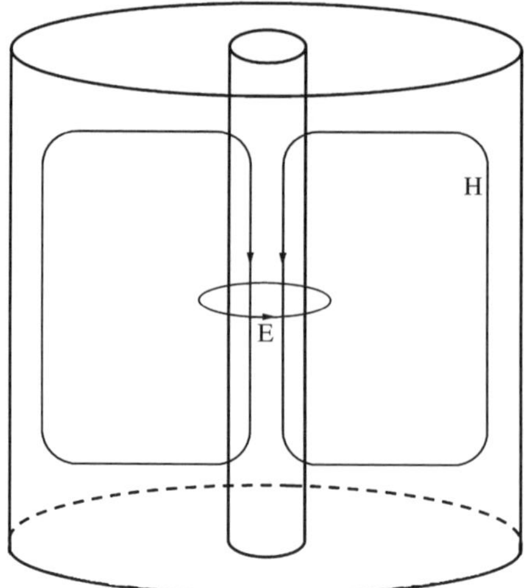

Figure 3.7   The TE_{011} mode cylindrical resonator.

(p) index is equal to zero. With $p=0$ the E field is distributed in a radial direction and has no variations along the z axis.

The second interesting circular waveguide-based resonator mode, named TE_{011}, may deliver the H field component, which along the axis is very similar to that of an ICP inductor coil (Figure 3.7). Unfortunately, in this case the index $p=0$ is not allowed and the cavity must have the axial length somewhat longer than free space halfway, which would be 7–8 cm at 2.45 GHz frequency. Trials of argon plasma generation in this cavity at reduced pressures revealed the formation of separate moving rings of plasma which did not form a con-solidated discharge.

Excitation of an atmospheric pressure oxygen plasma using the TE_{011} mode was quite successful at power levels in excess of 1500 W at 2.45 GHz.[27] However, the plasma ignition was difficult and the plasma (*ca.* 6 cm long) was not very stable. Perhaps the idea of a MIP should still be taken into account by applying high-power MWPs at twice the higher frequency (5.8 GHz) and thus having the expected plasma heating section half as short.

3.2.1   Requirements for an Ideal Microwave Cavity

An ideal microwave cavity should have several important features:

- First is the ability to initiate the discharge and maintain the plasma per-turbed by the sample inflow. The discharge must possess a strong ability to evaporate, excite and/or ionize a portion of the sample material. However, in contrast to industrial plasmas, the power efficiency of the source may not be the most important factor, although some non-critical energy matching may always be provided for a better performance.

- The second feature is to have the plasma shape and excitation region unchanged from sample to sample and have the loaded and unloaded plasma contained within a desired volume.
- Thirdly, the dimensions of the plasma cavity should be acceptably small, with the torch length (and thus cavity thickness) as short as possible.
- The lifetime of the torch as well as the other high-temperature-contacting materials in the torch-less designs should exceed several hundreds of hours (the more the better).

From a more careful look at different plasma cavities it is apparent that an expected "good plasma" comparable to ICPs must do the following:

1. Allow a sufficient amount of sample through the cavity. The plasma loading should not change the plasma shape, *i.e.* the excitation (ionization) region should always be predictable.
2. Interact with the sample in a sufficient period of time. The residence time of a sample in the plasma must be long enough to evaporate, atomize and excite the sample.
3. The stability of the discharge must be good and stay unaltered by the injection of samples. One has to remember that the smaller the plasma and its maintenance power, the smaller the sample it can take and the smaller the gas flow necessary to carry the sample.
4. The plasma does not have to be large. From the point of view of analytical applications the only parameter which finally counts is the signal-to-noise ratio (S/N), which must be high.

For a plasma which would guarantee the fulfilment of the above conditions one can expect that a reasonably low operating power will be consumed when both the plasma and the path of the sample introduction have approximately one common axis of symmetry.

3.2.2 What Makes a Good Microwave Plasma?

The first important parameter is the discharge symmetry, which should be introduced during the design stage. The plasma must not "see" the direction from which the MW power is coupled to the discharge. For instance, in the simplest and very popular rectangular waveguide-based cavity there exists a non-symmetrical distribution of the electric field and therefore heating of the plasma is always stronger from the side of the energy inflow and much weaker from the opposite side. The arrangement of a rectangular waveguide non-symmetrically excited plasma is acceptable when the sample is in a gas form. This was the case of the plasma environmental monitor known from the work of Woskov *et al.*[28] Nevertheless, when running such a plasma a swirl of stabilizing gas flow should be introduced in order to keep the plasma at the centre of the torch. Otherwise the torch will melt from the side of the energy inflow from the magnetron toward the plasma simply because much more power can be absorbed on one side.

Applying a really symmetrical cavity construction would allow the non-flowing plasma to fill up the torch without any melt heating it up to red heat at high power levels. Introducing good symmetry allows for saving that portion of energy that is necessary to heat the cold stream of shroud gas, which in these situations must be used just for plasma stabilization.

The symmetry of energy coupling is generally a difficult issue. However, some good examples have been drawn from the well-elaborated high-power technological "plasmatrons" like the one using a waveguide and a mushroom-shaped waveguide-to-coaxial transition (see Figure 3.30 below),[29] both solutions exactly as those used years later in the Okamoto cavity.[30–32] In the high-power plasmatron design, however, the constructors applied one more symmetry-introducing element in the form of a radial line serving as a resonant mode purifier and at the same time as an impedance transformer. This specific design (a little too large to be adopted as an analytical MWP) has allowed for generation of powerful 100 kW air discharges[29] that did not burn the plasma tube. Another symmetry-introducing design known from the first patent of Kirjushin[33] has found its continuation in the SLAN plasma source,[34] the Cyrranus plasma[35,36] and many others used in the semiconductor industry. The principle of operation of that cavity is the existence of symmetrically—as regard to the plasma—placed slits which symmetrically couple MW energy to the discharge (see Figure 3.32 below). A general advantage of these plasma sources is homogenous plasma heating from all sides. In connection with symmetry, the authors of this book are familiar with the new planar discharges generated in multi-phase-energized cavities, including the three-phase-energized cavities described in Section 3.5.

3.2.3   Sample Introduction into a Microwave Plasma

The problem of sample introduction into a plasma is similar for MWPs and non-MWPs. A good tolerance to sample loading is one of the characteristics achieved in ICPs. That tolerance is still not the best characteristic of a MWP. In the world of different plasmas used in analytical practice, at least a few of them are known from their high loading resistance. In that number, the direct current plasma (DCP), and the ICP should be listed first. In a DCP an electric arc having more than two points of attachment is expanded by the sample-feeding gas, thus forming the colder channel for passing the analytical sample through; in the ICP case, a similar channel is allowed along the axis of the plasma column where heating currents and the E-field strength are minimal by definition. There exist a number of different plasmas that are in use as excitation sources: DCPs, ICPs, MIPs, CMPs and glow discharge (GD) devices with different supplies: GD-dc, GD-RF and GD-MW. An ICP was and perhaps even today is assumed to be the reference plasma that is still gaining attention and which has great popularity among analytical specialists. This may change as the new MWPs are improved by researchers, while ICP sources seem to have already reached their commercial optimization and they exhibit still more limitations in modern applications, including mass spectrometry.

3.3 General Classification of Possible Microwave Plasma Sources

3.3.1 E-type Microwave Plasma Sources

3.3.1.1 Capacitive Microwave Plasmas

The capacitive mode of coupling exists when the plasma in the cavity has at least one point of immediate contact with the electric conductors. This kind of plasma is specific for microwaves. Owing to the high frequency of microwaves, the contact usually exhibits a capacitive character. Nevertheless, at higher power levels and, correspondingly, at higher current densities at electrode surfaces, the etching of the electrode material may occur, leading to contamination of the plasma that can be detected in the optical spectrum.

3.3.1.1.1 Coaxial Single-Electrode CMPs. These have been known since the early 1950s as coaxial plasma torches in which a plasma is maintained in an extension of the inner conductor of the coaxial line (Figure 3.8).[9,37] The inner conductor is usually terminated with a heat-resistant tip, which may be water cooled. The outer conductor of the coaxial plasma torch can be cut short to form a nozzle or it can be left long to act as a screen along the plasma. The stability of the plasma is great but the plasma is constricted near the tip-end, making it difficult to feed any sample into it. The sample material always tends to bypass the plasma.

Even if it is not cooled, the tip does not melt, but the plasma is still likely to become constricted in the near-tip region and introduction of a sample into the discharge remains difficult. The discharge can be obtained at rf as well as MW frequencies. In the microwave range a capacitive coupling between the plasma and the tip is usually strong enough to transport the energy between the central conductor and the plasma. This limits electrode etching. However, the constriction of the plasma near the tip can promote etching of the electrode material, adding unwanted lines to the observed spectrum. The coaxial plasma at low power levels, up to 150 W, is used in commercial instruments.[38,39]

3.3.1.1.2 Microwave Plasma Torch. A microwave plasma torch (MPT) is a coaxial CMP equipped with a hollow central electrode in the form of a thin-walled tube, the round edge of which enables a circumferential distribution of the plasma attachment (Figure 3.9). As a consequence, the plasma takes a toroidal form.[40,41] This configuration of cavity can tolerate liquid samples easier

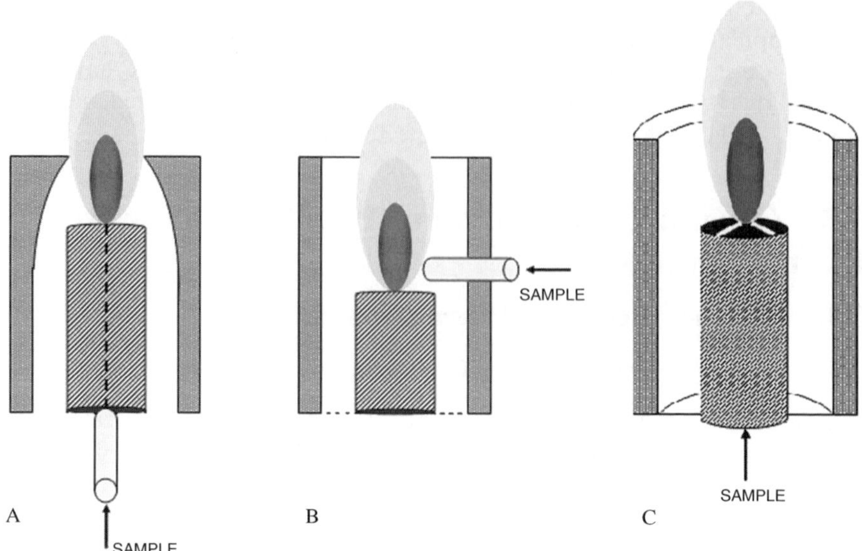

Figure 3.8 Coaxial torch with the sample fed through (A) a narrow bore in the central tube; (B) with a side-arm injector; (C) with a central tube with cross-mounting of the tip.

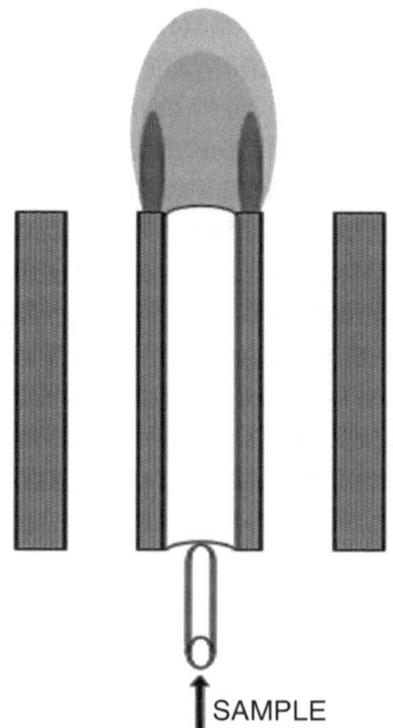

Figure 3.9 A MPT device.

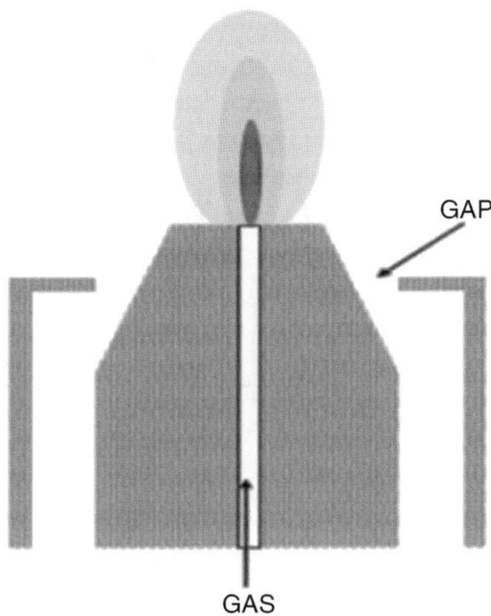

Figure 3.10   Torch injection axial plasma.

than when the plasma attachment occurs only as a single spot. A micro-MPT has been described by Stonies *et al.*[42] and more recently by Ray and Hieftje.[43]

The idea of a MPT was initially presented by Jin *et al.*[44] and later extended by others.[45] The inner conductor of the first MPT and the CMP plasma tip therein had the form of a thin-walled metal tube. Using such a central tube one could spread the plasma attachment around the tube edge, which created the discharge with a neutral hole at the axis. Thus, part of the central flow together with the sample could go along the plasma column axis. It is necessary to stress that traces of the tube material in MPTs, even made from copper, have not been detectable in plasma spectra. For this reason, one can assume that the MPTs are also electrodeless plasmas, like the MIPs, but in our classification they still belong to the CMP group since the plasma is in contact with metals immediately, *i.e.* without any solid dielectric barrier.

3.3.1.1.3   Torch Injection Axial Plasma. In the torch injection axial (TIA) plasma design the tip end of the inner conductor of the coaxial line sticks out beyond the coaxial outer shield (Figure 3.10). A narrow coaxial gap between the inner electrode and the shield is claimed to be important.[46,47] The gas passes axially through the tip and a stable plasma is formed in the extension of the tip end. The device may be screened to prevent radiation, but the screen should be distanced from the tip in order to keep the impedance high.

3.3.1.1.4   Microwave Plasma Jet. This design is similar to the TIA plasma, with the difference that the tip end is positioned close to the plane of

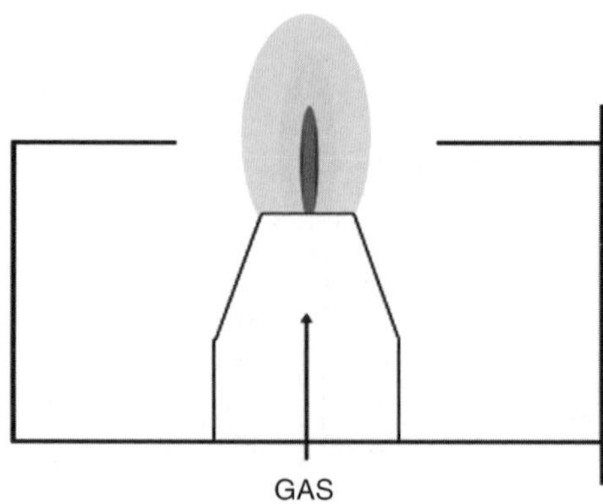

Figure 3.11   Microwave plasma jet cavity.

waveguide wall, where both form a nozzle that plays an important role in restricting the gas flow (Figure 3.11).[48,49]

The CMP-type plasma jets can also be realized with the use of a strip-line technique.[50] Particularly interesting may be a future application of CMPs in the microelectromechanical system environment which will accompany the lab-on-a-chip instrument solutions.

3.3.1.1.5   Cold Plasma Coaxial Torch.   A cold plasma coaxial torch consists of a thin (less than 1 mm) tip and has a long dielectric discharge tube, *e.g.* made of PTFE, and an external shield (Figure 3.12). Operating from just a few watts in argon, it generates a very cold (touchable) plasma discharge.[51]

3.3.1.1.6   Multi-electrode CMPs.   A multi-electrode assembly has been proposed with a form similar to an MPT in which a hollow central electrode is arranged as a plurality of single-tip electrodes distributed symmetrically on a circle (Figure 3.13).[6] Each of the electrodes preferentially has the length of a quarter of the wavelength and is simultaneously energized from the same inner conductor in the coaxial line. The latter should be hollow to allow the sample through. In a new design the tube-shaped central conductor in the MPT has been substituted by a multi-tip assembly, where instead of the round tube edge the plurality of the symmetrically placed single electrodes is applied. These electrodes can be made from appropriate high-temperature-resistant materials. In that configuration the energy density at the electrodes can be higher than at the tube edge of a regular MPT. For this reason the electrodes can be maintained at higher temperatures, thus importing more stability to the set of candle-like discharges. From the authors' experiments it is known that the crown of the discharges then collapses into a column of a larger plasma. However, the plasma obtained this way never completely collapses. This is because the charge in every candle-like component discharge is

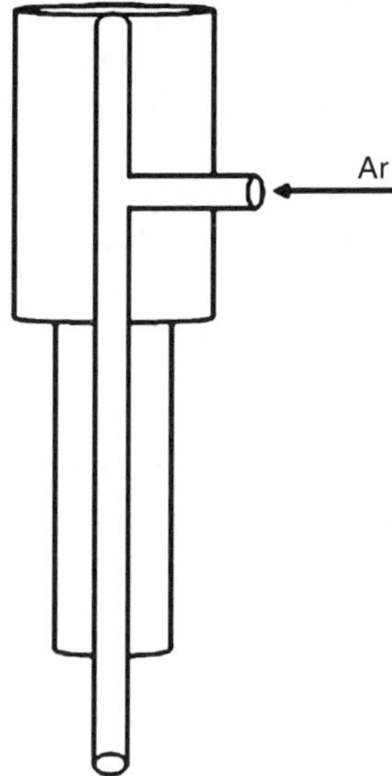

Figure 3.12 Cold plasma torch.

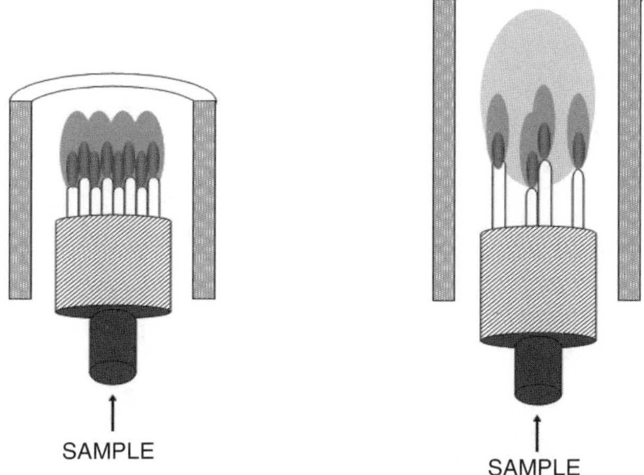

Figure 3.13 Multi-candle MPTs.

the same and the discharges repulse each other. Also the vicinity of an electric shield may pull the separate plasma candles toward the screen as they tend to close their currents along the lines of the electric field. The shield should therefore be sufficiently wide.

Observing the plasma along the axis reveals a division of the discharge by several candles, each corresponding to a given electrode. Therefore the sample can easily be injected and penetrate the plasma between the component discharges. This new version of a MPT can be optimized by choosing the electrode material and modifying along the axis-placed sample injection nozzle. However, the simplest implementation of the idea would be just cutting quarter-wave slits along the inner tube of a conventional MPT. It should be pointed out that the tube end divided by slits into sectors can be analyzed as elementary arms in the Wilkinson-type power splitter, which assures that the portions of the split power are equal and separated from each other.[52]

3.3.1.1.7 Multi-electrode MWPs. The multi-electrode assembly built into the resonant cavity of any kind (circular, rectangular, coaxial or strip-line) can be used as a means improving the plasma stability (Figures 3.14, 3.15). Thus the Beenakker cavity has been equipped with three electrodes grounded to the body of the cavity, as described by Horvath *et al.*[53]

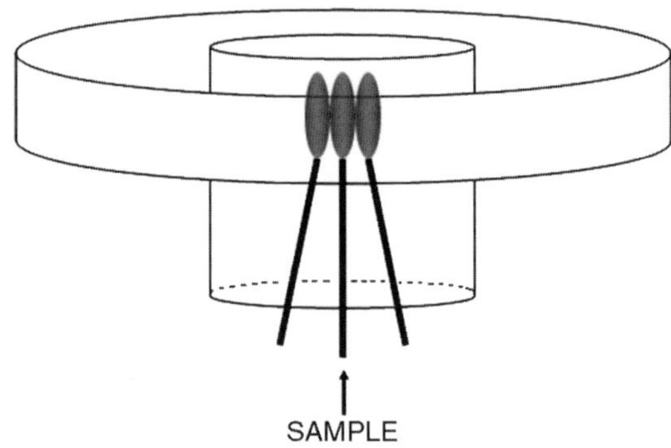

Figure 3.14   A multi-electrode CMP-MIP.

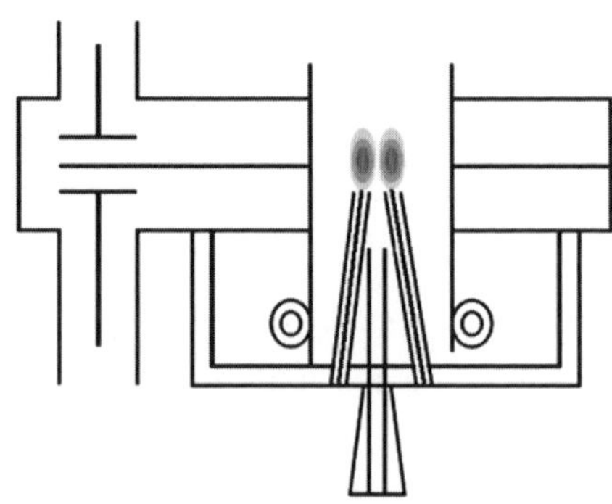

Figure 3.15   A multi-electrode CMP strip-line.

A portion of the MW power lights up the candles, which assures better stability by acting as pilot plasmas when the sample is introduced between the candles. As a result, the use of a hydraulic high pressure nebulizer could successfully be used.[53]

3.3.1.1.8 Multiphase CMP Cavities.

These plasmas are sustained with the electric field vector rotating in the plane of the discharge (Figure 3.16). Separate phases of the field are delivered to the coupling electrodes distributed circumferentially and symmetrically whose number is equal to the number of phases. Especially the three- to six-phase systems seem to be applicable, but a larger number of phases may also be of interest. Interaction of the discharges generated between the electrodes results in the formation of a hollow plasma, which is generally planar without any gas flow and can be expanded along with the sample containing the gas to form a doughnut-shaped jet in which excitation takes place. The system can also be configured as a plasma processing reactor equipped with a number of plasma coaxial torches, each working with pure gas. The thus combined plasma reactor can easily be fed with samples or substrates. Even the larger plasma systems consisting of several plasma sources can be combined in one chamber.[6]

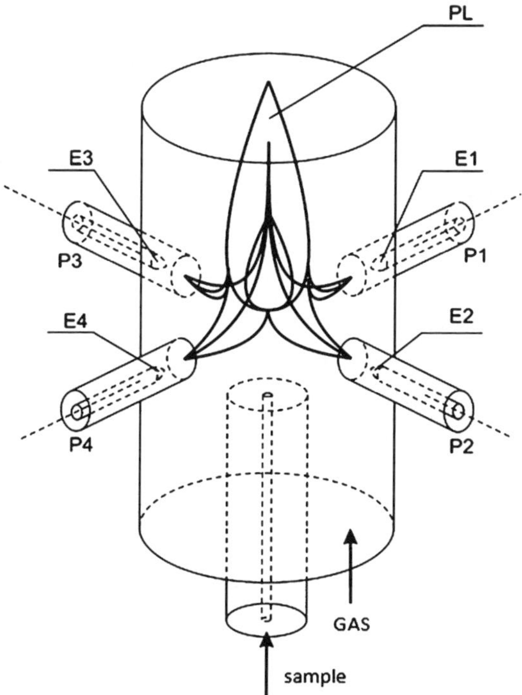

Figure 3.16 Multi-torch design comprising a plurality of four plasma torches: E_1-E_4 are electrodes, P_1-P_4 are power inputs, PL is the plasma.

3.3.1.1.9 Microdischarges with Micro-gaps. Microplasmas are generated in narrow working gaps below 1 mm. Different versions of microdischarges have been described.[54–58] One of them consists of a strip-line ring-based balloon which delivers two symmetrical TEM waves to a 100 μm working gap filled with microplasma.[54–56] One can suggest also a coaxial design of a micro-gap cavity,[57,58] which in its embodiment with scaled-up dimensions has already been widely used in contemporary chemical vapour deposition (CVD) microwave plasma reactors.

An interesting micro-gap plasma source has been described,[50,59] where a plurality of micro-gaps placed at the edge of a sapphire plate is supplied from strip-line splitters, enabling in-line operation. A number of gaps can be oriented along the edge of the plate or can be set up in the form of a stack. One of the expected applications of multi-gap systems is the generation of intense UV radiation.

It is worth adding that, using the new technique of multiphase supply, a similar long gap assembly can be obtained by applying a plurality of in-sequence running microwave phases connected to the sequence of gaps. Generally the concept of multiphase supply can readily widen the scope of possible constructions. For example, using strip-lines one can arrange a three-phase microwave microplasma source with a symmetrically energized discharge delivering a plasma with improved stability and geometry (Figure 3.17).

The multi-phase approach can be applied not only to symmetrical devices. It may be found useful in designing in-line as well as matrix geometries by applying the sequence of phases to neighbouring points of the power feeding system. Moreover, the tendency to reduce sizes of the chips should result in a vanishing of differences between the MW and RF circuits and, when only TEM lines are applied, also between the RF and audio frequencies. In some cases one may find these similarities with frequency going down even to the dc level. For

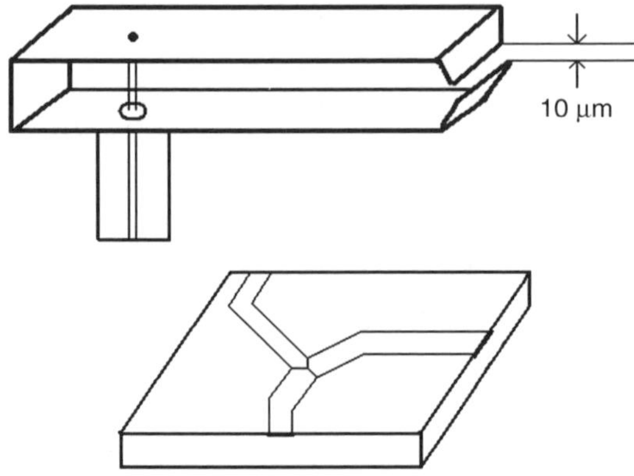

Figure 3.17 Micro-gap CMP plasma couplers: a single and a suggested three-phase coupler.

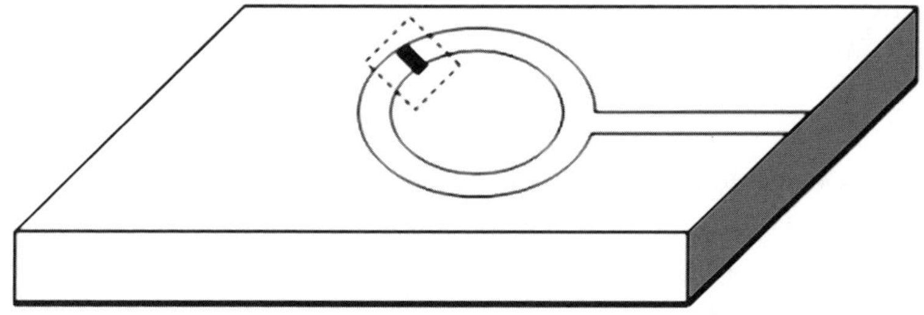

Figure 3.18 Micro-gap ring resonator cavity.

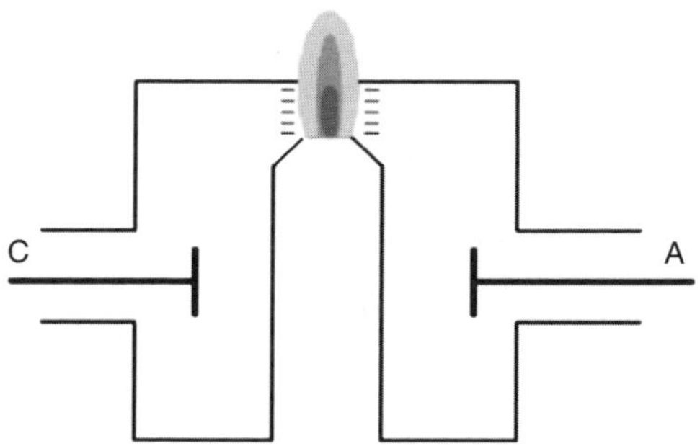

Figure 3.19 Multi-gap coupling in the form of a leaky wall (distributed gap). A is the antenna, C is the capacitive tuner.

example, a DCP with one cathode and two anodes is indeed similar to the rotating three-phase RF plasma arrangement.

Very good energy matching of a micro-gap excited plasma can be obtained with the use of wave splitting into two arms feeding both sides of the gap. Applying different lengths of the arms such that the waves reflected from the plasma within the gap going back to the splitter have a phase difference 180° should result in a self-matching feature of such a system (Figure 3.18).

The micro-gap concept can be adapted to a surfatron-like coaxial device and applied in the form of a "leaky wall" which in fact can act as a set of multi-gaps featuring the symmetry in plasma coupling (Figure 3.19). In the experiment performed by the authors, a leaky wall was formed by aluminum washers separated by mica dielectric spacers, but it seems that a dense spiral would do the same job as well.

3.3.1.2 Microwave Induced Plasmas

Microwave induced plasmas are obtained when the plasma in the cavity does not have any immediate contact with electric conductors. Contact may occur only through a dielectric barrier (a layer of gas, a wall of dielectric, discharge tube wall, *etc.*)

3.3.1.2.1 Waveguide-based Plasma Cavity. The discharge is isolated by a dielectric tube which crosses the rectangular waveguide through wider walls. In Russian papers this device is called a rectangular waveguide-based plasmatron (Figure 3.20).[2,60] At the low-pressure range and with a reduced waveguide height the design with a rectangular waveguide was introduced as the Surfaguide (Figure 3.21).[61]

All waveguide-based cavities equipped with standard flanges are well suited for integration with commercial power generators. Normally one port of the cavity is connected to the power generator and the second port is equipped with a regulated short (plunger).[62] For analytical applications it is convenient to make a compact device combining a microwave power generator with a plasma

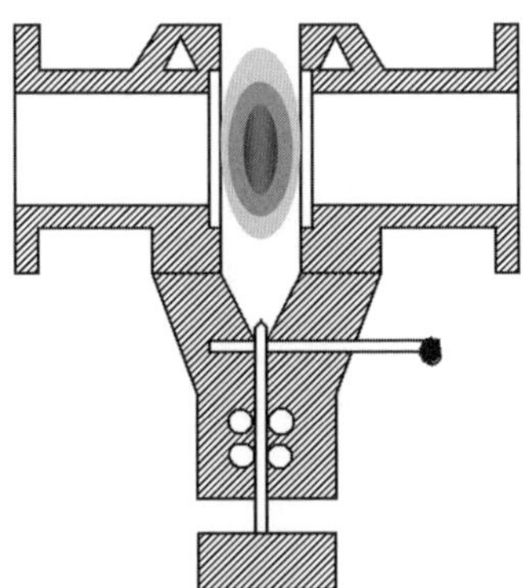

Figure 3.20   Typical waveguide plasmatron.

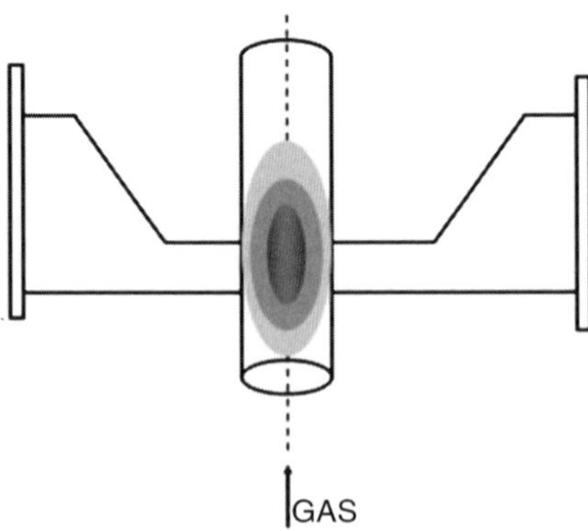

Figure 3.21   The Surfaguide.

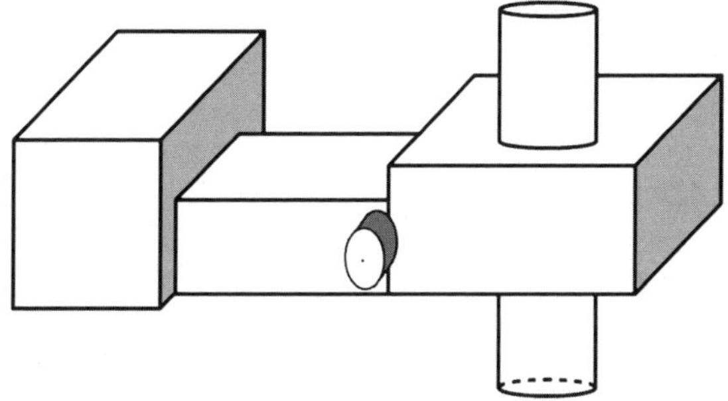

Figure 3.22   Integrated waveguide cavity.

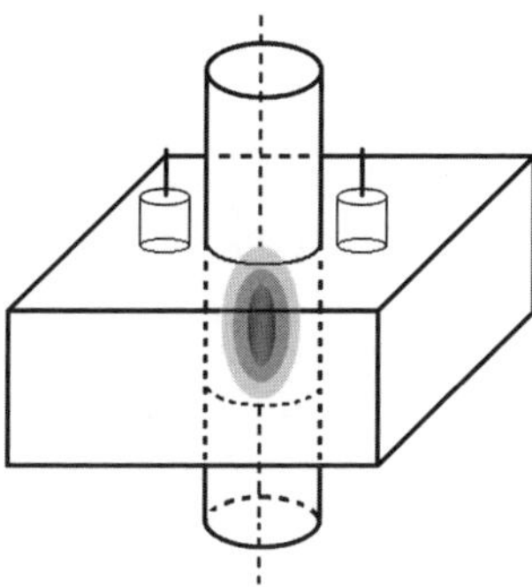

Figure 3.23   Square-shaped cavity.

cavity (Figure 3.22). The rectangular cavity described by Matusiewicz[24] and the one published by Jankowski *et al.*[26] as an improved version are both integrated with magnetron power sources and equipped with a resonant length tuning plunger, as well as with an efficient iris-type diaphragm for tuning the coupling coefficient.

As already mentioned, by choosing a proper dimension of the waveguide one can make a square cavity with an E field distribution similar to a cylindrical TM_{010} mode (Figure 3.23). Indeed, the fields in both cavities, whether cylindrical or rectangular, are almost the same, especially in close vicinity to the plasma column.

At present the popular rectangular waveguide-based plasmatron can be found under different commercial makes,[63] but the spectroscopic plasma parameters seem to be unchanged compared to classical devices. By splitting the incident wave into two waveguide arms, one can energize the plasma from both sides, as shown in the book by Toumanov.[64] Tapered sections in the surfaguide device (see Figure 3.21) are essential from the point of view of impedance matching. Several non-analytical applications of such structures have successfully been used in plasma chemistry.[64]

3.3.1.2.2   Circular Resonator with a TM$_{010}$ Mode (Beenakker Cavity). The TM$_{010}$ mode Beenakker cavity can also be classified as a TEM mode radial cavity, as both have exactly the same field distribution. Plasma is excited in a quartz tube at the resonator axis. In recent years there have been several modifications introduced in both the means of energy coupling and frequency tuning. The originally proposed loop coupling and dielectric rod inserts for frequency tuning were found insufficient to provide energy matching at high powers. Nowadays, plate antennae and disk tuners are common solutions.[65] Generally, two elements are needed for tuning the plasma cavity: one to tune the resonant frequency and the second to adjust the coupling coefficient, and these two elements have finally been realized in a Beenakker cavity by introducing a capacitive disk frequency tuner and an adjustable disk antenna coupler (Figure 3.24).

3.3.1.2.3   Beenakker Cavity with a Surfatron Tuner. One of the new modifications of the Beenakker cavity includes the use of a surfatron as a tuner to couple the energy in a symmetrical manner. The task of symmetrical energy supply can be fulfilled if all tuning elements are axially symmetrical or placed sufficiently far away from the plasma. The key design factor is the Q of the cavity. With a low Q-factor, when the cavity oscillations are strongly damped in the presence of lossy plasma, one of the ways towards obtaining symmetrical heating and good plasma stability is to avoid coupling or tuning elements which would perturb the symmetry. One of the best solutions may

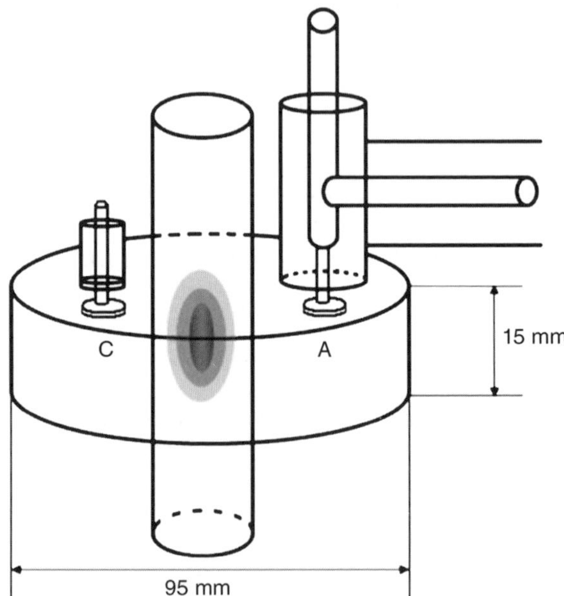

Figure 3.24   Disk antenna (A) and a capacitive disk tuner (C) in a Beenakker cavity.

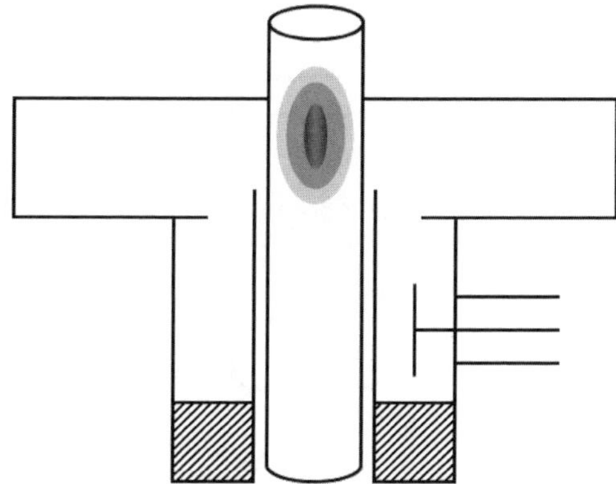

Figure 3.25   Beenakker cavity coupled symmetrically via a surfatron tuner.

be the adaptation of a well-designed tuner from a surfatron,[46,61] as shown in Figure 3.25.

3.3.1.2.4   Circular TM$_{020}$ Mode Resonator. As present, this cavity has only been suggested for heating liquid samples.[13] It may enable a generation of plasmas even larger than those obtained in the Beenakker cavity. This cavity must have a much greater diameter, but fortunately it can be made folded. Because there are three half-ways that fit in a cavity cross-section, the tuner and coupler can be placed at a distance from the plasma. Further separation of the coupling and tuning elements shown in Figure 3.26 can be accomplished using a low impedance section. Double folding of this cavity is shown in Figure 3.27.

3.3.1.2.5   Circular Resonators with Non-zero Longitudinal Indexes. These resonators, *e.g.* TM$_{011}$ and TM$_{013}$, are mostly used to construct high-power "plasmatrons" for plasma processing and for fusion experiments (Figures 3.28, 3.29).[21,66,67]

Among the sources where symmetrical coupling with microwaves has been well engineered, a device called a radial resonator-based plasmatron is worth mentioning. The device, known from a paper by Galliker[29] and originally devoted to the generation of plasma streams at power levels up to 100 kW, is shown in Figure 3.30. As can be seen, after the waveguide-to-coaxial transition, one more mode of filtering was introduced in the form of a radial line resonator, which plays the role of a symmetrical coupler of MW energy to the plasma. One can assume that this cavity, although relatively complex, can be adopted as a good MWP launcher. Note that the Okamoto cavity (see Figure 3.36) can be obtained by simply cutting the radial line out from the original cavity shown in Figure 3.30.

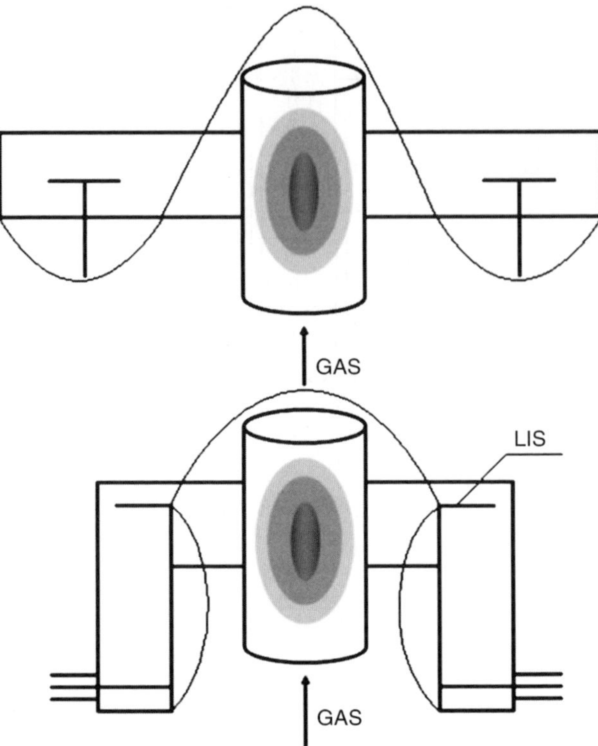

Figure 3.26 The TM_{020} mode cavity (*top*) and its folded version (*bottom*) with a low impedance section as a means for purification of the circular mode symmetry.

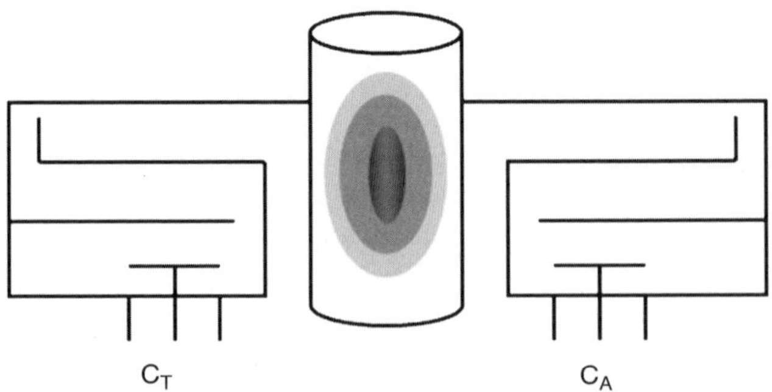

Figure 3.27 Another double folded version of the TM_{020} cavity: C_T is the capacitive disk tuner, C_A is the disk antenna.

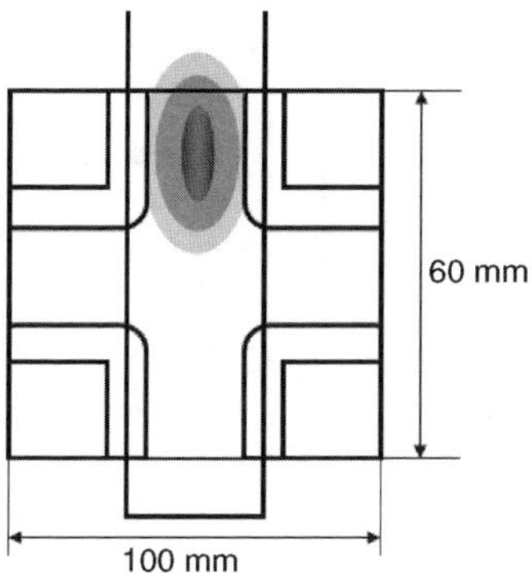

Figure 3.28 A TM_{011}-mode plasmatron.

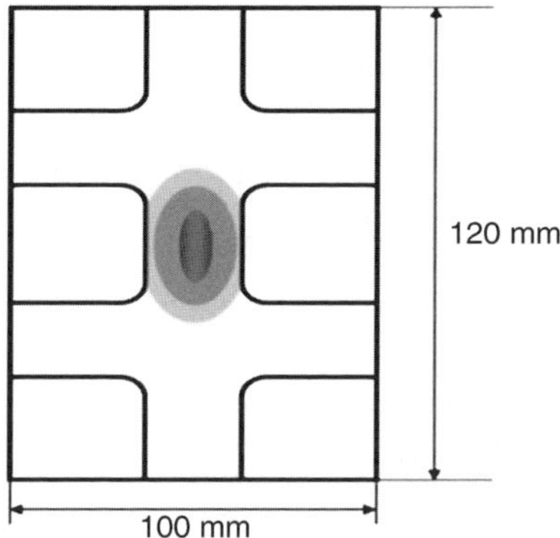

Figure 3.29 A deuterium gas plasmatron.[21]

The use of a flat dielectric barrier (Figure 3.31) may enable the creation of a planar slow wave discharge (SWD) plasma at low pressures.[61] As a matter of fact, SWD plasmas are better suited to low-pressure applications, when the plasma can propagate MWs with less attenuation.

3.3.1.2.6 Resonators with Gradient-type Coupling. The power can be delivered to the resonator through the side wall *via* slits or holes placed symmetrically to couple the microwaves from the waveguide which is wound around the cavity, as suggested by Kirjushin (Figure 3.32).[33] Different uses of this idea, known as SLAN,[34] Cyrranus,[35] *etc.*, have already been mentioned. Another similar construction could be suggested in the form of a circular resonator with common-phase excitation by a plurality of symmetrically distributed antennae supplied from the common power divider. Although the benefit of such devices lies in a very good symmetry of the

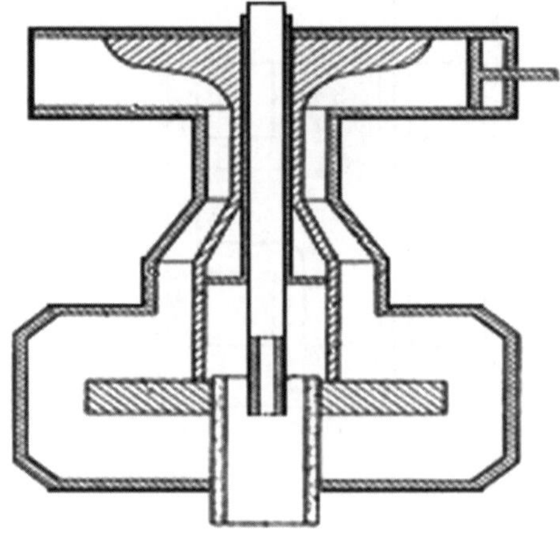

Figure 3.30   Plasmatron with radial line filter.

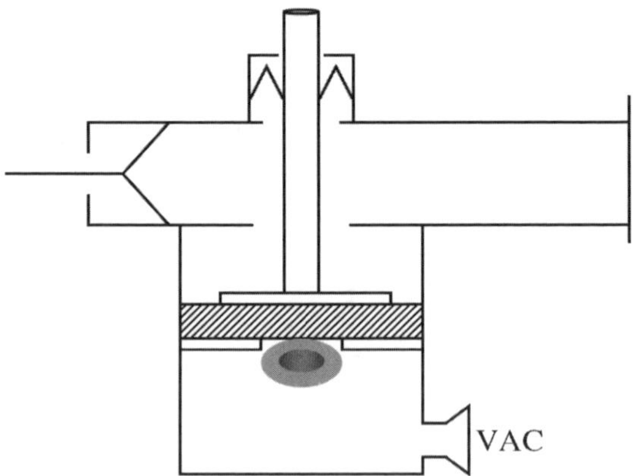

Figure 3.31   Flat discharge SWD coupler.

microwave field, they really seem to be geometrically too large for analytical use. For a more complete illustration, a kind of gradient-type coupling is shown in Figure 3.33. No coupling would occur if the plasma-exciting gap was placed in the middle. Therefore the gap has been shifted down towards the plane of the narrow wall of the rectangular waveguide. As a result, better impedance matching is achieved and better symmetry of the MW excitation.

3.3.1.2.7   Coaxial MIP Resonators. Coaxial MIP resonators may be considered as folded TM_{010} or TM_{012} cavities with the inner conductor taking

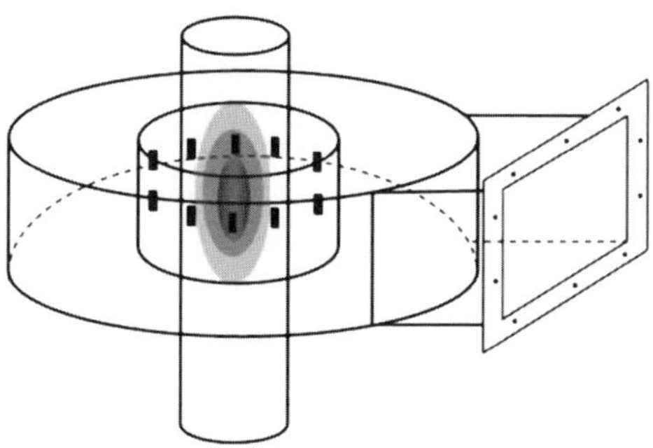

Figure 3.32 A plasmatron.[33]

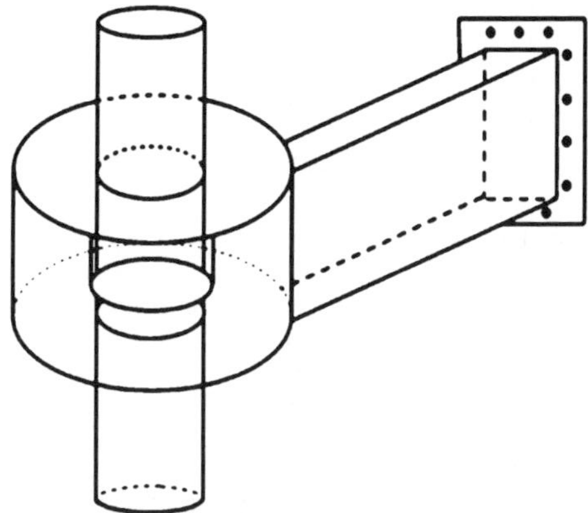

Figure 3.33 Design with gradient-type coupling.

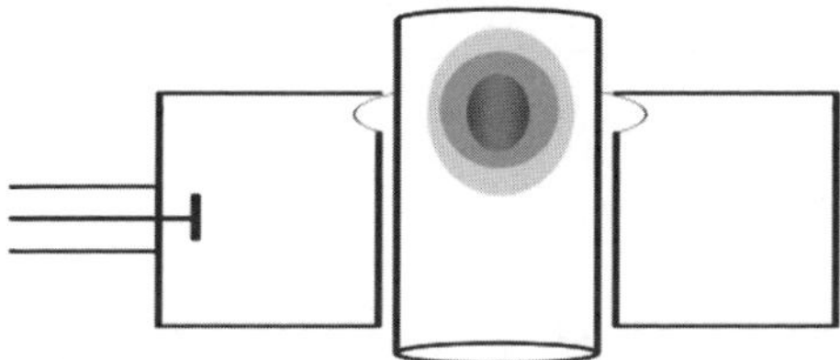

Figure 3.34 Thick radial resonator (sufatron).

the form of a tube and with the discharge tube passing along its axis. Coaxial resonators may include a circular gap to promote excitation of the surface waves. The device called a surfatron has a thin wall around the plasma tube, which promotes excitation of axially symmetrical surface waves (Figure 3.34). There exist a lot of possible constructions (*e.g.* see Figure 3.35), but

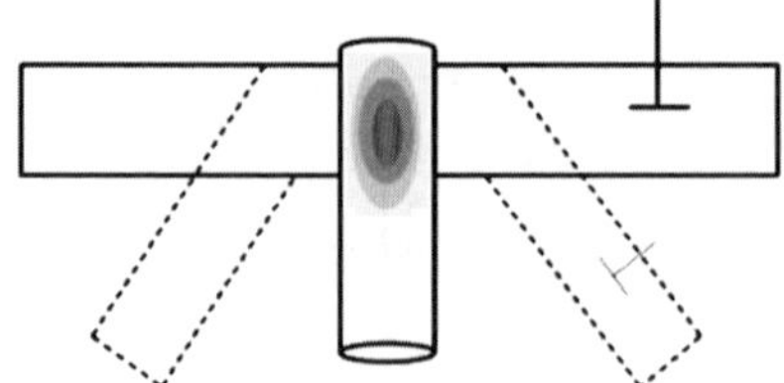

Figure 3.35   Radial-to-coaxial transformation.

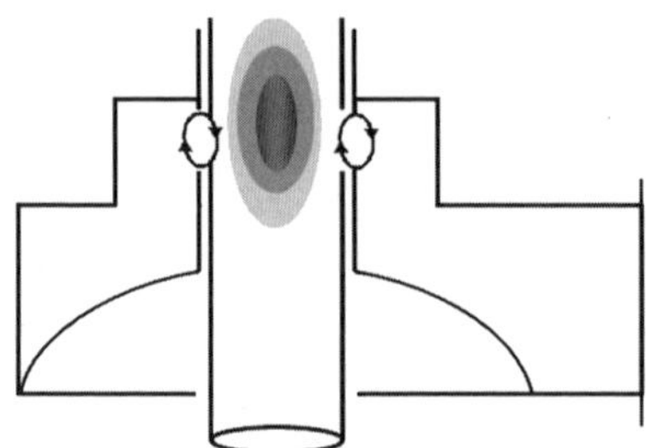

Figure 3.36   An Okamoto cavity.[22]

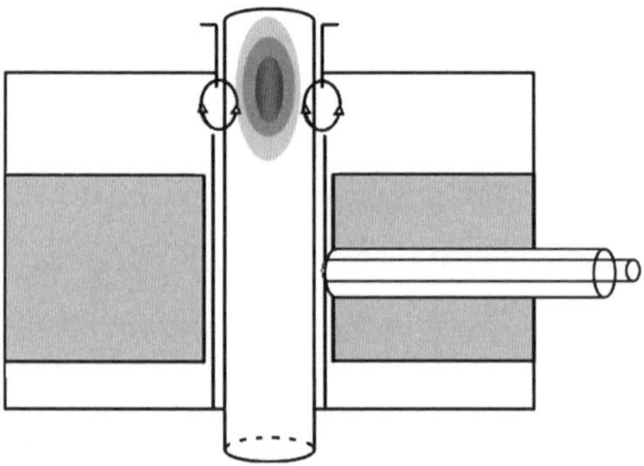

Figure 3.37   A TEM cavity.

the real use of surface waves can be exploited at low gas pressures when the low attenuation in surface wave propagation allows for generation of long plasma columns.[61] In terms of the present classification, the Okamoto cavity (Figure 3.36) belongs to the same coaxial group. In the Okamoto cavity the shape of the circular gap is crucial in generation of the electric field, which, owing to the use of well-shaped fringing field, can be symmetrical and at the same time denser at the peripheries of the plasma. Application of multiple gaps ("leaky walls") can further improve the stability of the plasma. The symmetry of the MW field excitation is very important. Jankowski *et al.*[68] described a TEM-mode cavity in which the symmetry was achieved by

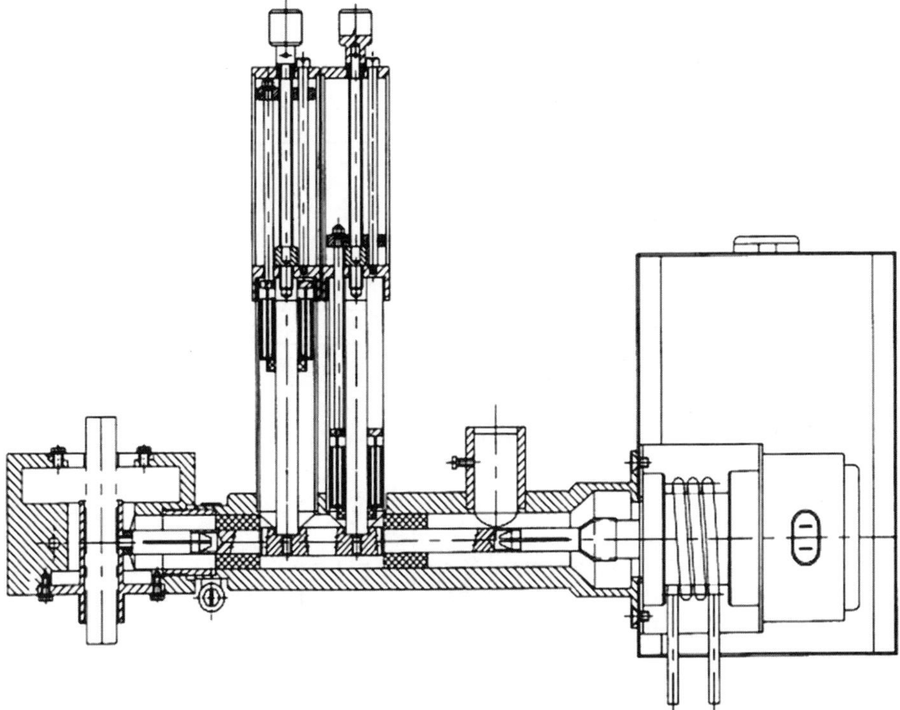

Figure 3.38　A TEM cavity cross-section (the open "short chimney" was for placing a reflectometer; note all coaxial connections to the magnetron antenna).

applying a coaxial coupler and a front ring tuner, both symmetrically placed at the axis of the radial cavity (Figure 3.37). There are several ways to obtain the radially symmetrical excitation.

In the front of the cavity shown in Figure 3.38 a ring tuner has additionally been installed to enable tuning of the resonant frequency of the resonator. In the rear side a secondary radial resonator has been used, which assures that the electrical length from the rear short to the point of coaxial feeder line is $3/8\ \lambda$. The electrical length from the feeder line to the plane of the front cavity is *ca.* $1/4\ \lambda$. The feeder line is equipped with a double stub coaxial tuner, which is energized immediately from the antenna of the CW magnetron. The magnetron box has a built-in filament transformer (not shown). The body of the TEM cavity has provisions for water cooling, which is necessary as the maximum operating power of the cavity is in excess of 800 W.

3.3.1.2.8　Folded Coaxial Resonators. The application of folded constructions allows for a reduction in resonator dimensions, as well as reducing the perturbation introduced by the coupling and tuning elements. In the cavity shown in Figure 3.39 the inner conductor of the coaxial line is folded at a

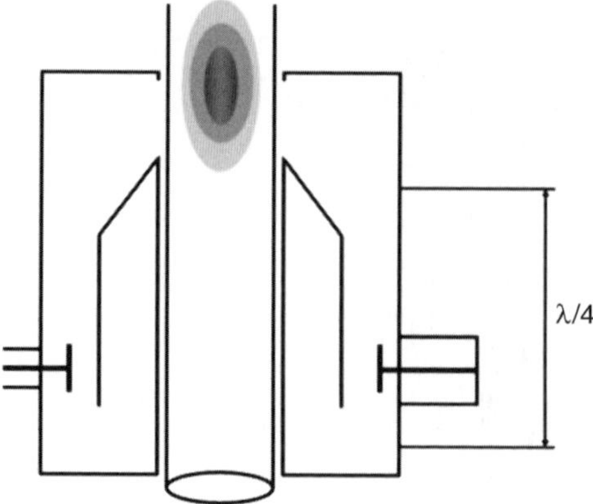

Figure 3.39 Folded version of a coaxial TEM cavity.

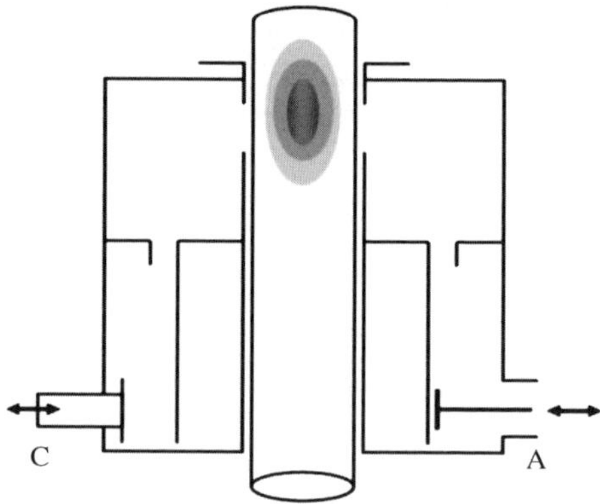

Figure 3.40 Another TEM with a low impedance section.

distance of 1/4 λ. As a result, both the tuner and coupling antenna are oriented in the high impedance region generated from the tapered short.[69]

A novel construction of a TEM cavity is shown in Figure 3.40, which has been additionally equipped with a low impedance section and a front ring tuner and this should deliver the field distribution even more symmetrically.

The last example (Figure 3.41) has antenna and tuner capacitors hidden behind the folded shield, which forms a tapered section with low impedance at the end. One can expect that this cavity should focus the attention of potential users.

3.3.1.2.9 MW-driven Glow Discharges. A cavity with a folded dielectric tube has especially been devoted to generate electrodeless glow-discharge plasmas. Single and double assemblies have been claimed in patent

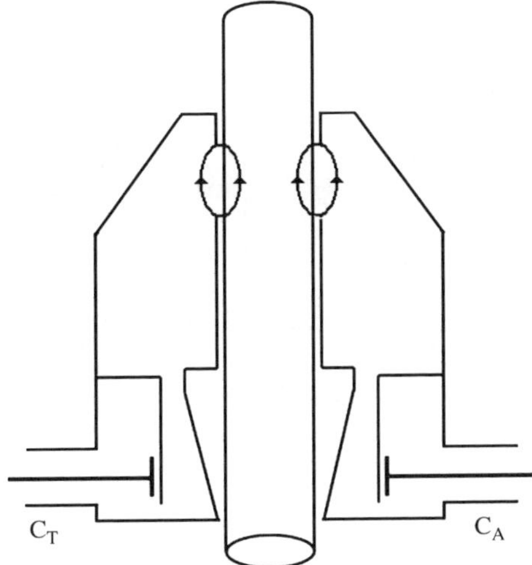

Figure 3.41   A new TEM cavity.

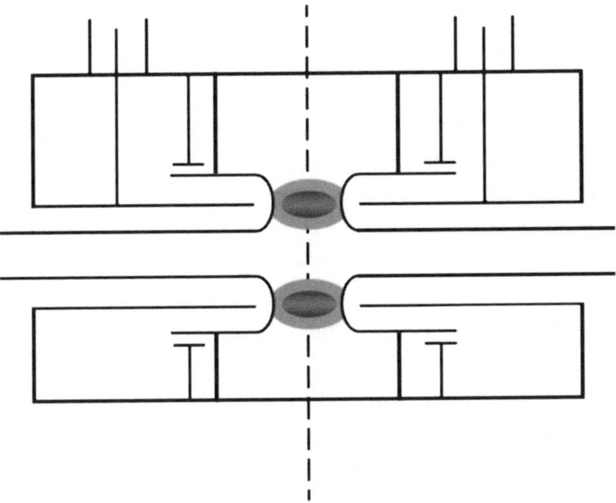

Figure 3.42   Double glow-discharge source with folded quartz tube applicators.

applications (Figure 3.42).[70] The assembly of a folded tube can easily be realized by simultaneously welding edges of two quartz tubes with different diameters. The discharge is very stable in air, oxygen, nitrogen and helium gases at pressure ranges of 100 Pa to 10 kPa and power levels from 10 to 250 W.

3.3.1.2.10   Double Loop Cavity. This cavity[12,60] was built very similar to the so-called stabilized capacitive plasma (SCP) reported by Knapp,[10] where the

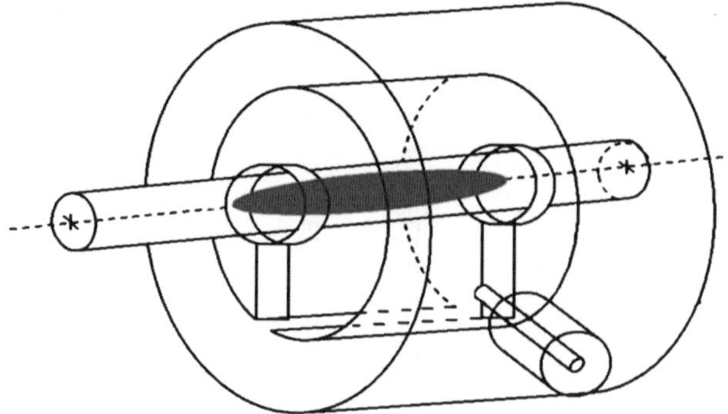

Figure 3.43   The cavity by Piotrowski *et al.*[60]

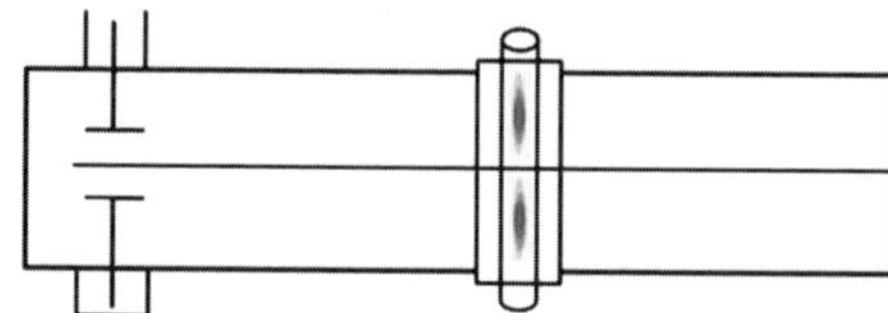

Figure 3.44   Strip-line cavity.

excitation was provided by two collars attached to the discharge tube and energized from an RF power source. Note that in this classification the MW-driven SCP with a dielectric tube barrier (Figure 3.43) is a MIP and not a CMP as is frequently suggested.

3.3.1.2.11 Strip-line Resonators. The strip-line resonator, originally patented by Barnes and Reszke,[71] is equipped with a dielectric tube with argon or helium gas and can deliver a double plasma placed on both sides of the strip (Figure 3.44). On the other hand, when the gap height on one side is reduced the device will promote propagation of surface waves along the plasma and, by analogy to the Surfaguide, it could obviously be named a "Surfastrip" (Figure 3.45).

Contemporary strip-line resonators with a dielectric tube cut in the sapphire ceramics and a variety of possible strip-line assemblies with dielectric barriers have been reported (Figures 3.46, 3.47).[54–56]

3.3.1.2.12 Slab-line Cavities. A single-electrode design used by Outred[72] in a slab-line cavity brings together the simple idea of using a 1/4 wavelength antenna coupler and a low impedance section (LIS) in order to improve matching of the electrodeless discharge lamps (EDLs) (Figure 3.48). This kind of coupling

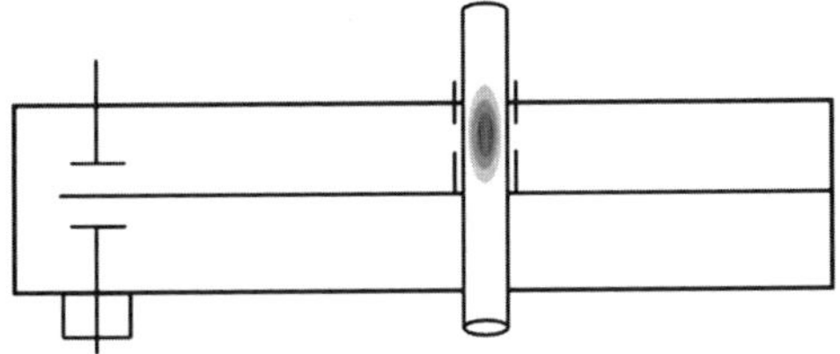

Figure 3.45 A "Surfastrip" cavity.

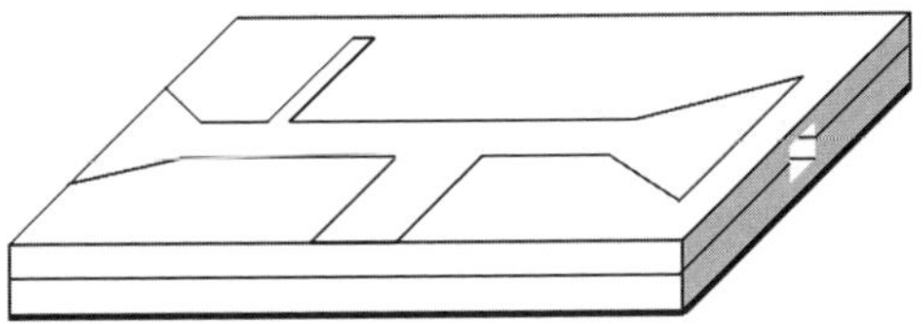

Figure 3.46 Strip resonator-based plasma generator.

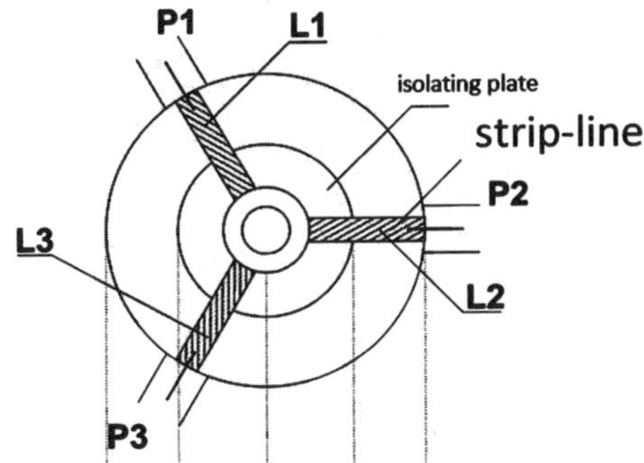

Figure 3.47 A three-phase strip-line resonator: L1–L3 are striplines, PA–P3 are power inputs.

employing 1/4 λ long sections has been adopted in multi-phase cavities described in this book.

3.3.1.2.13 Circular waveguide-based Cavities with an Orthogonal Rotating Field.

Two TE_{11} modes can be excited at the same time in two orthogonal planes, resulting in a rotating field propagating in the circular waveguide (Figure 3.49). This kind of cavity will generate a stable plasma which, however, may be too large for analytical applications. This idea has been implemented to deliver power to efficient microwave-excited lamps.

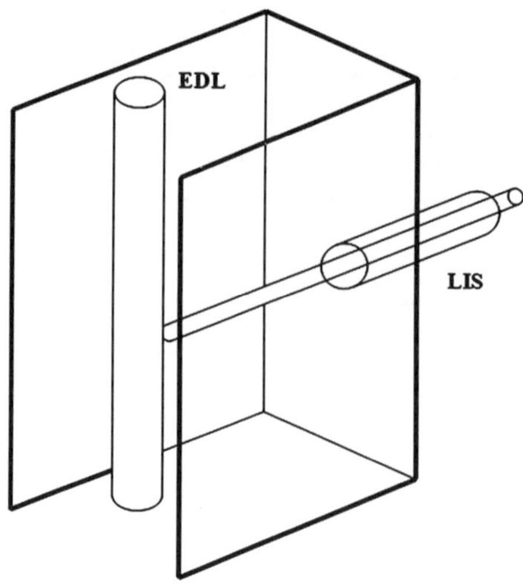

Figure 3.48   Quarter wavelength coupler in a slab-line cavity.

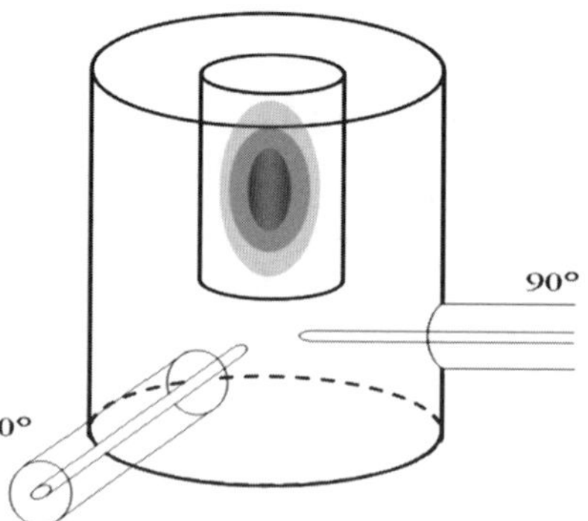

Figure 3.49   The TE$_{111}$ rotating-mode plasmatron.

Another rotating field system for energizing high-intensity discharge lamps in the form of closed gas-filled bulbs has been patented.[73] Four phases of RF supply have been obtained with the use of hybrid couplers. It may be interesting to add that one of the first systems with a rotating field used for energizing the plasma was designed and implemented in Russia. A 500 kW (!) plasmatron working at a 915 MHz frequency was applied for plasma chemistry processing, *e.g.* for reforming of H$_2$S.[74]

3.3.1.2.14   Rotating Field Assemblies with Strip-lines and Dielectric Barriers. An example of a cavity with a dielectric barrier and a rotating MW field is shown

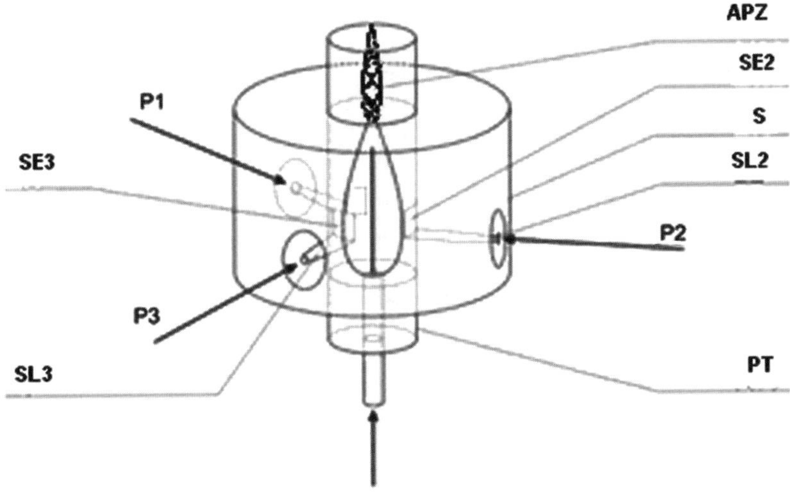

Figure 3.50 A three-phase MIP: P1–P3 = power inputs, L1–L3 = striplines, SE1–SE3 = strip electrodes, PT = plasma torch, APZ = analytical plasma zone, S = screen.

in Figure 3.50, where the energy is supplied through the coaxial ports P1, P2 and P3, each differing in phase by 120°. This energy is coupled to the plasma with the use of capacitive plates placed close to the external surface of the discharge tube. A similar coupling employing a multi-helix coil comprising four helixes is shown in Figure 3.51.

It is interesting to note that the rotating field approach has also been found useful in generating large-volume glow discharges using a multiphase mains supply. A 12-phase system has been described.[83]

3.3.2 H-type Microwave Plasma Sources

3.3.2.1 *A TE$_{011}$-mode Resonator with a Solenoid-like Distribution of H-lines*

Using this resonator for plasma generation has a long history. There were several obstacles, for instance the difficulty of resonant tuning and efficient single TE$_{011}$-mode coupling (Figure 3.52). A low-pressure operation has been achieved by Asmussen.[75] An atmospheric pressure oxygen plasma has been obtained by the authors using 1500 W at 2.45 GHz.[27]

3.3.2.2 *H-type Cavities with Lumped Inductances*

3.3.2.2.1 The Planar Coil ICP-like Discharge. Here the discharge is found to be limited to low pressures.[2] A 144 MHz design with a serpentine or

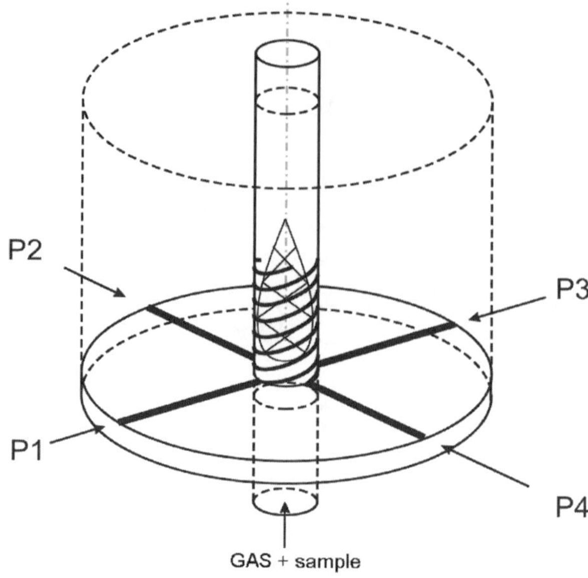

Figure 3.51   A multi-helix MIP: P1–P4 are power inputs.

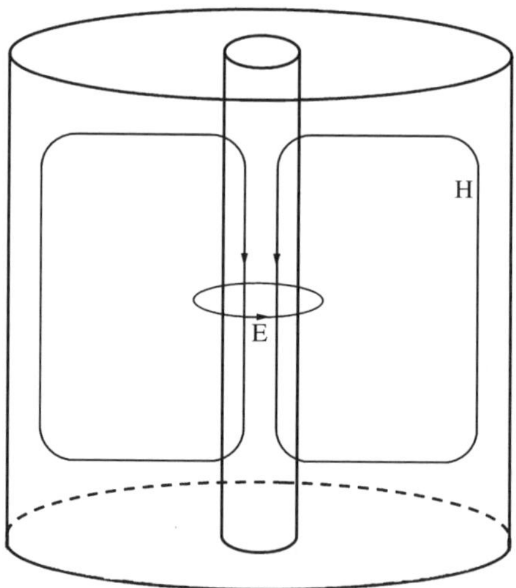

Figure 3.52   A microwave ICP with a TE_{011}-mode resonator.

Archimedes coil can generate an atmospheric pressure plasma using 50 W of power drawn from a regular RF transmitter (Figures 3.53, 3.54).[16]

Single- and multi-turn microfabricated ICPs have been studied by Hopwood *et al.*,[76] enabling work with a low gas flow.

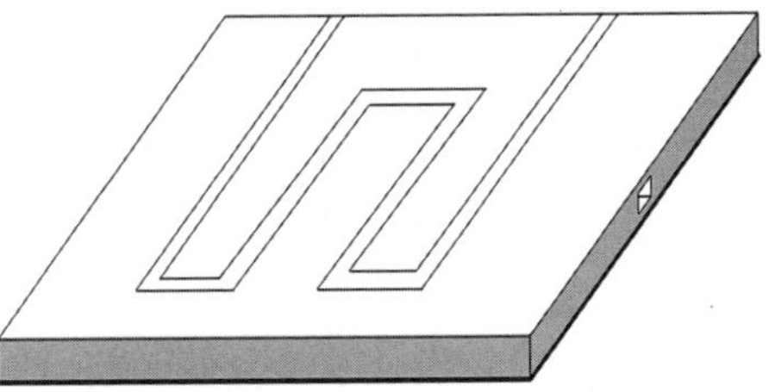

Figure 3.53 Serpentine coil coupler.

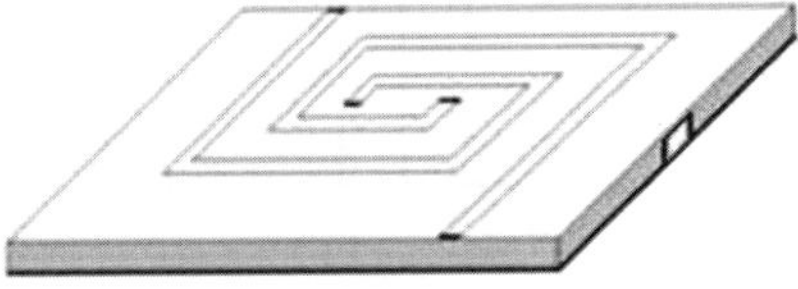

Figure 3.54 Archimedes coil for a flat ICP.

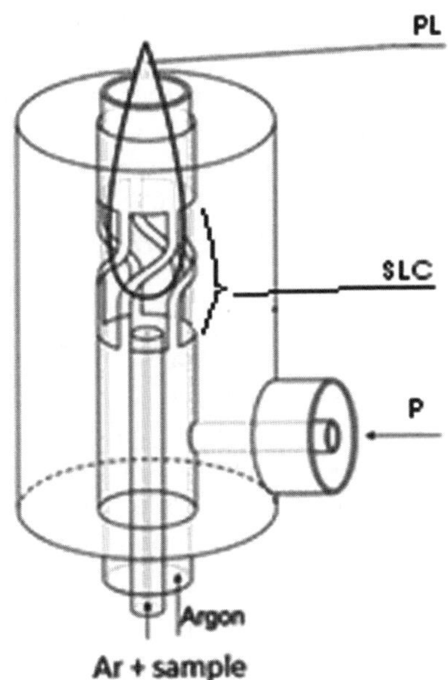

Figure 3.55 MW semi-lumped coil (SLC): PL = plasma, P = power input.

3.3.2.2.2 The Semi-lumped Multi-helix MW-ICP. A new idea of using multi-helix coils, each having just a fragment of a turn and thus having a length small compared to the working wavelength, led to the development of ICP-like MW-driven H-type plasma generators (Figures 3.55, 3.56).[6] This

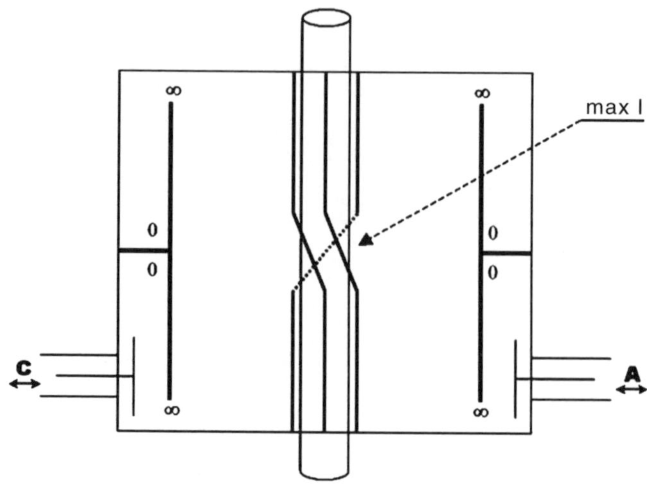

Figure 3.56   A practical coupler for a MW-ICP.

new idea, proposed by the authors, can be beneficial not only for MW frequencies when a single full turn of an ICP coil would be too long to play the role of a lumped inductor. It can also serve as the means to improve the symmetry of classical RF designs to further minimize the E field that is always associated with any regular RF coils.

In the MW range this new approach enables H-type conditions to be accomplished by combining fields generated by partial turn semi-lumped "inductors", giving as a result a fairly clean H field inside the plasma tube.[6,27]

3.3.2.2.3   Multi-loop Split Power Design. A multi-loop split power design giving a prevailing H field can be arranged as in Figure 3.57. This assembly can also be utilized as a base for explanation of the operating principle of the partial-turn helices described above. Showing them separately may be a didactic issue, as instead of power splitters one can introduce a plurality of power amplifiers each working with a common phase or with a desired phase shift (Figure 3.58).

3.3.2.2.4   H-field Coupling in a Single Rectangular Waveguide. In H-field coupling in a single rectangular waveguide the tube crosses through the waveguide perpendicularly to the narrow waveguide walls (Figure 3.59).[7] The advantage of this kind of coupling, compared to the cylindrical TE_{011} resonator, is the ease of impedance tuning in a rectangular waveguide where the standard EH or multi-stub tuners can be applied. The disadvantage is that the magnetic field is not exactly symmetrical around the axis of the plasma torch. However, the analytical results reported are encouraging.

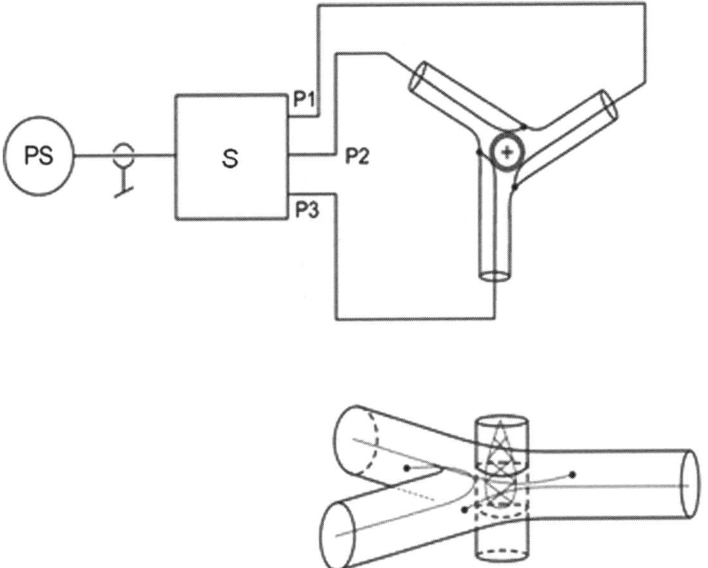

Figure 3.57 A multi-loop coupling: PS = power source, S = power splitter, P1–P3 = equiphase power inputs.

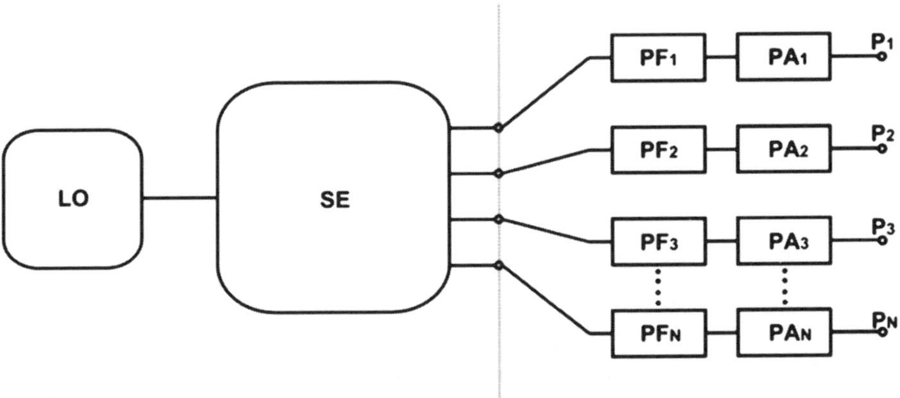

Figure 3.58 Application of a plurality of power amplifiers (LO = local oscillator; SE = splitter; PF_1–PF_N = phase shifters; PA_1–PA_N = power amplifiers).

3.3.2.2.5 H-field Coupling with Multiple Rectangular Waveguides. According to the deduction concluded from the present classification, a very symmetrical H-type plasma heating arrangement could be obtained using a microwave circuit formed by a plurality of rectangular waveguides (Figure 3.60). This idea of an improved Hammer-like assembly comes immediately from the idea of a

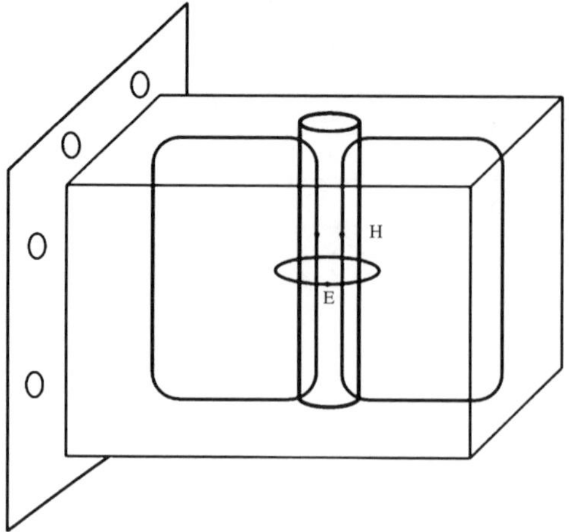

Figure 3.59   H-type waveguide coupling as disclosed by Hammer.[7]

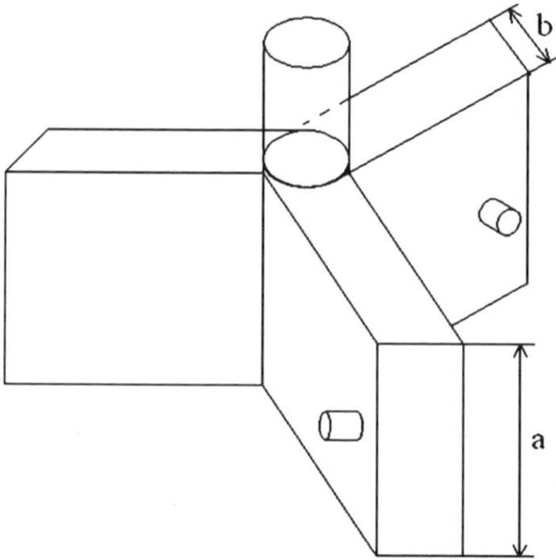

Figure 3.60   Star connection of rectangular H-type plasma couplers which can be energized by means of a power splitter or from separate MW power sources.

directionless split-MW power supply of plasma. This idea was introduced by the authors for the first time in 2007.[77] Possible applications of three phases for this case may include the use of three rectangular waveguides, but also three fin-line concentrators or three strip-line couplers. The involvement of a larger number of phases and the use of other types of MW lines, *e.g.* strip-lines of regular or co-planar geometries would also be feasible.

3.3.3 Hybrid EH-types of Microwave Plasma Sources

This type of cavity is an obvious consequence of the classified E- and H-types. Both single and double helices have been tested, confirming their applicability as plasma couplers that can excite atmospheric and reduced pressure discharges. The advantage of these cavities is good impedance matching, which seems to be independent of the plasma gas and can be attained without any operator's manipulated tuners.

3.3.3.1 Single-helix Single-phase Designs

This design is similar to the RF helical plasma reactor[64] and is illustrated in Figure 3.61.

3.3.3.2 Double-helix Two-phase Designs

The double-helix two-phase assembly can be positioned where the field between the helices differs by 90° or 180° (Figure 3.62). In case of a 90° difference, one can use the principle of self-matching, which is based upon the fact that the phase difference for the reflected waves is 180° and they cannot be summed in the splitter and must travel back to meet and heat the plasma again.

3.3.3.3 Triple or Quadruple Helices

Triple or quadruple helices may involve three or four phases (Figure 3.63). Even more phases can be used if desired. A characteristic feature of multiphase assemblies is very symmetrical heating, which should occur as a result of the rotating field, and the possibility of self-matching of the plasma impedance to the split microwave source. The greater number of phases, the more symmetrical the heating of the plasma and the easier generation of doughnut-shaped discharges.

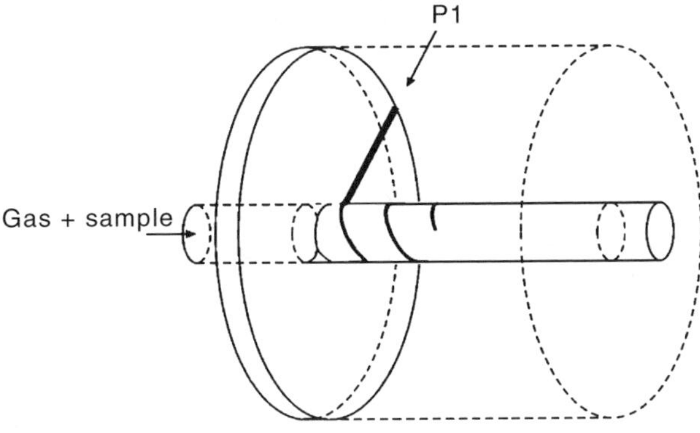

Figure 3.61 Single-helix coupler.

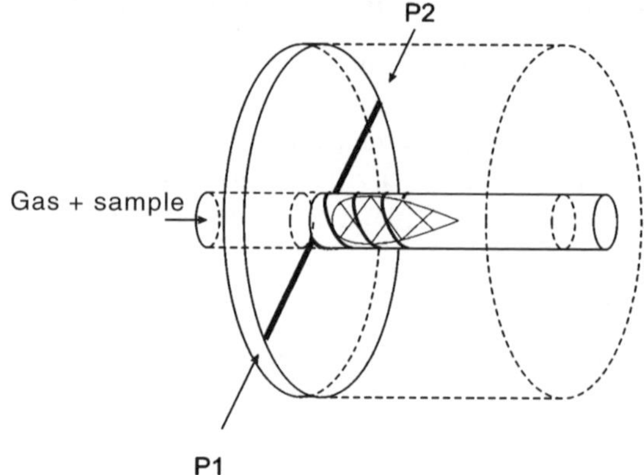

Figure 3.62 A double-helix coupler.

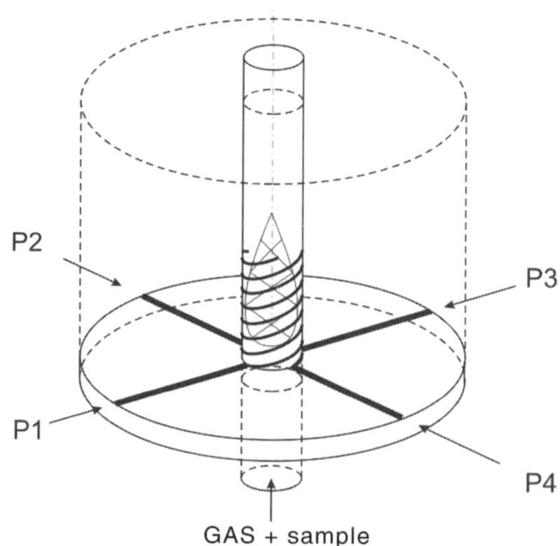

Figure 3.63 A four-helix four-phase-supplied plasma cavity.

3.3.3.4   *The Loop-gap Concept*

An interesting type of semi-lumped microwave circuit called a loop-gap resonator has been introduced by Froncisz and Hyde.[78] This type of resonator is
used in EPR spectrometers, where homogeneity of the magnetic field in the
sample is essential. A similar device with a gap length close to halfway has been
disclosed in a patent[79] and claimed as a means of guaranteeing that minimal
power levels for ignition of the discharge and for discharge maintenance are
essentially equal.

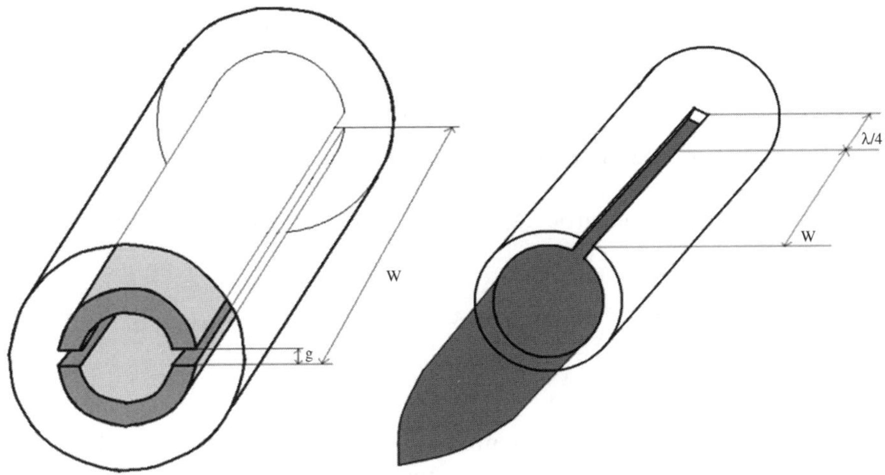

Figure 3.64　The loop-gap resonator[78] and a patented loop-gap plasma coupler.[79]

3.3.3.5　The Loop-gap–Waveguide Combination

The loop-gap–waveguide combination as a plasma applicator has been patented.[79] It comprises single or multiple loop-gap resonators as couplers that are energized using a rectangular waveguide (Figure 3.64). The so-arranged plasma sources enable the generation of stable oxygen, nitrogen or air plasmas at atmospheric pressure.

3.3.3.6　Discharges Generated with the Use of Dielectric Waveguides

Taking into account the infinite applications of silicon oxides around us, it may be useful to develop plasma sources in which the plasma torch plays the role of the MW-guiding structure and at the same time is the means for focusing the MW energy onto the discharge. Such a device has been registered as a compact electrodeless discharge lamp for lighting purposes[80] and in near future it may become a promising MIP.

3.4　Making Annular-shaped Microwave Plasmas

3.4.1　Introducing the Symmetry of Microwave Energy Coupling and Making a Doughnut-shaped Plasma

3.4.1.1　The Case of Stationary Fields

Attaining symmetrical modes of oscillation in high-power plasmatrons must incorporate special means to introduce a high degree of mode purity, for instance the above-mentioned filtering of mode purity is realized using a radial line resonator in series with a smooth waveguide-to-coaxial line transition (see Figure 3.30). Some methods involve the following.

1. Good symmetry can be obtained by applying symmetrical coaxial coupling of a circular resonator involving a low impedance section in the TEM-mode cavity, as presented in Figure 3.40, as well as the use of symmetrical concentric gaps such as those applied in the surfatron (see Figure 3.34).
2. A doughnut-shaped plasma can be obtained by applying a MPT in which the plasma attachment can be spread symmetrically around the axis, as in Figure 3.9.
3. Continuation of the MPT is used in common phase-split multi-candles, as in Figure 3.13.
4. Different TEM cavity assemblies in combination with coaxial coupling and low impedance sections, for instance the one shown in Figure 3.40.
5. Multipoint directionless coupling through the slits, as in the SLAN, Cyrranus and Kirjushin designs (see Figure 3.32) and also the leaky wall of Figure 3.19.
6. Multipoint plasma excitation by a plurality of symmetrically placed antennae, *e.g.* a plurality of short loops placed around the plasma (Figure 3.57). One should also include the technique of multi-helix coils shown in Figure 3.51.
7. Application of multi-waveguide structures, especially those promoting H-type discharges.

3.4.1.2 *The Case of Non-stationary Fields*

1. Application of rotating fields originating from splitting of the same power source or combined from a number of precisely synthesized and synchronized individual power sources.
2. Two orthogonal fundamental TE_{11} modes in a circular waveguide deliver the synchronically rotating field having its maximum on the axis. This system has been proposed for supplying MW-energized electrodeless discharge lamps.
3. Fields with random rotation may be obtained using two independent power sources which may differ in phase and frequency. The field rotates with a frequency that is equal to the difference between the frequencies of both power sources. A polarizer-combiner may be used to introduce isolation between the sources.
4. A plurality of electrodes connected to three or more phases with a phase shift equal to 360° divided by the number of phases can generate a synchronously rotating field having the minimum on the axis.
5. Commutated (switched) fields may be the technique of choice for energizing microdischarges using a plurality of small power sources. A

number of microwave switches (PIN diodes or mechanical switches) may be applied, generating a sequence of power pulses sent to coupling elements surrounding the plasma. This coupling may be arranged to cause the predicted plasma shapes.

3.4.2 Plasma-to-doughnut Shape Approaches

1. By taking advantage of the strong skin effect and applying the proper gas-flow shaping. The sample can be injected into the essentially field-free region along the discharge axis. This technique can be implemented for larger diameter plasmas, *i.e.* when a plasma is generated using relatively high-power microwaves and a large diameter torch.
2. By flow shaping only. Two separate gas flows in the torch are usual. This approach is usually less efficient and requires a precise division between the plasma and sample gas flow.
3. By fringing field shaping in the circular gap (Okamoto, surfatron). The field is partially arrested within the symmetrical gap to provide a stronger coupling at plasma boundaries than in the axis, leaving the plasma axis essentially free of currents. A plurality of gaps or a dense spiral may form a leaky wall (see Figure 3.19) acting the same way.
4. By using the MPT (as in Figure 3.9) or by introducing the multi-candle arrangement (as in Figure 3.13).
5. By applying a synchronously rotating field with three or more phases.
6. By controlling the commutation of the field pulses and making them rotate around the plasma axis.

3.4.3 Making the Annular-shaped Microwave Plasma

The swirl or laminar flow of shroud gas can prevent the plasma torch from melting; however, introducing symmetry of the MW supply can assure that the tendency of the discharge to move towards the wall is the same in all radial directions. Provided that the symmetry of the field is the ideal one, it is still necessary to have the gas flow shaped enough to maintain a stable plasma position. A new example of generating a field which is substantially symmetrical can be a multi-loop coupler having the form of a multi-turn helix, which leads to the concept of a microwave frequency-driven ICP-like discharge. The annular shape of a practical microwave-driven ICP has been obtained using the arrangement shown in Figure 3.65.

Symmetrical coupling of the MW energy to the plasma is realized in a new H-type plasma device using a multi-helix coupler that may become a building block for MW-ICPs. Application of common phase currents shows the way to reduce the electric field between the turns. This obvious feature can also bring some improvement to non-microwave traditions, *i.e.* an RF driven-ICP reducing the disturbances caused by the excess of the longitudinal component of the electric field and using the idea of having the RF energy supplied

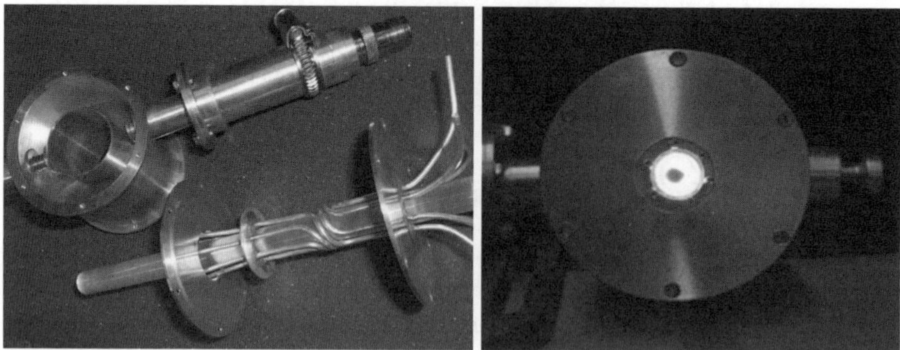

Figure 3.65 MW-ICP plasma at a power level of 200 W in a laminar flow of argon (*ca.* 1 L min^{-1}).

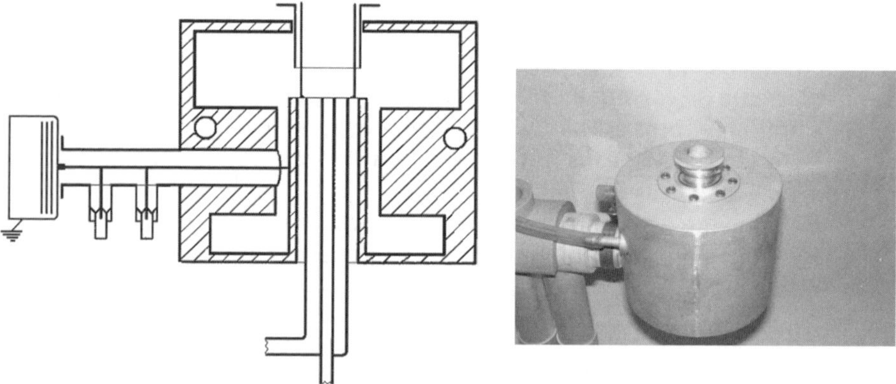

Figure 3.66 The TEM cavity cross-section and view.

simultaneously from a plurality of power amplifiers. There is great potential in the applications of H-type microwave discharges and their development seems just to be a question of time.

Another example of a doughnut-shaped plasma can be presented here, pointing to a good performance of the E-type TEM-mode circular cavity with a dual-flow torch (Figure 3.66). The low-flow annular-shaped helium MIP has been evaluated by the authors.[68,81] The new cavity is a good example of a combination of a coaxial coupler and a radial TEM-mode resonator. One of the aims in MIP development is the symmetrical delivery of MW energy to the plasma. The simplest symmetrical mode is the one propagated inside coaxial TEM-mode transmission lines. This very symmetrical mode is also found inside the TM_{010} resonators, but only in the case when their symmetry is not perturbed. The Beenakker cavity has coupling and tuning elements placed non-symmetrically and perturbation of the field is obvious. Placing the coaxially

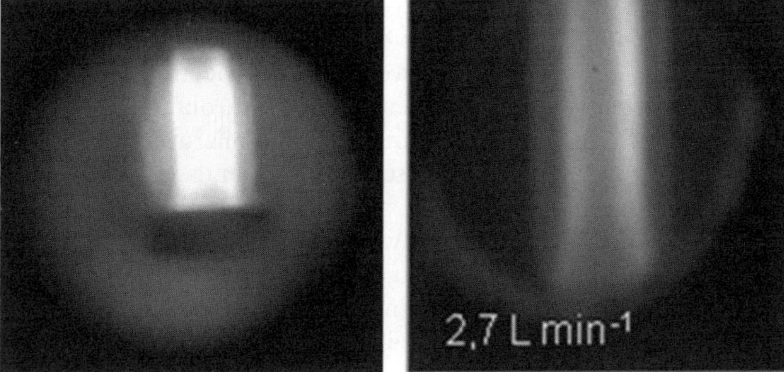

Figure 3.67   Low-flow helium plasmas.

shaped coupling and tuning elements on the axis of the Beenakker cavity, as shown in Figure 3.24, introduces the desired symmetry and in this way the authors obtained a doughnut-shaped helium plasma (Figure 3.67) at a low gas-flow rate $(1–3\,\text{L}\,\text{min}^{-1})$.[68]

A detailed idea of a symmetrical TEM-mode double (folded) MIP cavity has been described.[69] The construction of TEM cavities shown in Figures 3.37–3.41 can also be recalled as examples of how to couple the MW energy to the resonators without perturbing the symmetry of the plasma heating.

3.5   The Concept of Microwave Cavities with Rotating Microwave Fields

The operational principle of the multi-phase plasma is quite simple. It is assumed that the field is delivered by electrodes which are coupled to plasma immediately or through the dielectric barrier. When the number of phases is $N = 2$ and the difference between the phases is $360/N = 180°$, then the field vector between the electrodes appears as a stationary field that does not move but in exchange it exhibits symmetry of both the electrodes with respect to the ground shield. However, when the phase difference is chosen as only $90°$, then the field can rotate one revolution per period. When two antennae are involved the field may be generated only in the case that it can be excited in the cavity, *i.e.* the cavity must have dimensions which allow for excitation of at least a fundamental mode. Without the cavity one should use four electrodes in two pairs, each generating the TEM-mode field with a $180°$ shift; for the second pair the field is also shifted $180°$ but in quadrature, *i.e.* it is orthogonal to the first pair. With three and more electrodes involved, the field can propagate in TEM modes rotating at least in the fringing vicinity of the electrodes whether contacting the plasma or acting through the barriers of the dielectric material. Each electrode involved, even in a case when it is in contact with plasma, can work without releasing electrode material due to etching or sputtering. If the frequency of the field is very low, for instance a line frequency of an ac mains, then depending on what is the actual sign of the electrode it periodically becomes an anode or a cathode when conducting the ac currents. Rotating fields of this kind may be excited not only at microwave frequencies. The first is a two-phase electric arc at the frequency of the mains. By altering different frequencies one can obtain discharges at the mains and at audio frequencies, RF frequencies and microwave frequencies. For higher RF and microwave cases the discharge surrounds the electrode tips and the capacitance between the electrode and plasma is usually high enough to conduct the current flow in a capacitive way. This means that at MW frequencies the electrodeless mode can be realized without introducing the real physical solid-state dielectric barrier. The rotating field-excited discharges have already been applied for electrodeless discharge lamps[73] as well for plasma processing.[74] The microwave version of such a plasma generated in a circular waveguide using a rotating MW field has been applied in an ultra-high-power plasmatron[64,82,86] and in a diamond growing plasma rig.[77]

The systems with multiphase-combined power can be arranged by a plurality of high-power oscillators or regular oscillators with power amplifiers where the oscillators are mutually synchronized in order to realize the required number of channels, each equal in magnitude and different in phase. From the point of view of the unambiguous experimental conditions, each channel should be terminated with a ferrite isolator or a three-port circulator with a matched load. In this case, whatever are the matching parameters of the plasma to the channel line, neither the power nor the phase shift would be affected by power reflections.[3] The use of stable power amplifiers can also assure the required rigidness of the multiphase system. A closer look at the problem of plasma matching indicates the possibility of self-compensation, which is likely to occur in a multiphase system. The representation of several impedances which are the same in the module, but different in phase when displayed in the Smith Chart,[3] clearly indicates this possibility. First of all the impedance resulting from the appropriate summation of individual channels has reduced to zero the imaginary part, which means the ability of the system to self-tune its resonance frequency. This feature can be deduced from a simple analysis of a multiphase system which comprises of a set of N sections of coaxial line, each having a different length to provide a different phase shift. Each section terminates with an arbitrary admittance that represents the plasma load and has an input admittance $Y_n(n)$ which results from transformation of Y_L through the length of section with characteristic wave admittance Y_0. The overall value of the sum of all admittances can be expressed as:

$$Y_{\text{in}} = \sum_{n=1}^{N} Y_n = \sum_{n=1}^{N} (G_n + jB_n)$$

where

$$Y_n = G_n + jB_n = \frac{Y_L + jY_0 \tan(\beta l_n)}{Y_0 + jY_L \tan(\beta l_n)}$$

and is a complex input admittance of the nth section which can be calculated according to general formulae.[3]

Using the above formula, one can confirm, for example, that when the number of electrodes $N \geq 3$, the imaginary part of the input admittance (*i.e.* the input susceptance B_{in}) is vanishing, leaving only the real value G_{in} which has to be matched to the output admittance of the MW power source. However, the equal splitting of the energy can be achieved only when all the real values of Y_n (*i.e.* the conductances G_n) have the same equal values as is the case, for instance, when the number of phases $N = 4$ and the phase angle of Y_1 is 45°, which corresponds to a line section of $\lambda/8$. For this case, one may obtain the simplest plasma cavity shown in Figure 3.68 in which $l_1 = \lambda/8 + n\lambda/2$, $l_2 = 3/8\ \lambda + n\lambda/2$ and $l_3 = 5/8\ \lambda + n\lambda/2$. Another simple structure for a similar plasma cavity is shown in Figure 3.69, where four quarter-wave-long antennae are connected to a common point by $\frac{1}{4}\lambda$-long sections of coaxial lines with different characteristic impedances.

In a more general case when an arbitrary number of phases is to be involved, separation of the channels may be necessary. This can be accomplished using ferrite circulators. For experimental purposes, when one wants to control the flow of power in each channel of the system it is convenient to use ferrite isolators which guarantee that the power reflected will not influence the equal partition of MW energy between the electrodes. The isolators can be substituted with ferrite circulators having the third arm connected to dummy loads, as shown in Figure 3.70.

However, microwave circulators can play the role of non-reciprocal phase shifters, the use of which makes possible the real compensation of power reflected from the plasma, as shown in Figure 3.71. As a matter of fact, this original idea is about how to build a system that does not need too much tuning. In each channel (see Figure 3.71) the waves reflected from the plasma have phase differences of 120° and they cannot combine as their vector sum is always zero and therefore they must return to the plasma again. Of course, a larger number of phases can also be applied, having the same self-tuning feature.

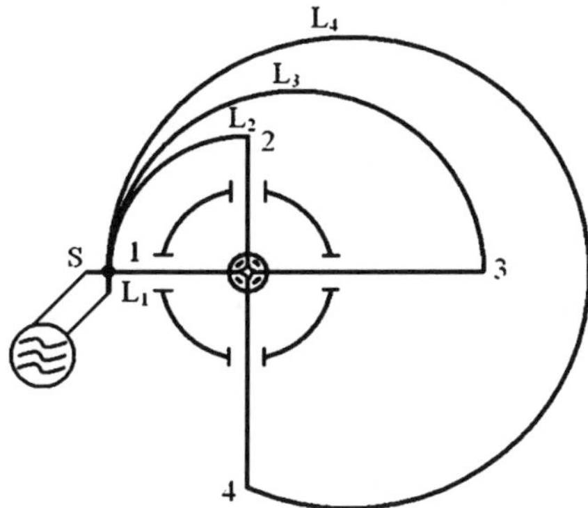

Figure 3.68 Simple microwave cavity with four-phase electrodes.

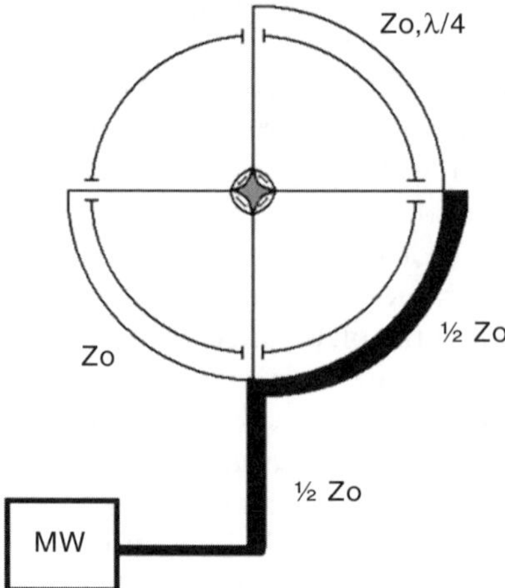

Figure 3.69   A cavity based on the design of a commonly used TV antenna splitter.

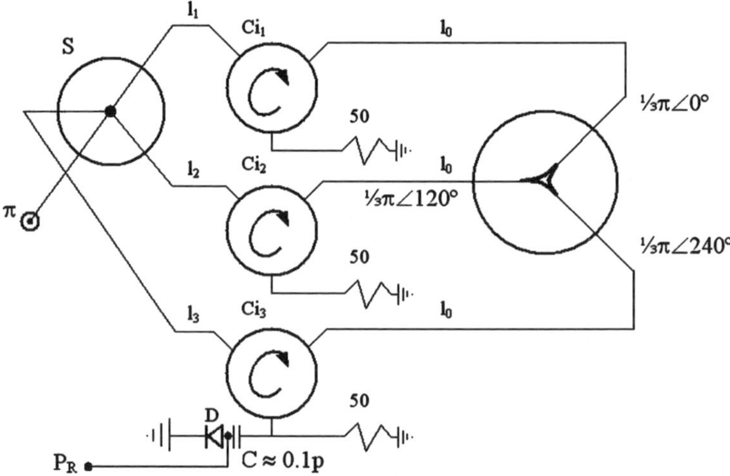

Figure 3.70   A typical experimental system with a power splitter (S), three isolators (circulators) and a power-reflected monitoring diode (D).

The role of the channel separator can be fulfilled by introducing power amplifiers separately for each channel. The new system with power distributed by channels is especially suitable, even beneficial, to solid-state applications. In Figure 3.72 a set of five power amplifiers has been used for plasma generation.

Even a relatively high-power operation (1–3 kW) can be arranged in a 10-channel system using regular power amplifiers and the vector arithmetic necessary for each phase can be done at the low-power side.

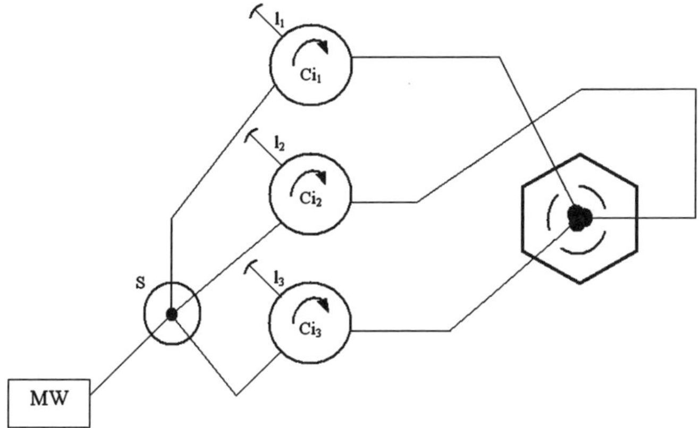

Figure 3.71 A plasma system with self-compensation of power reflected (a three-phase example).

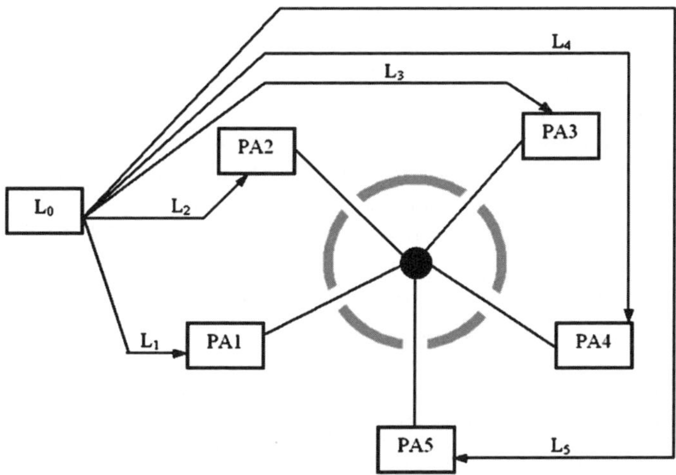

Figure 3.72 A plasma system comprising power amplifiers PA1–PA5 driven by a single local oscillator (LO).

3.5.1 Comments on Plasma Contamination in the New Capacitive Microwave Plasma Systems

The rotating plasma electrode system that has all the CMP features may contaminate the plasma with electrode material that can be released by sputtering, melting or by chemical reactions between the sample, plasma gas, the liquid and the gas carrying the sample, *etc.* In order to protect the electrodes, one can apply a special flow of the electrode shroud at a flow rate suitable to

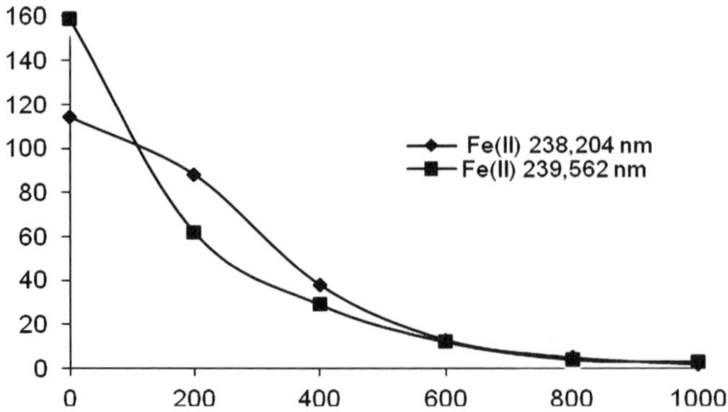

Figure 3.73   Reduction of Fe lines emission *vs.* protective gas flow (ml min^{-1}).

stop the electrode wear. The results of the corresponding experiments are shown in Figure 3.73.

Other non-contaminating systems have been realized by the introduction of platinum tips covering the electrodes. Moreover, successful trials have been done by changing the CMP system into the MIP system. This aim has easily been achieved by having the electrode tips embedded in high-temperature-resistant sapphire ceramic shells.

Similar no-tube arrangements may also be managed using circular resonator-based MIPs with a set of additional CMP dielectric-embedded electrodes used in order to improve the plasma stability, enhancing the sample passing through the plasma and improving the symmetry of the MW field.

There are several interesting features of the new CMP multi-electrode cavities. Firstly, the plasma is generated in a planar geometry and the currents between the electrodes and shield are not important or can be eliminated. Secondly, the new cavity can operate at MW but also at much lower frequencies. This concerns designs with a common phase (multi-candle) as well as with multiphase rotating fields. In the case of lower frequencies the common phase 1/4 λ Wilkinson-like splitter may be arranged by applying series inductances with a resonant capacitors at their ends.

A rotating plasma working at the frequency of the mains and dedicated to analytical applications was described in the 1980s.[84] Such a plasma can also be energized at audio and RF frequencies. Using higher frequencies seems to be advantageous. A 27 MHz case arranged from the sections of a TV feeder has already been positively tested in one of our experiments. Moreover, having in mind atmospheric pressure glow discharge applications, the tests have been extended down to audio frequencies and a very stable cold plasma has been found feasible well below 1 kHz frequency, which can be a good argument that this technology is completely aperiodic. For instance, a 600 Hz (audio) frequency plasma system based on a 12-phase supply derived from three-phase mains has found application in the production of carbon nanotubes.[83] The three-phase rotating-field excited plasma has found industrial applications in specialty fibre splicing,[84] where the three-phase field was used at a working frequency of 30 kHz. An even more powerful 500 kW plasma heated in a rotating microwave field at a frequency of 915 MHz has been described.[85]

It is worth stressing that the use of multiphase techniques may also be applied to generate other than cylindrical shapes of plasma, *e.g.* a planar shape can be generated at the surface of a micro electro-mechanical system (MEMS) by applying a plurality of phase electrodes having the form of a strip-line ladder arranged in-line or dual-in-line.

3.6 Final Remarks: Thinking of the Future

In light of the present classification, one needs to comment on some terminological difficulties. The first difficulty is connected with a definition of CMPs that was introduced many years ago for a single electrode coaxial cavity in which, despite a plasma–metal electrode contact, there was no contamination in the spectrum. It was due to a specific microwave feature that the energy could be coupled efficiently even through a very small capacitance formed within the plasma-to-metal gap. At RF frequencies the term capacitively coupled plasma usually means that there exists a solid dielectric layer separating the plasma from the electrode. If there is no solid dielectric, the discharge is called an RF arc. The next difficulty is connected with MIPs which are specifically microwave not RF devices. Going down toward even lower frequencies, one can notice that the use of the term "dielectric barrier" is in fact more general and could be extended to other capacitively coupled RF or ac plasmas and, by analogy, to all MIPs. However, it has to be noticed that the dielectric barrier has been developed not to separate electric potentials but, before all, to limit the RF or ac currents per unit surface (current density) and in this way eliminate any tendencies toward the formation of electric arcs. The same current density limiting role is played by the resistive barrier or any other current-limiting means, for instance metallic or dielectric plates with a network of small holes which cause plasma constriction and make the plasma look like a gas burner.

The known H and EH plasma configurations at MW frequencies are mostly MIPs, but perhaps one can imagine also some possible realizations of CMPs.

The present classification clearly demonstrates its usefulness in generating new MIP cavities and to point out what may be important factors in the design of an efficient microwave plasma excitation or ionization source. Namely, at least five new constructions were deduced from the classification and assigned as those which were worth trials. In that number there were three TEM symmetrical cavities, two new H-type cavities and three helical hybrid EH cavities, along with different non-stationary field arrangements including multiphase excitations as well as the possible amplitude modulation or switching of the plasma heating fields.

The development of analytical instruments is probably a never-ending story. Every year one notices something exceptionally improved or completely novel. This will also be the case for MWPs. Each novelty should pass extensive examination and become accepted by researchers in many laboratories and finally by equipment manufacturers. The new classification has certainly widened the scope of possible designs and may help to find new solutions for the future.

References

1. G. I. Babat, *News of Electro-industry*, 1942, **3**, 47–51 (in Russian).
2. *Theory of the Electric Arc*, ed. M. F. Zhukov, Nauka, Novosibirsk, 1977 (in Russian).
3. D. M. Pozar, *Microwave Engineering*, Wiley, New York, 2005.
4. E. H. Evans, J. J. Giglio, T. M. Castillano and J. A. Caruso, *Inductively Coupled and Microwave Induced Plasma Sources for Mass Spectrometry*, RSC, Cambridge, 2005.
5. E. Reszke *et al.*, *Microwave H-type Discharge Trials*, Plazmatronika report 2001/2.
6. E. Reszke, K. Jankowski and A. Ramsza, *Pol. Pat. Appl.*, P-385484, 2008.
7. M. R. Hammer, *US Pat.*, 7 030 979, 2006; *US Pat.*, 6 683 272, 2004.
8. M. R. Hammer, *Spectrochim. Acta, Part B*, 2008, **63**, 456–464.
9. J. D. Cobine and D. A. Wilbur, *J. Appl. Phys.*, 1951, **22**, 835–841.
10. G. Knapp, E. Leitner, M. Michaelis, B. Platzer and A. Schalk, *Int. J. Environ. Anal. Chem.*, 1990, **38**, 369–378.
11. A. Simon, T. Frentiu, S. D. Anghel and S. Simon, *J. Anal. At. Spectrom.*, 2000, **20**, 957–965.
12. A. Piotrowski, R. Parosa, E. Reszke and J. Rzepka, *Proceedings of the XVIII ICPIG*, Swansea, 1987, **vol. 4**, pp. 860–861.
13. *Advances in Microwave and Radio Frequency Processing*, ed. M. Willert-Porada, Springer, Berlin, 2006.
14. A. C. Metaxas, *Foundations of Electroheat*, Wiley, New York, 1996.
15. Gerling Applied Engineering; www.2450MHz.com.
16. T. Ichiki, T. Koidesawa and Y. Horiike, *Plasma Sources Sci. Technol.*, 2003, **12**, S16–S20.
17. M. Miclea and J. Franzke, *Plasma Chem. Plasma Process.*, 2007, **27**, 205–224.
18. J. A. C. Broekaert, V. Siemens and N. H. Bings, *IEEE Trans. Plasma Sci.*, 2005, **33**, 560–561.
19. C. I. M. Beenakker, *Spectrochim. Acta, Part B*, 1976, **31**, 483–486.
20. J. Asmussen, R. Mallavarpu, J. R. Hamann and H. C. Park, *Proc. IEEE*, 1974, **62**, 109–117.
21. P. L. Kapica, *J. Exp. Theor. Phys.*, 1969, **57**, 1801 (in Russian).
22. E. Reszke *et al.*, *Square Shaped Microwave Plasma Cavity*, Plazmatronika report 2001/1.
23. H. Feuerbacher, *Microwave Plasma Emission Spectroscopy*, commercial leaflet, AHF, Tübingen.
24. H. Matusiewicz, *Spectrochim. Acta, Part B*, 1992, **47**, 1221–1227.

25. J. Yang, J. Zhang, C. Schickling and J. A. C. Broekaert, *Spectrochim. Acta, Part B*, 1996, **51**, 551–562.
26. K. Jankowski, R. Parosa, A. Ramsza and E. Reszke, *Spectrochim. Acta, Part B*, 1999, **54**, 515–525.
27. E. Reszke, *H-type Microwave Plasmatron*, Ertec report, 2002/3.
28. P. P. Woskov, D. Y. Rhee, P. Thomas, D. R. Cohn, J. E. Surma and C. H. Titus, *Rev. Sci. Instrum.*, 1996, **67**, 3700–3707.
29. J. Galliker, *Bull. d'Information, Union Internationale d'Electrothermie, Paris*, 1972, **11**, 1–12as cited in M. Hering, *Foundations of Electroheat, Part II*, WNT, Warsaw, 1998 (in Polish).
30. Y. Okamoto, *Anal. Sci.*, 1991, **7**, 283–288.
31. K. Oishi, T. Okumoto, T. Iino, M. Koga, T. Shirasaki and N. Furuta, *Spectrochim. Acta, Part B*, 1994, **49**, 901–914.
32. Y. Okamoto, M. Yasuda and S. Murayama, *Jpn. J. Appl. Phys.*, 1990, **29**, L670–L672.
33. V. P. Kirjushin, *US Pat.*, 3 577 207, 1971.
34. H. Conrads and M. Schmidt, *Plasma Sources Sci. Technol.*, 2000, **9**, 441–454 see also www.plasmaconsult.com.
35. R. Spitzl, B. Aschermann and M. Walter, *US Pat.*, 6 198 224, 2001; see also www.cyrannus.com/pages/welcome.html.
36. L. Liang, Z. Guixin, L. Yinan, Z. Zhijie, W. Xinxin and L. Chengmu, *Plasma Sci. Technol.*, 2008, **10**, 83–88.
37. R. K. Skogerboe and G. N. Coleman, *Anal. Chem.*, 1976, **48**, 611A–622A.
38. S. Murayama, *Spectrochim. Acta, Part B*, 1970, **25**, 191–200.
39. G. F. Larson and V. A. Fassel, *Anal. Chem.*, 1976, **48**, 1161–1166.
40. Q. Jin, C. Zhu, W. Borer and G. M. Hieftje, *Spectrochim. Acta, Part B*, 1991, **46**, 417–430.
41. Y. S. Bae, W. C. Lee, K. B. Ko, Y. H. Lee, W. Namkung and M. H. Cho, *J. Korean Phys. Soc.*, 2006, **48**, 67–74.
42. R. Stonies, S. Schermer, E. Voges and J. A. C. Broekaert, *Plasma Sources Sci. Technol.*, 2004, **13**, 604–611.
43. S. J. Ray and G. M. Hieftje, presented at the Winter Conference on Plasma Spectrochemistry, Florida, 2010.
44. Q. Jin, G. Yang, A. Yu, J. Liu, H. Zhang and Y. Ben, *J. Natl. Sci. Jilin Univ.*, 1985, **1**, 90 (in Chinese).
45. S. J. Ray and G. M. Hieftje, *Anal. Chim. Acta*, 2001, **445**, 35–45.
46. M. Moisan, G. Sauve, Z. Zakrzewski and J. Hubert, *Plasma Sources Sci. Technol.*, 1994, **3**, 584–592.
47. J. Torres, E. Iordanova, E. Benova, J. J. A. M. van der Mullen, A. Gamero and A. Sola, *J. Phys.: Conf. Ser.*, 2006, **44**, 179–184.
48. J. D. Yan, C. F. Pau, S. R. Wylie and M. T. C. Fang, *J. Phys. D: Appl. Phys.*, 2002, **35**, 2594–2604.
49. J. Kim and K. Terashima, *Jpn. Pat.*, 2007, **29**, 9720.
50. J. Kim, M. Katsurai, D. Kim and H. Ohsaki, *Appl. Phys. Lett.*, 2008, **93**, 191505.

51. M. Goch, M. Jasinski and Z. Zakrzewski J. Mizeraczyk, *Czech. J. Phys.*, 2006, **56**(suppl. 2), B795–B802.
52. E. J. Wilkinson, in *Classic Works in RF Engineering*, J. L. B. Walker, D. Myer, F. H. Raab and C. Trask, (ed.), Artech House, Boston, 2006, pp. 311–313.
53. M. Horvath, E. Reszke and G. Heltai, presented at the Colloquium Spectroscopicum Internationale XXXVI, Budapest, Hungary, 2009.
54. J. Gregorio, O. Leroy, P. Leprince, L. Alves and C. Boisse-Laporte, *IEEE Trans. Plasma Sci.*, 2009, **37**, 797–808.
55. F. Iza and J. A. Hopwood, *IEEE Trans. Plasma Sci.*, 2003, **31**, 782–787.
56. A. M. Bilgic, E. Voges, U. Engel and J. A. C. Broekaert, *J. Anal. At. Spectrom.*, 2000, **15**, 579–580.
57. M. Nagai, M. Hori and T. Goto, *J. Vac. Sci. Technol. A*, 2005, **23**, 221–225.
58. T. Goto, M. Hori, S. Den and M. Nagai, *US Pat.*, 2008 00 29 030.
59. J. Kim and K. Terashima, *Appl. Phys. Lett.*, 2005, **86**, 191504.
60. R. Parosa and E. Reszke, in *Advances in Low-Temperature Plasma Chemistry, Technology, Applications*, Technomic Publishing, Lancaster, PA, 1991, **vol. 3**, pp. 264–274E. Reszke and R. Parosa, *ibid.*, vol. 1, pp. 1–7.
61. M. Moisan and Z. Zakrzewski, *J. Phys. D: Appl. Phys.*, 1991, **24**, 1025.
62. M. Jasinski, M. Dors and J. Mizeraczyk, *J. Power Sources*, 2008, **181**, 41–45.
63. Z. Machala, M. Janda, K. Hensel, I. Jedlovsky, L. Leštinska, V. Foltin, V. Martišovitš and M. Morvova, *J. Mol. Spectrosc.*, 2007, **243**, 194–201.
64. I. N. Toumanov, *Plasma and High Frequency Processes for Obtaining and Processing Materials*, Nova Publishers, New York, 2002.
65. K. A. Forbes, E. E. Reszke, P. C. Uden and R. M. Barnes, *J. Anal. At. Spectrom.*, 1991, **6**, 57–71.
66. W. F. Burov and Y. W. Strizhko, *Microwave Plasmatron with Freely Floating Plasmoid*, Publication U.D.K. 621.181.662.9 (in Russian).
67. W. F. Burov, *WO Pat. Appl.*, 2009, **128**, 741.
68. K. Jankowski, A. Jackowska, A. P. Ramsza and E. Reszke, *J. Anal. At. Spectrom.*, 2008, **23**, 1234–1238.
69. E. Reszke *et al.*, *Folded Coaxial Cavities*, Plazmatronika report 2001/5 (in Polish).
70. K. Jankowski, A. Ramsza and E. Reszke, *Pol. Pat. Appl.*, P-385512, 2008.
71. R. M. Barnes and E. Reszke, *US Pat.*, 5 049 843, 1991.
72. M. Outred and C. B. Hammond, *Phys. Scr.*, 1976, **14**, 81.
73. J. E. Simpson and M. Kamarehi, *US Pat.*, 5 227 698, 1993.
74. A. Z. Bogautdinov, V. K. Zhivotov and V. D. Rusanov, in *Problems of Nuclear Science and Engineering*, 1987, vol. 2, pp. 57–58.
75. J. Asmussen, H. H. Lin, B. Manring and R. Fritz, *Rev. Sci. Instrum.*, 1987, **58**, 1477–1486.
76. J. A. Hopwood, *J. Microelectromech. Syst.*, 2000, **9**, 309–313.
77. K. Jankowski, A. Ramsza and E. Reszke, presented at the East European Symposium on Plasma Chemistry, Brno, 2007; also K. Jankowski, A.

Ramsza and E. Reszke, presented at the XIXth Slovak-Czech Spectroscopic Conference, Casta-Papiernicka, Slovakia, 2008.

78. W. Froncisz and J. S. Hyde, *J. Magn. Reson.*, 1982, **47**, 515–521.
79. A. Yonesu, *US Pat.*, 2009 0260972.
80. F. M. Espiau and Y. Chang, *US Pat.*, 2009 243488.
81. K. Jankowski, A. Ramsza and E. Reszke, presented at the 2010 Winter Conference on Plasma Spectrochemistry, Fort Meyers, FL, 2010.
82. M. A. Lieberman and A. J. Lichtenberg, *Principles of Plasma Discharges and Materials Processing*, Wiley, New York, 1994.
83. T. Matsuura, K. Taniguchi and T. Watanabe, *Thin Solid Films*, 2007, **515**, 4240–4246.
84. T. R. Mattoon and E. H. Plepmeier, *Anal. Chem.*, 1983, **55**, 1045–1050.
85. R. Wiley and B. Clark, *Area Isothermic Plasma for Large Diameter and Specialty Fiber Splicing* , 3SAE Technologies, Franklin, TN, Optical Fiber Communications, National Fiber Optic Engineers Conference, 2008. See also US Pat. 2008187273 (A1), 2008.
86. V. Ivanov, B. Nikitin, S. Brykov, G. Eilenkrig, V. Rusanov, V. Zhivotov, S. Dresvin, D. Ivanov and N. K. Shi, *Highly Efficient Plasma Technologies Using Powerful High-Frequency and Microwave Sources*, 2010; available on the web: leonardo-energy.org/webfm_send/80

CHAPTER 4
Microwave Safety

4.1 Introduction

A microwave plasma system can be assembled by connecting several functional blocks, such as a microwave power supply, a circulator or isolator with water load or water cooling impedance tuners and a cavity applicator. The last is equipped with a gas supply, an optical pick-up assembly or ion sucking skimmer equipped with a mass flow control system for working with gases. Moreover, the system should be equipped with a microwave cable or waveguide sections, tuners and plasma ignition tools, a forced air and/or water cooling assembly, a sample introduction system with drainage and venting and an instruction manual (usually considered as too long). General laboratory rules must be obeyed when operating a microwave plasma (MWP) system, but there is one task that is new for a chemist who is not necessarily experienced in power microwaves: a microwave hazard which must always conform to the required rules and standards. If the equipment is supplied by a company, then that company is required to provide all the necessary certificates which should confirm that the equipment was fabricated conforming to the prescribed local laws.[1,2] It is common to deal with home-made apparatus that has been fabricated at laboratory facilities or modified therein. In such cases, in order to import the minimal safety requirements, basic measurements must be executed "in situ", *i.e.* at the place where the modifications were done and just after they have been implemented.[3,4]

4.2 Microwave Frequencies Permitted to be Used in Analytical Instrumentation

There are at least three possible microwave frequencies which may be assumed to be "legal" for microwave heating: 915 MHz (L-band), 2.45 GHz (S-band) and 5.8 GHz, among which the 2.45 GHz is the most popular. The 2.45 GHz in the S-band has already been attained by high-power transistors and microwave

RSC Analytical Spectroscopy Monographs No. 12
Microwave Induced Plasma Analytical Spectrometry
By Krzysztof J. Jankowski and Edward Reszke
© The Royal Society of Chemistry 2011
Published by the Royal Society of Chemistry, www.rsc.org

induced plasma (MIP) devices may be energized using solid-state power sources. Until now the cheaper magnetrons from microwave ovens are still used as building blocks of MWP systems since there is no real benefit from using transistors. The situation may change for the new microwave cavities described in Chapter 3, when a plurality of power sources may be applied synchronously. Assembling such a system composed of a plurality of relatively small solid-state sources will be easier and may also be cheaper and more reliable than in the case of vacuum tubes. Nevertheless, it seems necessary to mention the use of higher frequency bands which have become available for modern microwave processing of materials. In the book by Willert-Porada,[5] among the standard frequencies are also listed the non-standard choices going up to sub-millimeter frequencies, along with references to the literature on high-power sources such as travelling wave tubes or gyrotrons.

4.3 Working with Microwave Plasmas

Every piece of microwave equipment must firstly be certified by the manu-facturer and, secondly, after installation in the laboratory it must be reported in writing as checked in the place of use. The problem is that most of the plasma rigs are treated as experimental instruments which are to be modified in the course of laboratory expertise. Therefore, as mentioned, some more careful experimental procedures must be implemented in order to protect the operators against microwave radiation. First of all, an elementary knowledge of the equipment by the user is necessary for knowing how to measure and how to deal with MW leakages. It is good if the group is supported by a microwave engineer who is equipped with at least one of the standard leakage testers, such as the very popular, although old, model 1501 tester from Holaday Instru-ments. Much simpler and cheaper meters can also be used. They may help to detect a leakage and learn whether there is a more serious problem or not. One can assume that a good microwave oven tester should detect the leakage radiated from any regular microwave oven showing about half of the allowed field density somewhere at a distance of 5 cm from the doors of the oven cavity when it is loaded with a glass of tap water. It is reasonable to expect that a professional instrument to be sealed at least twice to ten times better. The excessive increase in radiated field can be noticed during the ignition of the plasma and therefore special care must be undertaken when starting the dis-charge. When the tool for ignition is a simple piece of wire, the radiation during the ignition may be very high. First of all the ignition must not take a long time during which the operator may expose his hand but also his head, if not even more sensitive gonads. It is imperative that during plasma ignition the operator should be trying to move as much as possible, as the danger is even more severe from stationary than from non-stationary field exposures. Generally, this method of discharge ignition should be considered dangerous and the exposure time during the ignition should be limited to less than 1 second. Much more reasonable is automation of the ignition procedure by moving the wire with the use of an electromagnet and/or application of a high-voltage spark taken from a piezo generator or Tesla transformer.

4.4   General Rules and Methods

One has to keep the MW energy closed in metal boxes, which may have openings. These openings, however, must be tightly closed for MW energy. Paraphrasing the well-known Murphy's law, one can say that if a cavity with plasma can radiate microwaves, it will. Therefore proper shielding must always be applied and appropriate means should be provided to eliminate any possibilities of radiation.

The principle behind good microwave shielding is continuity of the shield. If there are any openings whose size is close compared with half of the wavelength, then tight closing would mean a good continuous contact of an additional shield, usually in the form of a metal mesh. Continuous contact (soldering or welding) may be replaced with riveting, providing that the distances between the rivets are maintained as less than 1/8th of the wavelength. Adhesive Alu-tape can be applied in addition in order to seal the junctions. Large openings can be blocked using contact-less chokes built into the microwave system, exactly like those chokes which are used in the doors of modern microwave ovens.

As a matter of fact, one can use an empty microwave oven cavity to play the role of a Faraday cage, which will always prevent leakage from a MIP system, but then another set of special feed-through chokes should be used to decouple all the cables and metal tubing.

In the construction of MW plasma cavities the use of non-contact chokes may be considered generally applicable in commercial apparatus. This statement has taken into account the fact that the chokes of microwave ovens need more optimization than the less complicated traditional contact assemblies.

When openings are very small compared with the wavelength (*e.g.* 1%), then one can leave them open. In order to allow viewing through the electrical screen, one can always use metal mesh, such as that used in microwave cavity doors. Also, openings comparable but smaller than half of a wavelength may be left open if they have the form of the so-called under-critical tubes and are long enough to strongly attenuate and reflect the wave back into the cavity.

The connection between the shield and the under-critical tube must be made continuous by soldering, welding or by using a flange with plurality of rivets, screws or bolts. In fact, the continuity of a shield must be assured only along the paths of the electric currents. For instance, a slotted line which is used for impedance measurements has the slot cut along the outer conductor of the coaxial line and that slot is non-radiating by definition, as all the currents in the coaxial line flow along the axis. However, to be able to design slotted screens one has to know the field distribution exactly.

Usually, the plasma emanating from the cavity plays the role of a MW antenna and the MW currents all flow along the plasma. Therefore a slot cut along the screen may be used to allow for plasma observation without releasing MW energy outside.

During the experimental work there can never be too many precautions against MW radiation. The power of the MW source should be set up step by step and if there is a leakage one can react early to define what may be the problem. It is always desirable to detect MW radiation at the place it appears and where it shows a local maximum. According to regulations, which vary by country, such a local maximum measured at a distance d_x from the place of the leak should not exceed S_{max}. The value of d_x is typically 5 cm while S_{max} differs in different countries, lying in the range of $1-10 \, \mathrm{mW \, cm^{-3}}$, although the common opinion of microwave power engineers is that S_{max} should be kept below $2 \, \mathrm{mW \, cm^{-3}}$ during the experiments and well below $0.2 \, \mathrm{mW \, cm^{-3}}$ in case of 24/24 working of commercial equipment.[3] In Europe one has to fulfil the safety requirements within the frame of the so-called Low Voltage Directive (LVD)[1] and the radiated distortions must conform to the levels prescribed by the rules of the Electromagnetic Compatibility (EMC).[2] The detailed European regulations concerning laboratory equipment have been grouped together.[6]

References

1. *CE Conformity Requirements in Frames of the LV Directive*, European Commission, *Electrical Safety: Low Voltage Directive*, 2006/95/EC.
2. *Electromagnetic Compatibility (EMC)*, European Parliament, Directive 2004/108/EC.
3. P. Vecchia, R. Matthes, G. Ziegelberger, J. Lin, R. Saunders and A. Swerdlow, *Review of the Scientific Evidence on Dosimetry, Biological Effects, Epidemiological Observations, and Health Consequences concerning Exposure to High Frequency Electromagnetic Fields (100 kHz to 300 GHz)*, International Commission on Non-Ionizing Radiation Protection, ICNIRP 16/2009.
4. *NRL for Non-ionizing Electromagnetic Fields and Radiation; Electromagnetic Field (0 Hz–300 GHz) – Status as of 1 January 2009*, ICNIRP Information No. 16/2009.
5. W. Van Loock, *European Regulations, Safety Issues in RF and Microwave Power*, in *Advances in Microwave and Radio Frequency Processing*, ed. M. Willert-Porada, report from the 8th International Conference on Microwave and High Frequency Heating, Bayreuth, Germany, 2001, part II, pp. 85–91.
6. Selected contemporary European documents: (a) CISPR11 – *Industrial, Scientific and Medical (ISM) Radio-frequency Equipment – Electromagnetic Disturbance Characteristics – Limits and Methods of Measurement*; (b) EN 61010 – *Safety Requirements for Electrical Equipment, Control and Laboratory Use*; (c) EN 60519 – *Safety in Electroheat Installation, Part 6: Specifications for Safety in Industrial Microwave Heating Equipment, Control and Laboratory Use*.

CHAPTER 5

Optical Emission Spectrometry with Microwave Plasmas

5.1   Origins of Atomic Spectra

The emission of characteristic radiation by a sample requires the use of an excitation source at a temperature sufficiently high to cause thermal decomposition of molecules as well as collisional or radiative excitation and ionization of the atoms formed. Once the atoms or ions are in their excited states, characterized by very short lifetimes (typically 10^{-8} s), they can return to lower states through radiative energy transitions. All of the excited atoms and ions can than emit their characteristic radiation. Spectrochemical sources can populate a large number of different energy levels for several different elements at the same time. The emission line produced by a transition from the first excited state of the lowest energy to the ground state is called resonance line. It is worth noticing that a great number of prominent spectral lines for analytical optical emission spectrometry (OES) with microwave plasmas (MWPs) are atomic (or ionic) resonance lines (see Section 5.5).[1] The UV/Vis region (160–800 nm) of the electromagnetic spectrum is the region most commonly used for analytical atomic spectrometry. The essential advantage of OES lies in the multi-element determination capability as well as in the flexibility to choose from several different emission wavelengths for one element. However, when the number of emission wavelengths increases, the probability also increases for interferences that may arise from emission lines that are too close in wavelength to be measured separately.

Since, in general, MWPs can operate in a wide range of experimental conditions, it is difficult to propose a universal model to explain the spectroscopic features of the MWP. Argon microwave induced plasma (MIP) at atmospheric pressure bears features of recombining plasma and can be described by means of a collision–radiative model proposed by Boumans.[2] The division of atomic emission lines into *soft* and *hard* ones proposed for the inductively coupled

RSC Analytical Spectroscopy Monographs No. 12
Microwave Induced Plasma Analytical Spectrometry
By Krzysztof J. Jankowski and Edward Reszke

Published by the Royal Society of Chemistry, www.rsc.org

plasma (ICP) method seems to be adequate also for MWP spectrometry.[3] On the other hand, it was stated that the MWP has underpopulated electrons in the whole discharge and the plasma remains in an ionizing mode.[4] Brassem *et al.*[5] proposed for a low-pressure MIP the radiative ionization recombination (RIR) model, earlier elaborated by Schlüter[6] for discharges in hydrogen. Essentially the RIR model implies a partial thermodynamic equilibrium with the energy level populations near the ionization limit being governed by the local thermal equilibrium model, while the populations near the ground state are governed by radiative recombination of the corona model. The intermediate region called "thermal limit" seems to be of low importance for MWPs. This model considers, in the excitation process of analyte atoms and ions, collisions with high-energy electrons, radiative ionization and recombination processes. It also assumes that there exist two different groups of electrons in the discharge, a major group at low energy and an additional group at high energy. Ionization by the high-energy electrons is balanced by recombination with the low-energy electrons. The temperature of these latter ones reaches only a few thousand degrees kelvin, and thus they do not participate in collisional excitation processes.

In low-energy fields the electrons are the active species and their kinetic energies follow a maxwellian distribution. Thomson scattering spectra obtained for microwave plasma torches (MPTs) and torch à injection axiale (TIA) show a serious distortion from a gaussian profile of energy distribution.[7–9] The energy distribution function of electrons of atmospheric pressure MPTs show significant deviations from maxwellian distribution over the entire wavelength range (equivalent to electron energies of 0.1–0.6 eV). Thus, the occurrence of two types of electrons in plasma was postulated: high- and low-energy ones, while the number density of the high-energy electrons would be more than three orders of magnitude lower than those of the low-energy electrons, but their temperature would be more than one order of magnitude higher.

A detailed description of the mechanisms of analyte excitation in MWPs is beyond the scope of this book. Here we only mention some essential processes.[3,10–13] The most often proposed excitation mechanism is the Penning ionization described by the expression:

$$X + G^m \rightarrow X^+ + G + e^-$$

where X denotes the analyte neutral atom, X^+ the analyte ion, G the plasma gas atom and m the metastable state (or other energetic state).

This is followed by radiative ion–electron recombination:

$$X^+ + e^- \rightarrow X^* + h\nu_c$$

or ion excitation:

$$X^+ + G^m \rightarrow X^{+*} + G$$

where * denotes the excited state.

Direct analyte atom or ion excitation by metastable states is also possible:

$$X + G^m \rightarrow X^* + G$$

$$X + G^m \rightarrow X^{+*} + G + e^-$$

These processes are accompanied by energetic electron impact processes, especially in excitation of analytes of high ionization potential:

$$X + e^- \rightarrow X^* + e^-$$

$$X^+ + e^- \rightarrow X^{+*} + e^-$$

$$X + e^- \rightarrow X^+ + 2e^-$$

$$X + e^- \rightarrow X^{+*} + 2e^-$$

The low-energy electrons can participate in a three-body ion–electron recombination:

$$X^+ + e^- + e^- \rightarrow X^* + e^-$$

$$X^+ + G + e^- \rightarrow X^* + G$$

The existence of charge-transfer processes is suggested in order to explain the intense emission of the ionic lines for Cl, Br, I, S and P, especially in helium plasma:[4,14]

$$X + G^+ + \rightarrow X^{+*} + G$$

Additionally, in helium plasma, metastable helium molecules can be involved in the following excitation process:

$$X^+ + He_2^m \rightarrow X^{+*} + 2He$$

The excitation processes are supported by energy or mass transfer processes in the plasma, including radiative energy transfer *via* the plasma gas or matrix element resonance level, shift of the ionization equilibrium and ambipolar diffusion, increased analyte penetration and resonance collisions involving the matrix element.[3,15,16] These processes are of particular importance for the efficient excitation of elements of relatively low ionization potential in filament-type MIPs, owing to the problems of analyte penetration into the hot plasma zone.

5.2 Basic Spectroscopy Practice

5.2.1 Spectral Line Intensity

In OES, both quantitative and qualitative information about a sample can be obtained from the spectra. The qualitative information is related to the wavelength at which the radiation is emitted. The intensity of the light emitted at a specific wavelength is measured and used to determine the concentration of the element of interest. Spectral line intensities depend on the relative populations of the ground or lower electronic state and the upper excited state. The relative populations of the atoms in the ground and excited states can be expressed in terms of the Boltzmann distribution law, as follows:

$$N_1/N_0 = g_1/g_0 \exp(-\Delta E/kT)$$

where N_1/N_0 is the ratio of the number of excited atoms to the number of atoms in the ground or lower state, g_1/g_0 is the ratio of the number of energy levels having the same energies for the upper and lower energy states, respectively, ΔE is the difference in energy between the upper and lower energy states, k is Boltzmann's constant and T is the temperature.

Analyte line intensities vary with changes in plasma parameters. The intensity decrease at higher gas flow rates may be caused by a lower excitation temperature or by the shorter analyte residence time in the plasma that might result in lower atomization and excitation efficiencies. Analyte emissions also exhibit a generally increasing trend with increased sample flow rate. However, this is restricted by a limited plasma tolerance to sample loading. Higher analyte emission intensities might be observed with increases in applied microwave power; however, it is accompanied by an increase of the background.

5.2.2 Background Correction

In MWPs, strong emission from many elements appears on a weak, relatively unstructured plasma background, especially in the UV and NIR regions. In general, capacitive microwave plasma (CMP) spectra are characterized by higher background levels than those of MIP spectra. Background signals arising from the emission of non-analyte species, scattered radiation and the detector characteristics might be corrected with the aid of commonly used background correction techniques and strategies for OES.[17] Usually, background signals differ from one spectral region to another, and therefore the same region should be used for both analysis and background correction. The measurement of background is usually carried out at a wavelength adjacent to the analytical line. When the background emission is stable for the blank, standards and samples, then background correction can be effected by measurements at the analytical wavelength alone. The samples, standards and blank are measured in turn and then the blank is subsequently subtracted. If the background is a simple continuum and stable in the neighbourhood of the line, a single measurement is sufficient; when it is steadily changing with wavelength, measurements on both sides of the line may be required. When the background is highly structured, the precise location of that measurement should be selected with great care. The background may be measured and subsequently removed in the data processing system or it can be simultaneously subtracted electronically.

The principal difficulty is that of identifying and accurately measuring the background to be subtracted. When the background is large and unstable compared with the analyte signal, *i.e.* near the detection limit, then sophisticated signal processing or even chemical separation of the analyte from its matrix may be necessary. When the background is variable from sample to sample, accurate background correction requires measurement at additional wavelengths. In the case of a line overlap from an interfering element, the relative intensities of the overlapping and reference lines of the interferent could be measured to estimate the contribution of the former to the background at the analytical wavelength.

If the background is small and reproducible and the analyte signal is relatively large, the background correction may be effected simply by subtraction of a previously measured blank or by "zeroing" the instrument using a blank sample. The multi-point background correction procedure has been developed in MIP-based atomic emission detectors, leading to a substantial improvement in selectivity.[18] The spectrum acquisition mode performing the background subtraction over a wide spectral range led to an enhancement of the sensitivity of the charge-coupled device (CCD)-based optical detection systems of MPT-based instruments.[19,20] Wavelength modulation techniques have been successfully used for the background correction of steady-state as well as transient signals in some MIP-OES techniques.[21,22]

5.2.3 Transient Signal Measurement

In various spectroscopic techniques, including chromatography and electro-thermal vaporization coupled with MWP-OES, a transient (intensity *vs.* time) signal is registered for the analyte quantification. However, it is prone to spectral interferences and background variations and the sample matrix should be separated or time resolved. In the MWP-based techniques an increase of the background is usually observed; sometimes a sharp peak is formed which could overlap with the analyte peak.[23,24] The occurrence of this peak is explained as a result of a substantial influence of pressure changes during the vaporization stage on the small volume of plasma or by the sharp change in plasma emission due to the transient passage of the easily ionized matrix components. Background correction allows these shifts to be eliminated. In the case of multi-wavelength detection, the intra-element multiplication of the transient signals at several spectral lines of an analyte may improve the signal-to-noise of the background ratio and, as a consequence, the detection limit.[25]

When the MWP-OES system is coupled to a GC system, the analyte is temporally separated from the matrix; hence the background emission from that source is minimized. However, contributions to the background emission can arise from some components of GC effluents. It was found, for example, that background emission from hydrogen derived from hydrogen-containing compounds restricted the detection of deuterium in GC effluent at the 656.28 nm line owing to incomplete resolution of the hydrogen and deuterium lines.[26] The detection of oxygen using the line at 777.194 nm was found to be complicated by the presence of water vapour present in the plasma gas.[27]

5.3   Instrumentation

5.3.1   Spectrometer Configurations

Despite some technical limitations related to the limited volume of MIPs as mentioned in Chapter 1, various optical and detection systems have been successfully applied to MWP-OES techniques. The presently available MWP instruments utilize both the sequential slew-scanning single-channel mono-chromators and the direct-reading multichannel spectrometers.[28,29] The first instrument type offers flexibility in the choice of possible analytical emission lines and background correction strategy. The second type provides true simultaneous measurement capability and higher sample throughput, but is expensive. A sequential spectrometer typically consists of an entrance slit, a plane grating, either one (Ebert-type) or two (Czerny–Turner) mirrors for collimating and focusing the light, and a single exit slit. In side-on configured MPTs and CMPs, the external focusing optic is used in conjunction with mirrors that allow measurement of emission at different heights within the plasma.[30] In multichannel spectrometers, CCD or photodiode array detectors are usually used. It has been shown that by measuring the background signals at the same time as the analyte line, called simultaneous background correction, detection limits may be enhanced for certain elements. This simultaneous collection of background and analyte intensities removes noise originating from the sample introduction system. Conventional spectrometric systems operate in the 200–800 nm spectral region. To observe atomic emission at wavelengths less than 200 nm, an argon or nitrogen purged monochromator should be used. Alternatively, the optical system may be connected with a vacuum system. Various optical systems that can be used for MWP-OES are shown in Figure 5.1.

An alternative method of simultaneous observation involves using a rapid-scanning spectrometer based on the galvanometer-driven oscillating mirror. Movement of this mirror scans a given spectral region across the exit slit, with repetition of several tens of spectra per second. The increase of scan number and data smoothing enhances signal-to-noise ratios. Reports show that such a rapid-scanning spectrometer has potential as a multi-element detector when used in conjunction with electrothermal atomizers or chromatography and MWP.[31,32] Disadvantages include a relatively narrow wavelength window and

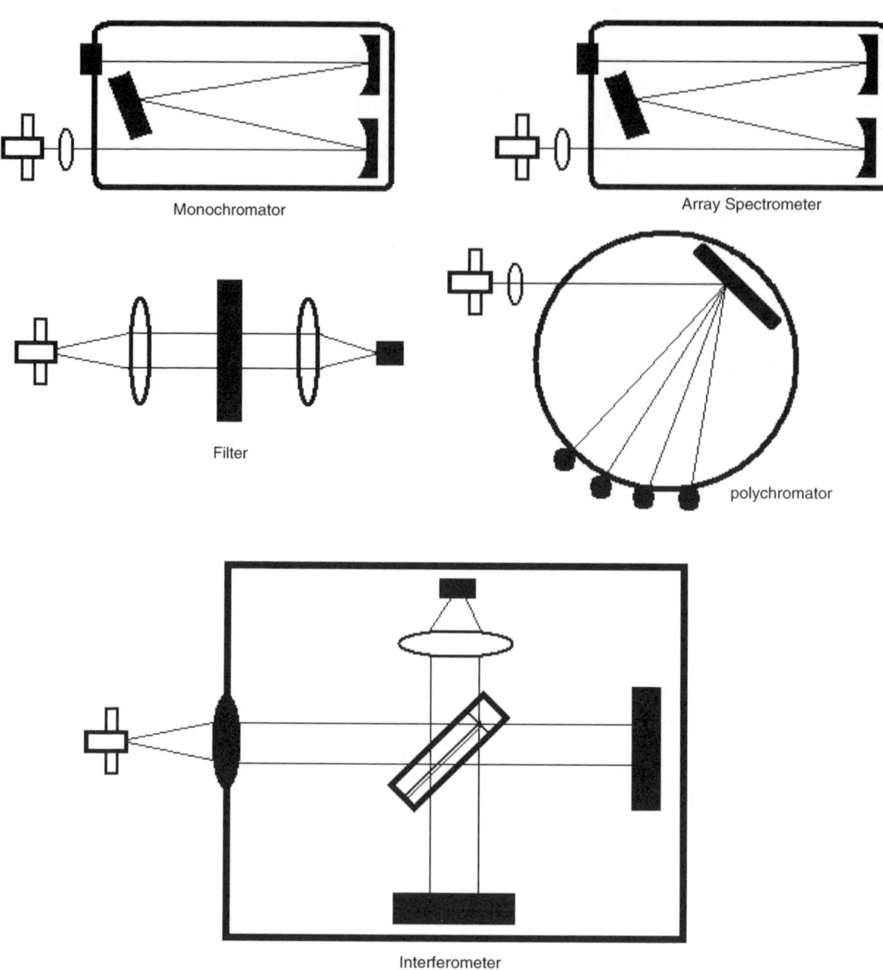

Figure 5.1 Optical systems useful for MWP-OES.

degradation of the detection limit compared with a single-channel monitor, owing to the lower integration time. Attempts have been made to couple the MPT source and an acousto-optic tunable filter spectrometer to attain high repetition scanning.[33]

5.3.2   The Use of Echelle Optics to Observe the Emission from Microwave Plasmas

The next possibility for simultaneous multi-element analysis is the use of echelle optics. The major component difference of the echelle spectrometer is the grating, which utilizes the spectral order for maximum wavelength coverage. Echelle grating-based spectrometers offer some advantages over conventional spectrometers, but have also some disadvantages. Firstly, the optics result in good efficiency in each of the spectral orders. Because of the use of higher orders with better resolution, the physical size of the instrument may be reduced. However, in order to separate spectral orders perpendicular to the diffraction grating, a secondary dispersion is carried out, typically by using a prism or second grating. As a result, a two-dimensional spectral "map" can be obtained where the spectral order is sorted vertically and the wavelength horizontally. However, this leads to worsening of detection limits due to the higher aperture number as a consequence of using of the double dispersion system that produces lower radiation energy inquiring into the detector.

An echelle emission spectrometer with CCD detection has been used in the element-specific MIP-OES detector for GC, providing a multi-wavelength simultaneous detection or even a complete fingerprint of a sample. A specific near-IR echelle system could be used for the determination of GC-separated organic compounds such as halogenated and sulfurous hydrocarbons, under the use of H-, C-, F-, Cl-, Br-, I- and S-selective signals.[34] A high-resolution spectrometer with a MPT source, echelle optics and UV-intensified CCD has been announced recently.[35] For spatial emission studies the monochromatic imaging spectrometer equipped with a silicon-intensified target vidicon tube was employed,[36] providing a 110 µm spatial resolution of the system.

5.3.3 Interference Filters

The relatively simple spectra emitted by MWPs are composed primarily of lines originating from ground state transitions. The continuum background is also low, so that high resolution is not necessary to minimize the background contribution at the analytical wavelength. Optical systems utilized for specific spectroscopic techniques [*i.e.* MWPs coupled with chromatography, hydride generation or electrothermal vaporization (ETV)], especially in cases where the spectral selectivity is high, could efficiently operate with low-cost, low-resolution, high optical aperture spectrometers. Some applications, such as the determination of the chemical composition of the separated sample components or even individual particles, can be accurately and rapidly achieved with optical systems consisting a set of interference filters[22] or low-resolution monochromators[37] that provide simultaneous monitoring of a few wavelengths. The basic idea behind the use of the oscillating interference filters is that the exact centre bandpass wavelength of such a filter depends on the light incidence angle. If this angle is less than 90° the centre bandpass wavelength is shifted towards shorter wavelengths. This property can be used for background correction in plasma emission detectors.[38,39]

Generally, high-wavelength resolution instruments are required to solve background correction problems when matrices are analyzed or a mobile phase used for chromatographic separation generates a structured background.[40] Selectivity may be significantly enhanced using high-resolution spectrometers by minimizing the effects of spectral overlap. A high wavelength resolution of the optical system is also useful for developing miniaturized spectrometers.[20,41]

5.3.4   Instruments Based on Fibre Optics

Fibre optics-based systems are often used to transmit the emitted light from the MWP to the optical system.[22,34,38,41] They offer a flexible arrangement of the experimental setup; however, the wavelength range available is restricted, depending on the fibre material, and the sensitivity of the measurement is usually lower at wavelengths below 240 nm. The other problem is focusing the optical beam from the plasma into the fibre optics to increase light collection efficiency. A double lens focusing assembly has been developed, resulting in an enhancement of the system sensitivity and signal stability.[19]

5.3.5   Detection Systems

Three different types of detectors are predominantly used in OES: photo-emissive (photomultiplier tubes, PMTs), photoconducting (photodiode arrays, PDAs) and photocharge storage [charge-transfer devices (CTDs), including charge-coupled devices (CCDs) and charge-injection devices (CIDs)]. Until the 1990s, the "working horse" of OES was the PMTs and they are still the best detectors in common use; however, more recently, multi-channel detectors based on CCD technology have gained substantial acceptance. The major advantages of the PMTs are a relatively wide wavelength coverage, high sensitivity and wide range of response to over nine orders of magnitude in light intensity measured. Advanced CCDs offer the flexibility in analytical wavelength selection, the possibility of the use of several lines (wavelengths) for the same element in order to extend the linear dynamic range and the potential for qualitative analysis, including some improvements in the identification of spectral interferences. Additionally, they are capable of measuring light fluxes over a relatively wide dynamic range with high quantum efficiency (up to 70%), which, however, is still worse than that for a PMT. CCDs can be interfaced directly to signal acquisition boards and data acquisition systems to provide a full spectrum coverage in a given spectral range. However, CTDs also suffer from some severe limitations, including a "blooming" effect, when individual pixels are saturated, a read noise (introduced when a single pixel is read out) much higher than that of a PMT and a poor response in the UV region of the spectrum.

Either PDAs or CCDs have significant read noises and dark currents at room temperature, resulting in the generation of a relatively large background signal. To overcome this problem, CCDs usually operate at low temperatures (down to 125 K) to reduce the dark current. However, this generates some technical problems, including detector dewing.

CCDs are preferably used as detectors for echelle spectrometers, although linear CCD arrays are also used in polychromator systems. A small, integrated spectrometer with a linear 2048 pixels-long CCD detector covering a spectral range of 40 nm has been used in a portable MPT-source instrument (EPD-1) for environmental analysis.[19,42] A linear UV-intensified CCD was used as a detector in a miniature MPT-OES spectrometer design.[20]

The ideal detection systems for OES should be able to accommodate simultaneous viewing of an analytical line and the adjacent background and offer a totally flexible choice of analytical lines. PDA detectors fulfil these two requirements. However, intensified PDAs should be used to attain the photometric sensitivity comparable with that of PMT and readout noise must be reduced to achieve relatively low detection limits. The PDA-based detection system, readily interfaced with computer data processing, gives access to a spectral range of 10–200 nm and allows the simultaneous measurement of up to ten elements. A flat focal plane spectrometer with a movable PDA was successfully used to construct a MWP-based detector for gas chromatography.[18]

The output signal of the detector requires electronic processing before being usable for analytical interpretation. Such processing may include gating, amplification, integration and analogue digital computation. The electronics used for signal processing in MWP-OES systems utilizing PMT detection are generally straightforward. The first step is to convert the anode current, which represents emission intensity, into a voltage signal that is subsequently converted into digital information *via* an A/D converter. This digital information can then be used by a computer for further processing, the end result being information passed on to the host computer or to the analyst in the form of a number representing either relative emission intensity or concentration. Thus, post-processing of the data is easily accomplished.

Attempts have been made to adopt Fourier transform spectroscopy (FTS) to the MWP-OES technique. FTS provides a very accurate and precise wavelength measurement, which is predetermined and requires no calibration, and simultaneous measurement in the whole observed spectral region; high resolution can be achieved while allowing both simultaneous qualitative and quantitative measurements due to minimizing line overlap and other spectral background problems. However, these advantages do not lead to improvements in signal-to-noise ratios, and hence in detection limits.

FTS was extensively studied for the detection of non-metal atomic emission from a MIP in the NIR spectral region.[43–46] However, systematic GC-MWP-OES studies proved that FTS offers the highest selectivity of the measurement but detection limits are about one order of magnitude poorer than those obtained with the use of a PMT or a PDA as a detector, even with the use of intra-element multiplication of several spectral lines.[25,47]

Some optical systems, including rapid scanning spectrometers, acousto-optic tunable-filter spectrometers and spectrometers equipped with a CCD or PDA detector, are capable of measuring the time-resolved spectra of a transient emission. This offers additional possibilities for analytical spectrometric studies of multi-element detection. Other possible applications of time-resolved spectra include advanced optimization of time-dependent sample introduction procedures such as ETV, GC or LC, and confirmation of elemental identity in speciation analysis or elemental ratio-based determination of chemical composition.[31,32,48–51] In chromatographic separations, the reliability of identification based on the retention time is highly dependent on the quality of the separation. An unknown compound can be confirmed by examining the spectra (as intensity–wavelength–time) collected during its elution. The presence of several characteristic lines (usually prominent lines) of the element determined, at the expected wavelengths and with the expected relative intensities, confirms the identity. When done together with the monitoring the isotopic pattern by mass spectrometry techniques, a more reliable proof of identity can be obtained.

5.4   The Microwave Induced Plasma Spectrum: General Description

The observed emission spectra from MWPs are rarely complex, consisting mostly of resonance transitions of neutral atoms (and also ions) from essentially all elements of the Periodic Table. Helium plasma, in particular, can produce intense atomic and ionic emission lines from non-metals, while in less energetic argon MWPs the band spectra of these elements are observed. Most of the analytically useful lines for MWP-OES are in the range of 190–450 nm; however, there are also some important lines between 160 and 190 nm and above 450 nm.

A MIP exhibits a remarkably simple and weak background spectrum. In the 190–250 nm region the spectrum exhibits a general rise in the intensity of the background continuum with increasing wavelength, with the carbon lines at 193.09 and 247.86 nm arising from the presence of carbon-containing species in the plasma gas. In the visible region the Balmer series hydrogen lines and the plasma gas atomic lines appear. The strongest lines of silicon and the alkaline earth elements may appear at low intensity as a result of erosion of the quartz discharge tube.

MWP emission spectra are further complicated by the presence of band spectra arising from the excitation of molecular species (ionic or neutral, and radicals), which are likely to cause spectral interference. Band spectra possess a strong intense head, which degrades in the intensity to either shorter or longer wavelengths depending on the nature of vibrations in the particular molecule. Two strong OH bands at 281 and 306 nm and the NH band at 336 nm seem to be more of a problem in MWPs. The major contributions to the background emission in the UV range are NO, OH and NH molecular emission bands, and the O_2 absorption bands below 200 nm. Other binary molecules (including ionized species) of potential interest are C_2, CH, C, CO, CS and SO, which result from various possible chemical reactions in the plasma involving organics, water or carbon dioxide. The major sources of interfering band emission are collected in Table 5.1.

The intensities of these molecular bands vary, depending on the sample composition as well as the sample introduction technique and the MWP used.[52–56] For aqueous solution nebulization MWP-OES, OH and NH are the dominant bands in the background emission, as shown in Figure 5.2. Compared with MIPs, the CMPs and MPTs suffer from a complexity of background emissions, because the plasma is directly exposed to air. As a result the

Table 5.1 Most prominent background spectral features in atmospheric pressure MWPs.

Originating species	Wavelength (nm)	Transition
NO		$[A^2\Sigma^+\!-\!x^3\pi]$
	205.24	2.0
	215.49	1.0
	226.94	0.0
	237.02	0.1
	247.87	0.2
	259.57	0.3
OH		$[^2\Sigma^+\!-\!^2\pi]$
	281.13	1.0
	287.53	2.1
	294.52	3.2
	306.36	0.0
NH		$[A^3\pi\,-\!x^3\Sigma^-]$
	336.00	0.01
	337.09	1.1
N_2		$[^3P_u\!-\!B^3P_g]$
	296.20	3.1
	311.67	3.2
	315.93	1.0
	337.13	0.0
	353.67	1.2
	357.69	0.1
	371.05	2.4
	375.54	1.3
	380.49	0.2
N_2^+		$[B^2\Sigma^+\!-\!x^2\pi]$
	391.44	0.0
CO^+	219.0	0.0
	221.5	1.1
	230.0	0.1
CN		$[B^2\Sigma\!-\!x^2\Sigma]$
	358.59	2.1
	359.04	1.0
	385.47	3.3
	386.19	2.2
	387.14	1.1
	388.34	0.0
	416.78	3.4
	419.72	1.2
	421.60	0.1
CH		$[A^2\Delta\!-\!x^2\pi]$
	431.42	0.0
C_2		$[A^3\pi\,-\!x^2\pi_u]$
	473.7	1.0
	512.9	1.1
	516.5	0.0

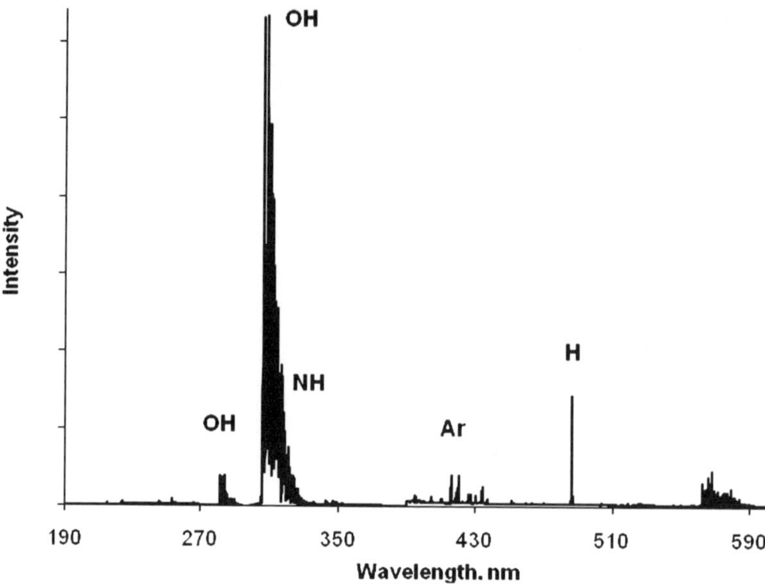

Figure 5.2   Spectrum of Ar MIP with aqueous solution nebulization.

molecular band emission, such as NO, NH, N_2 or N_2^+, is observed, which covers a wide wavelength range. A similar molecular emission is also present in the N_2-MWP spectrum. To solve this problem, an oxygen-sheathed MPT as well as a nitrogen/oxygen MIP-OES have been developed.[57,58] In spectra using GC-MIP-OES or supercritical fluid chromatography (SFC)-MIP-OES, the band emission from various carbon-containing molecules is of particular importance.

Excitation sources used in emission spectrometry differ with respect to the pattern of emitted spectra. A comparison of the emission spectra excited in typical electric discharges and plasmas reveals that the most complex spectra are emitted by a high-voltage spark. ICP is a very efficient excitation source and the spectrum exhibits a complex structure, including intense ionic lines of many elements. The dc arc and the ICP yield spectra of similar complexity. Compared with the ICP emission spectrum, a MWP gives a relatively simple spectrum that appears more "arc-like" than "spark-like", with an excess of atomic lines and a relatively small number of lines for particular elements, owing to the smaller excitation energy. Beenakker *et al.*[59] have reported that, in contrast to an ICP, a MIP does not exhibit the "ionic line advantage" in which the detection limits of the analyte ion lines exceed the respective values for the atomic lines by one to three orders of magnitude. By systemizing the excitation sources with respect to the sample ionization degree, the following order can be suggested:[15,60,61]

$$\text{spark} > \text{ICP} > \text{arc} \approx \text{direct current plasma (DCP)} > \text{MIP}$$

Hence, a MWP may be an excitation source predicted for qualitative analysis, since it shows a relatively smaller number of lines, which facilitates its interpretation.

Two essential features of MIP spectra different from ICP ones should be stressed. It is clear that the ICP spectrum is considerably more complex, in terms of the number of lines, than the MIP spectrum. Secondly, the most sensitive lines in the MIP spectrum cover a wide range of UV and Vis wavelengths. In particular, the prominent spectral lines emitted in the high-power N_2-MIP posses wavelengths greater than 300 nm.[58,62] This reduces the problem in the selection of lines for the direct-reading spectrometer. In ICP, however, the majority of the intensive emission lines are placed in the UV range. On the other hand, the fact that a certain number of intensive emission lines of some elements occur in the 306–320 nm range, where they are overlapped by the intensive OH band, is a limitation in the utilization of the MIP spectrum.

The distribution of intensities among atomic and ionic lines differs between the two plasma sources. To give an example, ionic lines of Tl and V, which are the most intensive lines of these elements in ICPs, are not found in the MIP spectrum, whereas their intensive atomic lines are identified. Also the intensities of the ionic lines of Cd, Co, Cr, Hg, In, Mn, Mo, Ni, Pb and Pt are much smaller than those for ICP spectra.[1,3]

5.5 Provisional Wavelength Tables Specific for Microwave Induced Plasma Spectra

The development of MWP-based spectroanalytical techniques has required the setting up of a spectral line atlas devoted especially to these specific techniques. It is expected that users of a MWP source will receive a comprehensive spectral line database. MWP studies have not yet received such an exhaustive treatment, mainly because of the non-existent wide-scale manufacturing of the instruments, as has been the case for ICP, which hinders the maximum utilization of MWP-based techniques in OES. Fragmentary treatment has been given in several works,[63–66] so the selection of appropriate wavelengths for trace analyses sometimes is difficult. The well-known tables of spectra from classical sources and ICP do not correlate well with the MWP intensity data.[67–70] This is not surprising, since the MWP method is qualitatively different from the ICP technique and other sources, and even certain prominent lines for ICP spectra are not present in MWP spectra. Our goals in the preparation of this provisional atlas were to obtain a comprehensive summary of the most prominent spectral lines of the elements useful for the determination of trace and ultratrace concentrations, as well as to provide the basic information concerning the analytical capabilities and the potential spectral interferences of the prominent spectral lines of the elements. Elemental lines over the whole operating range of the instrument (190–800 nm) were sought, and spectral regions of special interest were analyzed in detail. Studies carried out for many years allowed the collection of a large amount of data concerning the

Table 5.2 The most used optical emission lines in MIP-OES.

Element		Wavelength, nm	I_n/I_b	DL_{exp} ng mL^{-1}	Comments
Ag	I	338.29	100	9	
	I	328.07	96	14	OH band
	II	243.78	5		
Al	I	396.15	100	110	
	I	394.01	50	200	
	I	309.27	n.m.		OH band
	I	237.64	30		
Ar	I	415.86	100		
	I	420.07	98		
	I	419.83	44		
	I	419.03	42		
	I	427.22	42		
	I	425.94	39		
	I	430.01	34		
	I	433.37	30		
	I	404.44	29		
	I	426.63	28		
	I	394.90	27		
	I	668.44	27		
	I	451.07	21		
As	I	228.81	100	90	
	I	234.98	47	190	
	I	200.33	35		
	I	193.70	27		
	I	278.02	27		
	I	197.20	25		
	I	198.97	23		
Au	I	267.59	100	90	
	I	242.79	82	95	
	I	197.82	18		
	I	201.20	10		
	II	191.89	10		
B	I	249.77	100	110	
	I	249.68	84		
	I	208.96	80		
	I	208.89	73		
Ba	II	493.41	100	110	
	II	455.40	85	190	
	II	233.53	32		
	I	553.55	24		
Be	I	234.86	100	2	
	I	332.13	29		OH band
	I	249.47	21		
	II	313.04	18		OH band
Bi	I	223.06	100	80	
	I	472.24			
Br	II	470.49	100	20	
	II	478.55	65		
	I	635.07	45		
	II	481.67	40		
C	I	193.09	100	50	
	I	247.86	89		
	I	199.36	2		

Table 5.2 (*Continued*).

Element		Wavelength, nm	I_n/I_b	DL_{exp} ng mL^{-1}	Comments
Ca	II	393.37	100	3	
	I	422.67	62	4	
	II	396.85	47		
Cd	I	228.80	100	5	
	II	214.44	39	8	
	II	226.50	34		
	I	479.99	20		
	I	326.11	15		
Ce	II	413.77	100	80	
	II	422.26	95		
	II	399.92	70		
Cl	II	479.54	100	14	
	II	481.01	72		
	I	725.66	65		
	II	481.95	54		
Co	II	238.89	100	85	
	I	240.73	90	90	
	I	345.35	45		
Cr	I	357.87	100	45	
	I	359.35	90	55	
	I	425.43	85	60	
Cs	I	455.53	100	65	
	I	459.32	24		
	II	452.67	5		
Cu	I	324.75	100	20	OH band
	I	327.40	52	25	
	II	213.60	17		
	II	217.89	14		
	I	223.01	13		
	II	219.23	12		
	II	224.70	10		
Dy	II	364.54	100	36	
	II	396.84	90		
	II	353.17	65		
Er	II	390.63	100	44	
	II	369.26	80		
	II	323.06	75		OH band
Eu	II	420.51	100	12	
	II	381.97	80		
	II	412.97	70		
F	I	685.60	100	4000	
	I	623.96	80		
	I	634.85	60		
	I	690.25	48		
Fe	I	248.32	100	45	
	I	373.49	67		
	I	371.99	60		
	I	248.82	53		
	I	252.29	44		
	I	249.06	43		
	II	238.20	35	65	

Table 5.2 (*Continued*).

Element		Wavelength, nm	I_n/I_b	DL_{exp} ng mL^{-1}	Comments
Ga	I	417.21	100	16	
	I	403.30	53		
Gd	II	342.25	100	40	
	II	376.70	97		
	II	358.50	65		
Ge	I	265.12	100	65	
	I	265.16	62		
	I	275.46	47		
	I	303.91	41		
H	I	656.28	100		
	I	486.13	36		
	I	434.05	6		
	I	410.17	2		
He	I	587.60	100		
	I	706.57	28		
	I	388.87	26		
	I	667.82	24		
	I	501.57	18		
	I	447.15	12		
	I	492.19	8		
Hf	II	368.22	100	37	
	II	339.98	85		
	II	277.34	45		
Hg	I	253.65	100	23	
	I	365.02	90		
	I	435.83	65		
	I	546.07	62		
	II	194.23	43		
Ho	II	389.10	100	20	
	II	381.07	50		
I	I	206.24	100	60	
	II	516.12	71		
	II	546.46	35		
In	I	451.13	100	18	
	I	303.90	70		
	I	325.61	n.m.		OH band
Ir	I	380.01	100	130	
	I	208.88	65		
	I	322.08	n.m.		OH band
K	I	766.49	100	2	
	I	769.90	53	5	
	I	404.70	7		
La	II	408.67	100	60	
	II	398.85	60		
	II	379.48	55		
Li	I	670.78	100	0.3	
	I	610.36	21	2	
	I	460.30	3	11	
Lu	II	261.54	100	7	
	II	350.74	65		

Table 5.2 (*Continued*).

Element		Wavelength, nm	I_n/I_b	DL_{exp} ng mL^{-1}	Comments
Mg	II	279.55	100	7	OH band
	II	280.27	60	9	
	I	285.21	43	11	
	I	383.83	27		
	I	383.23	18		
Mn	I	403.08	100	25	
	II	257.61	98	30	
	I	403.30	83		
	II	259.37	81		
	I	279.48	70		
Mo	II	202.03	100	350	
	I	379.83	85	420	
	I	386.41	70		
N	I	746.88	100		
	I	744.26	61		
	I	742.39	31		
Na	I	588.99	100	0.9	
	I	589.59	55		
	I	330.29	2.3		
Nb	I	405.89	100	580	
	I	407.97	75		
	I	410.09	56		
	II	202.93	40		
Nd	II	410.95	100	110	
	II	430.36	99		
	II	401.22	80		
Ni	I	232.00	100	90	
	II	221.65	92		
	I	231.10	90		
	I	232.58	68		
O	I	777.19	100		
	I	777.41	70		
	I	777.54	49		
Os	I	201.81	100	600	
	I	202.02	95		
	I	305.86	60		
P	I	213.61	100	70	
	I	214.91	70		
	I	253.56	40		
	I	255.33	35		
Pb	I	405.78	100	60	
	I	368.35	41		
	I	261.42	32		
	I	363.96	31		
Pd	I	340.46	100	55	
	I	363.47	83		
	I	360.96	64		
	I	324.27	62		OH band
	I	344.14	38		
	I	342.12	36		

Table 5.2   (*Continued*).

Element		Wavelength, nm	I_n/I_b	DL_{exp} ng mL^{-1}	Comments
Pr	II	390.84	100	230	
	II	417.94	90		
	II	422.53	85		
	II	422.30	80		
Pt	I	265.95	100	160	
	I	217.47	84		
	I	292.98	56		
	II	214.42	44		
Rb	I	420.18	100	65	
	I	421.56	42		
Re	II	227.53	100	75	
	II	221.43	88		
	I	346.05	85		
	I	488.92	57		
Rh	I	343.49	100	60	
	I	369.24	88		
	I	350.25	75		
	I	352.80	71		
	I	339.68	69		
	I	248.33	67		
Ru	I	372.80	100	85	
	I	372.69	85		
	II	379.93	50		
S	I	469.41	100	70	
	I	190.03	54		
	II	545.39	54		
	II	481.55	50		
Sb	I	252.85	100	50	
	I	259.81	92		
	I	217.92	26		
	I	231.15	21		
Sc	I	391.18	100	27	
	I	402.04	90		
	I	361.38	75		
Se	I	203.99	100	47	
	I	196.03	76		
	I	206.28	36		
Si	I	251.61	100	75	
	I	250.69	42		
	I	252.85	42		
	I	252.41	37		
	I	221.67	30		
	I	251.43	30		
	I	212.41	29		
	I	221.09	20		
Sm	II	359.26	100	170	
	II	442.43	95		
	II	363.43	90		
Sn	I	235.48	100	190	
	I	242.95	88		
	I	270.65	88		
	I	224.61	68		

Table 5.2 (*Continued*).

Element		Wavelength, nm	I_n/I_b	DL_{exp} ng mL^{-1}	Comments
Sr	II	407.77	100	7	
	II	421.55	53	11	
	I	460.73	50		
Ta	II	268.51	100	750	
	II	263.56	70		
Tb	II	384.87	100	340	
	II	387.42	90		
	II	350.92	65		
Te	I	214.28	100	140	
	I	238.58	45		
	I	225.90	32		
	I	238.33	30		
Th	II	401.91	100	160	
	II	374.12	45		
Ti	II	334.94	100	70	
	II	336.12	70		
	II	337.28	45		
Tl	I	535.05	100	17	
	I	377.57	63		
	I	351.92	19		
	I	276.79	19		
Tm	II	384.80	100	23	
	II	376.13	60		
	II	370.03	30		
	II	424.22	30		
U	II	385.96	100	360	
	II	409.01	80		
	II	393.20	70		
V	I	437.92	100	90	
	I	438.47	75		
	I	411.18	70		
W	II	209.48	100	500	
	I	400.86	85		
Y	II	371.03	100	12	
	II	437.49	80		
	II	360.07	75		
Yb	II	369.42	100	17	
	II	328.94	70		
	I	398.80	45		
Zn	I	213.86	100	15	
	II	202.55	71	26	
	I	481.05	65		
	II	206.20	46		
	I	472.22	40		
Zr	II	339.20	100	320	
	II	343.82	85		
	II	349.62	70		
	II	360.12	47		

Explanation of Symbols and Abbreviations

The symbols I and II indicate that the spectral lines originate, respectively, from the neutral atom and singly ionized state.

DL_{exp} – experimental detection limit

n.m. – not measurable because of interfering line listed in comments column spectra of particular elements. A tentative listing of about 300 lines of 80 elements, including the lines generally considered to be the strongest atom and ion lines for each element, the best signal-to-background ratio values and the estimated detection limits of elements, are presented in Table 5.2. The spectra were obtained when aqueous samples were introduced into a low-power (200 W) argon MIP at atmospheric pressure with the use of the TE_{101} microwave cavity, excluding halogens, sulfur and helium, which were investigated by chemical vapour generation He-MIP-OES. The lines are listed alphabetically by element and the line of the highest signal-to-background ratio was taken as a reference intensity standard for each element separately. Argon lines in the range 300–700 nm and helium lines in the range 300–710 nm are included. The experimental detection limits were estimated according to Boumans from:

$$DL_{exp} = \frac{0.03 \times RSD_b \times C_{exp}}{I_n/I_b}$$

in which C_{exp} is the analyte concentration used in the experiment, RSD_b is the relative standard deviation of the background (1% as the typical value was applied generally to all analytical lines) and I_n/I_b is the signal-to-noise ratio obtained for a given element concentration.

References

1. K. Jankowski, *J. Anal. At. Spectrom.*, 1999, **14**, 1419–1423.
2. P. W. J. M. Boumans, *Spectrochim. Acta, Part B*, 1982, **37**, 75.
3. K. Jankowski and M. Dreger, *J. Anal. At. Spectrom.*, 2000, **15**, 269–276.
4. M. Huang, *Microchem. J.*, 1996, **53**, 79–87.
5. P. Brassem, F. J. M. J. Maessen and L. De Galan, *Spectrochim. Acta, Part B*, 1978, **33**, 753–764.
6. H. Schlüter, *Z. Naturforsch.*, 1963, **18A**, 439.
7. M. Huang, K. Warner, S. Lehn and G. M. Hieftje, *Spectrochim. Acta, Part B*, 2000, **55**, 1397–1410.
8. M. Huang, D. S. Hanselman, Q. Jin and G. M. Hieftje, *Spectrochim. Acta, Part B*, 1990, **45**, 1339–1352.
9. J. van der Mullen and J. Jonkers, *Spectrochim. Acta, Part B*, 1999, **54**, 1017–1044.
10. C. I. M. Beenakker, *Spectrochim. Acta, Part B*, 1977, **32**, 173–187.
11. J. P. Matousek, B. J. Orr and M. Selby, *Prog. Anal. At. Spectrosc.*, 1984, **7**, 275–314.
12. C. F. Bauer and R. K. Skogerboe, *Spectrochim. Acta, Part B*, 1983, **38**, 1125–1134.
13. K. Tanabe, H. Haraguchi and K. Fuwa, *Spectrochim. Acta, Part B*, 1983, **38**, 49–60.
14. P. G. Brandl and J. W. Carnahan, *Appl. Spectrosc.*, 1995, **49**, 1781–1788.
15. M. H. Miller, D. Eastwood and M. Schulz Hendrick, *Spectrochim. Acta, Part B*, 1984, **39**, 13–56.

16. J. P. Matousek, B. J. Orr and M. Selby, *Spectrochim. Acta, Part B*, 1986, **41**, 415–429.
17. J. B. Dawson, R. D. Snook and W. J. Price, *J. Anal. At. Spectrom.*, 1993, **8**, 517–537.
18. J. J. Sullivan and B. D. Quimby, *Anal. Chem.*, 1990, **62**, 1034–1043.
19. Y. Duan, Y. Su, Z. Jin and S. P. Abeln, *Rev. Sci Instrum.*, 2000, **71**, 1557–1563.
20. G. Feng, Y. Huan, Y. Cao, S. Wang, X. Wang, J. Jiang, A. Yu, Q. Jin and A. Yu, *Microchem. J.*, 2004, **76**, 17–22.
21. T. M. Spudich and J. W. Carnahan, *J. Anal. At. Spectrom.*, 2001, **16**, 56–61.
22. B. Rosenkranz and J. Bettmer, *Trends Anal. Chem.*, 2000, **19**, 138–156.
23. J. Yang, J. Zhang, C. Schickling and J. A. C. Broekaert, *Spectrochim. Acta, Part B*, 1996, **51**, 551–562.
24. M. Wu and J. W. Carnahan, *Appl. Spectrosc.*, 1990, **44**, 673–678.
25. C. Lauzon, K. C. Tran and J. Hubert, *J. Anal. At. Spectrom.*, 1988, **3**, 901–905.
26. S. R. Goode, J. J. Gemmil and B. E. Watt, *J. Anal. At. Spectrom.*, 1990, **5**, 483–486.
27. S. R. Goode and L. K. Kimbrough, *J. Anal. At. Spectrom.*, 1988, **3**, 915–918.
28. L. Zhao, D. Song, H. Zhang, Y. Fu, Z. Li, C. Chen and Q. Jin, *J. Anal. At. Spectrom.*, 2000, **15**, 973–978.
29. Y. Okamoto, *Anal. Sci.*, 1991, **7**, 283–288.
30. T. Maeda and K. Wagatsuma, *Microchem. J.*, 2004, **76**, 56–60.
31. K. J. Mulligan, M. Zerezhgi and J. A. Caruso, *Spectrochim. Acta, Part B*, 1983, **38**, 369–375.
32. K. J. Mulligan, M. Zerezhgi and J. A. Caruso, *Anal. Chim. Acta*, 1983, **154**, 219–226.
33. L. W. Zhao, Y. H. Zhang, M. J. Wang, D. Q. Song, H. Q. Zhang and Q. Jin, *Spectrosc. Spectr. Anal.*, 2002, **22**, 472–475.
34. J. Koch, M. Okruss, J. Franzke, S. V. Florek, K. Niemax and H. Becker-Ross, *Spectrochim. Acta, Part B*, 2004, **59**, 199–207.
35. J. Jiang, Y. F. Huan, W. Jin, G. D. Feng, Q. Fei, Y. B. Cao and Q. H. Jin, *Spectrosc. Spectr. Anal.*, 2007, **27**, 2375–2379.
36. M. Selby, R. Rezaaiyaan and G. M. Hieftje, *Appl. Spectrosc.*, 1987, **41**, 749–761.
37. K. Kobayashi, A. Sato, T. Homma and T. Nagatomo, *Jpn. J. Appl. Phys.*, 2005, **44**, 1027–1030.
38. B. Rosenkranz, C. B. Breer, W. Buscher, J. Bettmer and K. Camman, *J. Anal. At. Spectrom.*, 1997, **25**, 993–996.
39. T. Twiehaus, S. Evers, W. Buscher, J. Bettmer and K. Camman, *Fresenius' J. Anal. Chem.*, 2001, **371**, 614–620.
40. G. Heltai, B. Feher and M. Horvath, *Chem. Pap.*, 2007, **61**, 438–445.
41. P. Pohl, I. J. Zapata, M. A. Amberger, N. H. Bings and J. A. C. Broekaert, *Spectrochim. Acta, Part B*, 2008, **63**, 415–421.
42. Y. Duan, Y. Su, Z. Jin and S. P. Abeln, *Anal. Chem.*, 2000, **72**, 1672–1679.
43. J. E. Freeman and G. M. Hieftje, *Appl. Spectrosc.*, 1985, **39**, 211–214.
44. D. E. Pivonka, W. G. Fateley and R. C. Fry, *Appl. Spectrosc.*, 1986, **40**, 291–297.

45. J. Hubert, H. V. Tra, K. C. Tran and F. L. Baudais, *Appl. Spectrosc.*, 1986, **40**, 759–766.

46. D. E. Pivonka, A. J. J. Schleisman, W. G. Fateley and R. C. Fry, *Appl. Spectrosc.*, 1986, **40**, 766–772.

47. J. Hubert, S. Bordeleau, K. C. Tran, S. Michaud, B. Milette, R. Sing, J. Jalbert, D. Boudreau, M. Moisan and J. Margot, *Fresenius' J. Anal. Chem.*, 1996, **355**, 494–500.

48. J. P. Matousek, B. J. Orr and M. Selby, *Talanta*, 1986, **33**, 875–882.

49. J. A. Seeley, Y. Zeng, P. C. Uden, T. I. Eglinton and I. Ericson, *J. Anal. At. Spectrom.*, 1992, **7**, 979–985.

50. R. Łobiñski and F. C. Adams, *Trends Anal. Chem.*, 1993, **12**, 41–49.

51. R. Łobiñski and F. C. Adams, *Anal. Chim. Acta*, 1992, **262**, 285–297.

52. A. T. Zander and G. M. Hieftje, *Anal. Chem.*, 1978, **50**, 1257–1260.

53. L. J. Galante, M. Selby, D. R. Luffer, G. M. Hieftje and M. Novotny, *Anal. Chem.*, 1988, **60**, 1370–1376.

54. B. Riviere, J. M. Mermet and D. Deruaz, *J. Anal. At. Spectrom.*, 1988, **3**, 551–555.

55. M. Ohata, H. Ota, M. Fushimi and N. Furuta, *Spectrochim. Acta, Part B*, 2000, **55**, 1551–1564.

56. S. A. Estes, P. C. Uden and R. M. Barnes, *Anal. Chem.*, 1981, **53**, 1829–1837.

57. Q. Jin, W. Yang, F. Liang, H. Zhang, A. Yu, Y. Cao, J. Zhou and B. Xu, *J. Anal. At. Spectrom.*, 1998, **13**, 377–384.

58. M. Ohata and N. Furuta, *J. Anal. At. Spectrom.*, 1998, **13**, 447–453.

59. C. I. M. Beenakker, B. Bosman and P. W. J. M. Boumans, *Spectrochim. Acta, Part B*, 1978, **33**, 373–381.

60. K. Jankowski, *Chem. Anal. (Warsaw)*, 2001, **46**, 305–327.

61. R. K. Winge, E. L. DeKalb and V. A. Fassel, *Appl. Spectrosc.*, 1985, **39**, 673–676.

62. Z. Zhang and K. Wagatsuma, *J. Anal. At. Spectrom.*, 2002, **17**, 699–703.

63. A. E. Croslyn, B. W. Smith and J. D. Winefordner, *CRC Crit. Rev. Anal. Chem.*, 1997, **27**, 199–255.

64. Q. Jin, Y. Duan and J. A. Olivares, *Spectrochim. Acta, Part B*, 1997, **52**, 131–161.

65. W. Yang, H. Zhang, A. Yu and Q. Jin, *Microchem. J.*, 2000, **66**, 147–170.

66. K. Tanabe, H. Haraguchi and K. Fuwa, *Spectrochim. Acta, Part B*, 1981, **36**, 119–127.

67. *Massachusetts Institute of Technology Wavelength Tables*, MIT Press, Cambridge, 1969.

68. *Tables of Spectral Lines*, ed. A. N. Zaidel, 3rd edn., Plenum Press, New York, 1970.

69. C. C. Wohlers, *ICP Inform. Newslett.*, 1985, **10**, 593–688.

70. R. K. Winge, V. J. Peterson and V. A. Fassel, *Appl. Spectrosc* 1979, **33**, 206–219.

Introduction of Gases and Vapours into Microwave Plasmas

6.1 Introduction

In this chapter the techniques applied in cases when the determined substance is in the gaseous form at a temperature close to room temperature will be described. Gaseous samples can be directly introduced into the microwave plasma (MWP) without any pretreatment. They are delivered continuously after mixing with the plasma gas or injected as a portion of a microsample. Samples can be introduced directly into the plasma in their natural state, but more frequently they are transformed by a chemical or electrochemical process.[1–4] Especially, hydride generation (HG) has been much applied because the typical volatile hydride-generating elements can also be efficiently excited in argon and especially in helium MWPs. Moreover, for a number of elements, reactions other than HG could be efficiently used for generating volatile species that can be excited by MWPs. Using gas generation techniques, the sensitivity can be improved significantly because the transport efficiency of gas is much higher than that of liquid aerosols, and substantial matrix effects are reduced because of the use of the gas–liquid separation stage. Procedures including various trapping techniques can be used for preconcentration of analytes. Electrothermal atomization and graphite furnace atomization are sometimes classified as gas introduction techniques. However, in this book they will be discussed as solid sampling techniques, bearing in mind that a substantially higher temperature has to be used in these techniques to evolve the vapour phase for transport to the plasma, leading even to partial atomization of the sample. The coupled GC-MIP-OES technique is particularly discussed in this chapter, where even microsamples are injected into the carrier gas stream after conversion into

RSC Analytical Spectroscopy Monographs No. 12
Microwave Induced Plasma Analytical Spectrometry
By Krzysztof J. Jankowski and Edward Reszke
© The Royal Society of Chemistry 2011
Published by the Royal Society of Chemistry, www.rsc.org

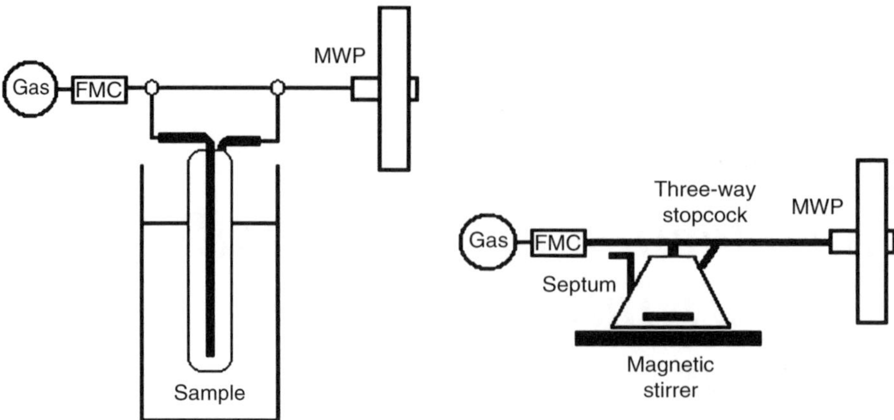

Figure 6.1 The schemes of DVS (left) and ED (right) gaseous sampling systems (FMC – flow mass controller).

the vapour state at a higher temperature and subjected to separation on a chromatographic column, before being introduced into the MIP.

Most often one of the following action modes is used for direct introduction of the determined sample in the form of vapour: direct vapour sampling (DVS) or exponential dilution (ED). The sample introduction systems are shown in Figure 6.1.

In the DVS technique a portion of liquid of the analyzed material is placed in a chamber, from where it is delivered as a vapour with the carrier gas to the plasma by washing the chamber with heated carrier gas flowing in the space above the liquid or by carrying out barbotage of the liquid sample with the carrier gas.[5–8] It is important to maintain a constant concentration of the vapour transported to the plasma during the measurement.

The ED technique consists of injection of a small sample to the flow chamber with a stirrer and recording its gradual dilution by the carrier gas. The sample is injected through a rubber septum located on the top of an exponential chamber. The sample vapour is mixed by a magnetic stir bar with the carrier gas in the chamber and the temperature is adjusted. Two three-way valves at the entrance and exit of the flask allow the flask to be isolated from the carrier gas to permit bypassing the flask during sample mixing, to measure background levels. Upon switching the bypass valves, the carrier gas purges the sample out of the flask into the plasma. The use of an ED flask has been shown to provide a means of continuously varying the analyte concentration as a well-defined function of time:

$$C_t = C_0 \exp(-Ft/V)$$

Where C_t is the concentration at time t, C_0 is the initial concentration at $t = 0$, F is the carrier gas flow rate (in mL s^{-1}), V is the exponential chamber volume (in mL) and t is the time (in s).

This technique has been elaborated for the needs of trace analysis to eliminate matrix effects. It is assumed that at certain sample dilution, when the signal originating from the analyte determined is still observed, the matrix concentration is small enough not to affect the measurement result. The signal measured under such conditions is extrapolated to the initial sample concentration on the basis of the recorded ED curve. The other applications of the ED technique include obtaining the calibration curve and determining the detection limit.[9–11]

As already stated, owing to the limited MWP energy amount, it is most convenient to introduce the sample in the form of a gas or vapour, since this ensures the best conditions of its atomization and excitation. However, a definite plasma tolerance to gaseous sample loading should be considered for choosing the appropriate sample amount, depending upon the sample matrix, the cavity type, the sample introduction method, the carrier gas flow rate, the operating pressure and the applied power used. Early MIPs could usually withstand the injection of a limited amount of sample material at about $1\,\mathrm{mg\,min^{-1}}$ or approximately $5\,\mu g$ absolute, without seriously degrading the signal intensity or stability.[12] However, today, MWP designs provide an enhanced sample loading plasma immunity.

It is well known that the addition of molecular gases to an argon or helium plasma affects the atomization and excitation efficiency. McKenna *et al.*[13] reported that adding more than 1% of oxygen, nitrogen or argon to a helium plasma sustained in a $\mathrm{TM_{010}}$ cavity caused a decrease of emission from their atomic species. However, the decrease was less remarkable in the case of a reduced-pressure helium plasma compared with that at atmospheric pressure. Olsen *et al.*[14] introduced $3.3\,\mathrm{mL\,min^{-1}}$ of offgas from an oil shale retort (nitrogen was used as a carrier gas) into a helium MIP to determine traces of selenium and arsenic, while Seravallo and Risby[15] observed that a reduced-pressure helium MIP can accept injection of air containing vinyl chloride vapour or methane with sample sizes at least as large as $100\,\mu L$ without affecting the stability. Ducatte and Long[16] found that non-metal emission in a helium MIP can be affected by the presence of carbon dioxide or hydrogen. Decreased signal intensity with carbon dioxide or hydrogen addition for ionic (and atomic to a lower extent) lines of sulfur, phosphorus, chlorine, bromine and iodine was observed, while the excitation and ionization temperature of the plasma remain virtually unchanged.

In order to compare three MWP sources towards the tolerance of an argon plasma to water vapour loading before being extinguished, Camuna-Aguilar *et al.*[17] introduced water vapour in a transient mode using a syringe. The amounts were about 50, 10 and 0.5 mg for the MPT, the Beenakker cavity and the surfatron, respectively. Similar trends exist with helium discharges to those obtained with argon.

MWPs are recognized to be susceptible to the introduction of excess hydrogen. However, detailed studies show that the MPT offers high tolerance to H_2 when the HG technique is directly coupled to MPT-OES without using any interface. The argon MPT plasma can work normally up to 30% volume of hydrogen. Moreover, the introduction of a suitable amount of H_2 to the MPT plasma proves to be beneficial for enhancing the excitation efficiency of the elements studied.[18] High robustness against hydrogen loading of an argon MIP sustained in a TE_{101} cavity coupled with a HG system without a hydrogen separator was stated, while the immunity of a helium plasma for this system was even better.[19] Also the nitrogen plasma in an Okamoto cavity exhibits a high tolerance for hydrogen loading when coupled with the HG technique.[1,20]

In another paper, Camuna-Aguilar *et al.*[9] conducted a comparison study of a Bennakker cavity, a surfatron and a MPT to evaluate atomization and excitation capabilities for chlorinated hydrocarbons introduced in the form of vapour into helium plasma. The amount of organic compounds withstood by the plasma prior to being extinguished was up to 100, 1 and 0.3 mg min^{-1} for the MPT, the TM_{010} and the surfatron, respectively. When monitoring the reforming of methane with the use of MIP-OES, about 7 mg min^{-1} of methane was continuously introduced into an argon plasma.[21] The analysis of trace impurities in a number of metal–organic vapours, including trimethylaluminum, trimethylgallium and trimethylindium, by direct introduction into the argon or helium plasma sustained with a TE_{101} cavity, was carried out at sample flow rates that ranged from 0.5 to 3 mg min^{-1} and a microwave power of 150 W.[22]

Despite the small volume of samples used in gas chromatography, a solvent passing through the microwave plasma region causes a disturbance in discharge, and sometimes even extinguishes the plasma. To assure stability of the plasma discharge, solvent venting systems were proposed.[23] However, studies of more advanced MWP sources, including the MPT used as a GC detector, showed that up to about 1 mg of organic sample could be injected without necessity of using a solvent venting system.[24–26]

6.2   Continuous Gas Introduction

In the case of gas analysis, samples should be introduced with direct injection sampling devices or delivered continuously from a gas flow. In our opinion, the latter method allows one to obtain more representative values. Moreover, on-line analysis minimizes the risks of contamination associated with discrete sampling. The sample gas as well as the gaseous standards are introduced continuously into the MIP directly from the gas or vapour container,[1,5,6,13,14,8–10,21,26–28] by dynamic mixing devices,[7,29] peristaltic or infusion pumps[30,31] or by injecting a small volume of the liquid sample into the mixing chamber where complete vaporization quickly takes place.[9–11] The discrete method consists of injection of a portion of sample directly into the plasma gas stream[15,32] using gas-tight syringes or sampling loops, or with the use of various trapping techniques as well as chromatographic separation.[23–24,33,34] Changes of the emission signal with time are recorded and integration of the peak obtained is performed. A similar procedure takes place with the signal obtained

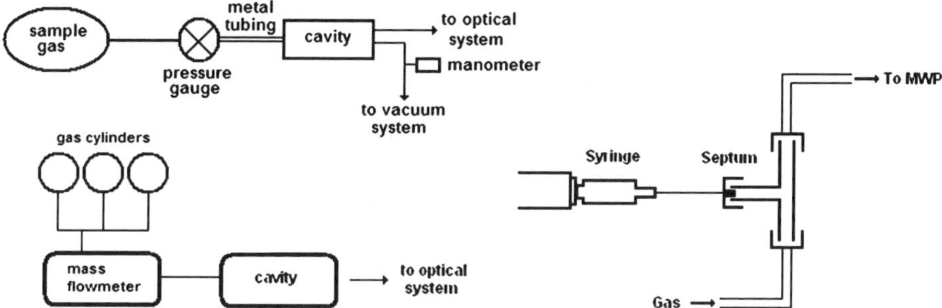

Figure 6.2 Gas sampling arrangements applicable for MWP-OES: for reduced pressure MWP – left top; for atmospheric pressure MWP – left bottom; discrete sample introduction – right.

using the quasi-continuous method, most often carried out in the system for flow injection analysis (FIA).[28,35]

Some schematic representations of the gas sampling arrangements are shown in Figure 6.2. The dynamic mixing of the two gases is performed with the use of a special type of mass flowmeter.[29] The direct determination of contaminants in hydrogen chloride using a contaminant-free gas injection system connected to a MIP-OES instrument has been developed.[31] In the reduced-pressure plasma system, gas is fed *via* a pressure gauge through a length of metal tubing to restrict the gas flow. The pressure in the plasma is measured with a manometer at the outlet of the gas line.[13]

The determination of nitrogen, oxygen, hydrogen, helium, krypton, ammonia, water vapour, carbon dioxide and organic gases in argon using discrete sample introduction can be especially mentioned.[30] The sample is introduced through the rubber septum located in the injection port. The detection limits permit direct analysis of argon of 5.5 N purity. Seravallo and Risby[15] determined the concentration of vinyl chloride in air by means of an He-MIP under reduced pressure on the basis of discrete injection and measuring the chlorine emission at the 479.45 nm ionic line. The determination of sulfur in the form of hydrogen sulfide or sulfur dioxide in air was studied by Taylor *et al.*[32]

6.3 Hydride Generation and Related Techniques

The HG technique is known as an effective sampling method, since it assures the introduction of a sample in gaseous form, without the unnecessary excess of solvent, which leads to a considerable increase in sensitivity and selectivity of the measurement. A MWP is an attractive excitation source for this type of sample, since due to low discharge energy a low background is observed under these conditions. Owing to nearly a 100% efficiency of the hydride formation reaction and natural separation of the analyte from the sample matrix, this technique is an attractive tool for the determination of about 10 elements that form stable hydrides (As, Ge, Pb, Sb, Se, Sn, Te, Tl, Bi). Recently, it has been shown that a number of other elements can be volatilized under HG conditions and MWPs are stated to be useful for the excitation of these species. Moreover, in a simple way the technique can be expanded with the pre-concentration stage, which still further increases the possibilities of trace analysis. The broad range of work on the HG and related techniques is reflected in reviews in this field.[2,3,36–38] Compared with chemical HG, electrochemical HG offers more stable production of hydrides and hydrogen, leading to a more stable plasma operation and reduced risk of contamination.[39] Electrochemical HG as well as chemical HG are successfully coupled with the MPT and TE_{101}-based MWP-OES operated with helium as well as with argon.[4,39]

The mercury cold-vapour atom generation technique (HgCVG) is noteworthy regarding exceptional sensitivity obtained, which is better than that required for baseline environmental mercury studies.[17,40–42] The use of HgCVG-MIP-OES is superior to the conventional atomic absorption approach. Mercury vapour can be generated from water samples by reduction with tin(II) chloride, purging with helium and subsequent trapping with a gold amalgamation trap.[42] Also, the HgCVG-MIP-OES technique has been applied to the indirect determination of iodide. A detection limit of $0.74\,ng\,mL^{-1}$ for iodine was reported.[43,44]

The HG or HgCVG processes can be carried out according to one of several methods: periodically, continuously or quasi-continuously. The periodic method consists of carrying out the hydride formation reaction in a batch vessel and transportation of the volatile part of the sample to the plasma. The approach consisting of the injection of microvolumes of the sample solution on a sodium borohydride pellet placed in an appropriate flow vessel is an interesting example of a HG discrete mode application.[45] Recently, a batch mode generation system for a slurry sampling HG has been developed by Matusiewicz and Ślachciński.[46] A flow-injection CVG technique has been successfully coupled with a microstrip plasma device for spectrometric determination of mercury.[28] An on-line HgCVG system was employed as the sample introduction system in a number of experimental setups with various MWPs.[17,41,47]

In a discontinuous mode of working, a trap interface can be used. Here, the hydrides are trapped in a graphite furnace (hot-trapping)[4,48,49] or in liquid nitrogen (cold-trapping),[32–33,50] both followed by evaporation and transport of the analyte into the plasma. Furthermore, the trap can act as a powerful pre-concentration step. The determination of analytes on a sub-$ng\,mL^{-1}$ level can be easily achieved. An argon or a helium plasma for HG with trapping followed by electrothermal atomization can be used, and again the differences with various MWPs are minimal.[48] Bulska *et al.*[51] trapped the hydrides *in situ* onto the inner wall of the graphite furnace and evaporated them subsequently after separation from hydrogen. An on-line amalgamation trap was constructed for the collection of mercury species separated by capillary gas chromatography (CGC) for detection by MIP-OES.[52] In the collection step, mercury species were amalgamated with gold–platinum wire. For the analysis step, the gold–platinum wire was heated to release mercury vapour.

The fact that the formation of hydrides is accompanied by hydrogen evolution, which in larger quantities can destabilize the plasma discharge, is a limitation in the application of the HG technique in combination with MIPs. The plasma in a Beenakker-type cavity and in the surfatron is susceptible to the excess hydrogen produced in chemical HG using sodium borohydride solution. Owing to this, design solutions and procedures are sought which would enable minimalization of the amount of hydrogen and water vapour reaching the plasma.[53] Tao and Miyazaki[54] and Gong *et al.*[55] applied a membrane separator for removing the excess hydrogen and water vapour. The use of a desiccator to remove water vapour from the gas phase is a common practice. These problems do not occur when using high-power plasmas such as the nitrogen high-power MIP operated in an Okamoto cavity. Matsumoto and Nakahara[56] do not report any problems with plasma instability when the excess hydrogen enters the plasma. Liang and Li[57] observed that the use of oxygen as a shielding gas in an argon MPT caused a higher tolerance to hydrogen loading. Low-pressure argon and helium MIPs proved to be easily maintained in the presence of hydrogen compared with the atmospheric pressure plasmas.[58,59] In the case of an MIP operated in a TE_{101} cavity at 200 W, a miniaturized cell enabled it to determine selenium down to $0.6\,\mu g\,L^{-1}$ without the need to remove the excess hydrogen.[60] Hence, it is concluded that applying higher microwave power levels and an improved symmetrical coupling of the microwave energy is beneficial for determining elements with a relatively high ionization potential (As, Se).[19]

6.4 Generation of Other Gaseous Species

Instead of hydrides, other volatile metallic species can be generated, both inorganic and organic. Volatile organometallic compounds are usually subjected to chromatographic separation (see Section 6.5). Skogerboe *et al.*[35] proposed a chemical vapour generation (CVG) system with hot trapping for determination of several metal chlorides by MIP-OES. Another interesting approach to gaseous sample introduction is determination of nickel as nickel tetracarbonyl with the use of FI-CVG-MIP-OES.[61] The detection limits for some important elements, *i.e.* Bi, Cd, Mo, Ni, Pb, Tl and Zn are in the range $0.5-3\,ng\,mL^{-1}$.

A distinctive feature of a helium MWP is the possibility of determining non-metals, including halogens, at low concentration levels. Interestingly, this high performance of MWPs results not only from the possibility of using helium as the plasma gas, but also from applying a microwave frequency. It has been shown that for radiofrequency-driven plasma sources the detection limits for non-metals are poorer than those for MWPs.[62,63] Most non-metal MWP-OES determinations have been performed with gaseous sample introduction.[1] The chemically produced volatile form of the analyte produced in continuous or batch mode is separated from the solution in a gas–liquid separator and swept into the carrier gas stream, usually helium. The sample introduction system is preferably supported with a desiccator to remove water vapour and a heated transfer line in the case of generation of relatively less volatile species, *i.e.* iodine and bromine, to avoid analyte losses due to condensation. Sometimes, silanization of internal walls of the generator, a desiccator and connector tubes are recommended to facilitate the efficient transfer of the analyte species.[64] The volatile species can be generated by an oxidation or reduction reaction and subsequently evolved by acidification of the sample solution. Halogens are commonly generated as molecular chlorine, bromine or iodine with the use of various chemical agents, including potassium permanganate, potassium dichromate, potassium bromate, hydrogen peroxide and others.[1,65,66] Interestingly, the generation of chlorine, bromine or iodine is also performed by on-column oxidation with solid lead dioxide.[67] Chloride and bromide can be evolved as their respective hydrides; however, poorer detection limits were obtained in comparison with the generation of molecular chlorine and bromine.[64] A limitation of the CVG-MWP-OES technique for halogen determination is a substantial influence of molecular gases in the helium plasma originating from helium impurities or evolved from the sample. The decrease of analyte emission in the presence of H_2, H_2O, CO_2 and N_2 is well documented.[9,16,55,57,68] Among other non-metallic analytes, trace ammonia, nitrite and nitrate can be converted into nitrogen by CVG and determined by MWP-OES, while sulfide, sulfite and carbonate can be evolved by acidification, with subsequent introduction into the MWP.[68,69] The CVG-MWP-OES technique can be utilized for speciation of inorganic nitrogen and iodine in sample solutions, owing to two differences in the chemical properties of various forms of the analytes.[68,70] Also, the CVG of non-metals has been coupled with MIP-MS.[71]

6.5 Microwave Induced Plasma Coupling with Gas Chromatographic Techniques

The combination of GC with MIP-OES is at present the most predominant and matured technique utilizing MWPs. MIP plays here the role of a typical element-selective flow-through detector and the effluent gas, containing components of the sample separated on a column, is continuously introduced into the plasma. Temporal separation of the sample components combined with selective elemental detection yields a great deal of qualitative as well as quantitative information. Since MIP-OES is a powerful technique for detection in GC, it is regularly reviewed[72–79] and updated in a number of books.[80–83] Hence, in this chapter only a short summary will be given. The microwave plasma detector (MPD) utilizing MIPs belongs to the most sensitive detectors used presently in GC.[79,84–88]

The MWP cavities are very easy to interface with capillary GC columns, providing minimal transfer volume. MWP sources coupled with GC have been examined at either reduced pressure or atmospheric pressure.[77] The use of reduced-pressure-doped plasmas can allow the solvent to enter the detector without extinguishing the discharge. Both the $\frac{1}{4}$ wave Evenson-type cavity and the surfatron are preferred for sustaining reduced-pressure helium or argon plasmas. In contrast, the TM_{010} Beenakker-type cavities have been easily adapted for atmospheric pressure GC-MIPs. Based on the comparative study of three MWPs, including the TM_{010}, surfatron and the MPT, Camuna-Aguilar *et al.*[9] concluded that for GC applications the MPT may offer the better performance. Nevertheless, only the TM_{010} approach succeeds on a commercial scale.

When the GC-MIP-OES setup consists in coupling of two independent instruments commercially available, the method of interfacing the column outlet with a detector in fact requires a more precise description. Relatively large dimensions of the MPD cause the necessity of prolongation of the route covered by the species detected, and hence lead to unfavourable peak broadening. To minimize this effect, transfer lines made of copper, nickel or steel, characterized by a minimal dead volume and heated to avoid analyte condensation, are used. Additionally, the makeup gas or scavenger gases can be introduced (Figure 6.3).

The early GC-MIP systems lacked ruggedness and reliability, as the plasma was extinguished during the solvent passage through the plasma and the discharge tube was prone to carbonization. The first problem has been resolved by introduction of the solvent-venting system,[23,89–92] an automatic device for plasma re-ignition[23] or by splitting of the effluent gas.[89] In order to avoid formation of carbonaceous deposits on the wall of the discharge tube, the addition of small quantities of a scavenger gas to the plasma (usually oxygen, rarely hydrogen or nitrogen) has been utilized.[11,93–95]

The MPD can co-operate with a chromatograph equipped with capillary as well as packed columns.[78] Capillary columns are readily used for analysis of reactive compounds, since it is easier to deactivate them on the immobile phase surface. Flexible fused silica capillary columns can be directly interfaced to within a few millimetres of the plasma by standard compression fittings. The possibility of introducing samples not larger than 1 μL is a limitation in applying capillary columns. Larger samples can be separated on packed

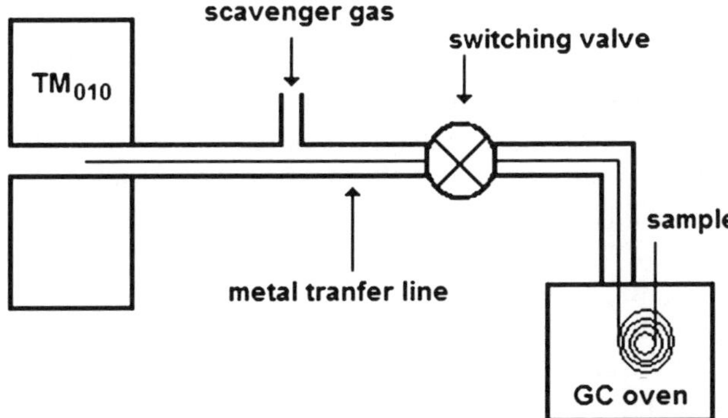

Figure 6.3 A GC-MIP interface with side-arm introduction of scavenger gas.

columns; however, greater broadening of the peaks should be taken into account. Some years ago the coupling of a MPD with multi-capillary chromatography (MC-GC)[96] was proposed. This allows shortening of the separation time while maintaining the possibility of introducing samples of a volume of several microlitres. Thus, a very effective analytical technique was obtained, combining the advantages of rapid and complete sample separation with high sensitivity, selectivity and universality of a MPD. The ideal fitting of gas flows used in MC-GC and MIPs is an additional advantage of this solution.

6.5.1   Atomic Emission Detector

The MPDs are a part of a wider class of atomic emission detectors (AEDs) for chromatography, but ones which have been particularly useful because of the advantages of compatibility with GC, compactness and cheapness. The most important plasma emission sources that have been used for GC detection have been the helium MIP, CMP, direct current argon plasma, alternating current plasma, and radiofrequency-induced plasma.[73,76,77,84,97] The MIP has been recognized as a more appropriate source than ICP for use as the excitation source in GC-AED; however, the latter is finding favour as a mass spectral ion source in the GC-ICP-MS mode. A comparison of the analytical capabilities of electron ionization mass spectrometry (EI-MS), MIP-OES and inductively coupled plasma time-of-flight mass spectrometry (ICP-TOFMS) for the speciation of organolead compounds in rainwater on the basis of sensitivity, selectivity and reliability has been made.[86] In general the detection limits obtained by ICP-TOFMS and MPD were similar.

To date, several MPDs have been commercialized, including MPD 850, HP2350 and its second-generation G2350A or 2370AA, and some others have been announced as a preserial. The MPD consists of a quartz capillary inlet which conducts the effluent gas from the GC column to the centre of a microwave cavity. The GC carrier gas can be argon or helium and the generally used flow rates are compatible with stable MIP operations. Argon/helium mixtures have also been used as both the carrier and discharge gases. Especially helium plasma is characterized by high sensitivity, both with respect to metals and non-metals (Table 6.1). The emitted spectrum is focused on the entrance slit of an optical spectrometer and the detector system typically consists of a photomultiplier tube with associated hardware. The discharge is initiated by a Tesla coil spark. It is sustained in a thin-walled silica discharge tube within a water-cooled microwave re-entrant cavity, which is a modification of the TM_{010} cavity.[23]

The major advantages of the GC-MPD system are high elemental sensitivity, high selectivity of detection of the element of interest over co-eluting elements and the tolerance to incomplete chromatographic resolution.

Despite the MPD, other microwave-driven detectors have been investigated and used, including photoionization detectors,[98,99] MIP reflected power detectors[100,101] and, more recently, MIP-MS detectors.[102–105]

Table 6.1 Analytical performance of a MPD.

Element	Wavelength (nm)	Detection limit (pg s^{-1})	Selectivity vs. carbon
Al	396.2	5.0	> 10 000
As	189.0	3.0	47 000
	228.8	6.5	47 000
B	249.8	3.6	9300
Be	234.9	(10 pg)	–
Br	470.5	10	11 400
	478.6	30	599
C	247.9	2.6	1
	193.1	0.2	1
^{13}C	171.0	10	–
Cl	479.5	39	25 000
	481.0	7	200
Co	240.7	6.2	182 000
Cr	267.7	7.5	108 000
F	685.6	8.5	3500
Fe	302.1	0.05	3 500 000
	259.9	0.28	280 000
Ga	294.3	*ca.* 200	> 10 000
Ge	265.1	1.3	7600
H	656.3	7.5	160
	486.1	2.2	variable
^{2}H	656.1	7.4	194
Hg	253.7	0.1	3 000 000
	301.2	1.0	–
I	206.2	21	5010
	516.1	50	400
Mn	257.6	0.25	1 900 000
Mo	281.6	5.5	24 200
N	174.2	7.0	–
	746.9	2900	6000
Nb	288.3	69	32 100
Ni	301.2	1.0	200 000
	231.6	2.6	6470
O	777.2	75	25 000
Os	225.6	6.3	50 000
P	177.5	1	5000
	185.9	1.0	–
	253.6	2.1	26 000
Pb	261	0.8	314 000
	283.3	0.17	25 000
	105.8	2.3	200 000
	406	0.2	286 000
Pd	340.4	5.0	> 10 000
Pt	405.8	(0.1 ng L^{-1})	–
	407.8	(1.1 pg)	–
Ru	240.3	7.8	134 000
S	180.7	1.7	150 000
	545.4	25	200
Sb	217.6	5.0	19 000
Se	196.1	2.3	135 000
	204.0	5.3	10 900

Table 6.1 (*Continued*).

Element	Wavelength (nm)	Detection limit (pg s^{-1})	Selectivity vs. carbon
Si	251.6	7.0	90 000
Sn	271	1.0	295 000
	284.0	1.6	36 000
	303.1	1.4	1 500 000
Ti	338.4	1.0	50 000
V	292.4	4.0	36 000
	268.8	10	56 900
W	255.5	51	5450

The MPD can be used either as a substitute for one of the more usual non-metal-selective detectors or it can be used as a metal-sensitive detector. Theoretically, any element can be detected with the MPD, providing that atomic emission can be monitored in the UV/Vis wavelength region. Depending on the plasma gas used and the element considered, absolute detection limits typically achieve low to hundreds of picogram levels.[80]

Theoretically, the MPD can operate in a number of modes: a simple element mode, a universal mode by monitoring carbon, or it can be used to determine empirical formulas by emission from all the various atoms in the molecule. In the first mode, the MPD exhibits selectivity which far exceeds all other selective detectors. In the universal mode, the MPD is the most sensitive GC detector, since organic molecules generally contain more carbon atoms than heteroatoms.

The MPD also provides a unique GC possibility of multi-element detection of the separated compounds. The linearity between response and the number of atoms of a particular element should enable the empirical formula determination for unknown compounds. Owing to this the selectivity of the detection increases, which is of great importance in the analysis of complicated materials.

Moreover, it is a universal device enabling the realization of various detection methods, *e.g.* identification of all organic compounds present in the sample on the basis of carbon concentration and individual retention times, identification of selected classes of compounds, *e.g.* halogen derivatives on the basis of the heteroatom emission, speciation studies or determination of the chemical composition of compounds on the basis of determination of the ratio of the concentrations of several basic elements, *e.g.* C, H, O, N and S.

The possibility of using the GC-MIP-OES technique for compound-independent calibration and the determination of molecular composition has been investigated for a long time. It has been shown for some elements and some groups of compounds that the MPD has nearly the same response for an element irrespective of the compound. Despite some unresolved questions, the results are promising and the studies will be continued.[106]

Despite high selectivity and sensitivity of MIP-OES, in several instances, preconcentration techniques are utilized for enhancing the analytical performance of the GC-MPD.[77] These include in-line preconcentration,[107] purge-and-trap thermal desorption[108–110,94] and different derivatization methods such as HG,[111,112] alkylation,[113,114] chelate formation[115,116] and others.

By employing a separation technique prior to atomic emission, valuable speciation information can be gained. Generally, the full benefit can be obtained from selective detectors when used in-parallel with nonselective detectors. This arrangement allows the maximum amount of information to be obtained for the sample, even though part of these data is not required. GC coupled with MIP-OES is the most common hyphenated microwave plasma technique. Expanded hyphenation techniques involve GC with parallel operation of both AED and MS detection.[117,118] The spectral information from the two techniques can be combined to enable the analyst to deduce much about unknown compounds in a complex mixture.

Finally, two specific and interesting applications of coupled or hyphenated techniques utilizing MPDs should be mentioned, namely pyrolysis GC (Py-GC)[119,120] and hot carrier gas extraction GC (HCGE-GC).[121] The former is widely utilized in organic geochemistry[122,123] and polymer science.[124,125]

6.6 Solid-phase Microextraction

Solid-phase microextraction (SPME) is a simple, solvent-free and efficient extraction technique. To date, SPME has been used successfully to analyze gaseous, liquid and solid samples. It can be easily coupled to GC-MWP-OES and with some modifications to the respective HPLC technique.[126,127] In general, two operational modes are available: headspace and direct liquid-phase sampling. The headspace (HS-SPME) technique consists of placing the fibre in the headspace of the sample container for analyte preconcentration and then, after the optimized sorption time, transferring the fibre to the hot injection port for desorption. Thus, it is in principle the sampling technique for gaseous analytes. For non-volatile species the direct liquid-phase sampling can be used.

In the most cases, SPME is coupled with chromatographic separation. However, the large selectivities normally observed for MIP-OES result in the possibility of injecting gaseous samples directly into the plasma, thereby obviating the need for GC. This may be useful in a number of applications, in particular when inorganic volatile species have to be determined. As a trapping technique, SPME play the role of a preconcentration technique and can be used to eliminate the load of interfering compounds to the plasma, *i.e.* hydrogen in the case of the HG technique or volatile acids when operating with high acid concentration in the liquid phase. Up to now, SPME-based sample introduction for ICP-MS has been reported as a promising analytical tool and here the application of MIP-OES will be interesting. Preliminary studies are in progress in our laboratory.

The SPME-GC-MPD coupled technique has been successfully used in a number of applications.[84,95,128–134] The use of flame photometric detection, pulsed flame photometric detection, MIP-OES and ICP-MS for the speciation of butyl- and phenyltin compounds after SPME and GC separation has been evaluated.[87] Compared with other detectors, MPD exhibited relatively poor sensitivity. For organoselenium compounds the versatility of SPME in combination with multicapillary gas chromatography was evaluated using different common detectors.[135] The detection methods were ICP-MS, MIP-OES and atomic fluorescence spectrometry (AFS). All detectors were found to be suitable, with the highest sensitivity being obtained for MIP-OES detection. The method has been applied to the determination of volatile alkyl selenides in biological samples.[136] The use of metal chelates was examined for the separation of Cu, Ni, Pd and V by CGC, with flame ionization and MIP-OES detection.[116]

6.7 Quantitative Analysis of Gases

In the case of gas analysis, sampling is one of the main problems and is therefore an important source of systematic errors. To determine the content of analytes in a gas, calibration in the same matrix is desirable. Gaseous calibration mixtures can be prepared on a high-pressure mixing line which allows the addition of individual pure gases to be monitored accurately by weight. The procedure is valid for mixtures with a percent content of components, then diluted gravimetrically. If the influence of the matrix gas on the plasma is well examined and low, matrix matching can be neglected. Standard addition of a small flow of calibration gases through the direct inlet connection allows calibration of selected analytes for which volatile components are available. A number of different vinyl chloride/air mixtures have been prepared containing different amounts of vinyl chloride monomer. Linear calibration curves were obtained by injecting the same volume of each standard.[15]

In continuous introduction techniques, calibration curves for the various species are obtained by controlled feed-rate direct injection of suitable gases into the plasma gas at the injection port.[30] The diffusion principle can be easily used for the dynamic generation of a known amount of mercury vapour in various carrier gases over a wide concentration range by only changing the operating parameters of the diffusion cell.[7] The sample gas as well as the standard gas mixture are injected continuously into the MIP by peristaltic pumps, permitting standard additions. During continuous analysis, the analyte emission reaches a steady state approximately 10 minutes after opening the sample cylinder. For calibration of the analytical system towards iron determination, a certified sample of standard iron pentacarbonyl in argon is used. To obtain different concentrations the standard gas has to be diluted with argon.[31]

An exponential dilution flask is evaluated for automated calibration curve preparation and determination of the detection limit by spectroscopic methods. The transient signal is registered at the wavelength corresponding to the element determined. The exponential decay curve is fitted by a liner regression to the semi-logarithmic plot of net signal against time. Solving the resulting linear equation at zero time gives the initial signal, which is directly proportional to the initial concentration of the analyte in the sample. Chlorinated compounds have been introduced into the plasma in gaseous form using the exponential decay method.[9]

References

1. T. Nakahara, *Anal. Sci.*, 2005, **21**, 477–484.
2. P. Pohl, *Trends Anal. Chem.*, 2004, **23**, 21–27.
3. R. E. Sturgeon and Z. Mester, *Appl. Spectrosc.*, 2002, **56**, 202A–213A.
4. B. Özmen, F. M. Matysik, N. H. Bings and J. A. C. Broekaert, *Spectrochim. Acta, Part B*, 2004, **59**, 941–950.
5. K. Tanabe, H. Haraguchi and K. Fuwa, *Spectrochim. Acta, Part B*, 1981, **36**, 119–127.
6. K. B. Starowieyski, A. Chwojnowski, K. Jankowski, J. Lewiñski and J. Zachara, *Appl. Organomet. Chem.*, 2000, **14**, 616–622.
7. V. Siemens, T. Harju, T. Laitinen, K. Laryava and J. A. C. Broekaert, *Fresenius' J. Anal. Chem.*, 1995, **351**, 11–18.
8. S. Schermer, N. H. Bings, A. M. Bilgic, R. Stonies, E. Voges and J. A. C. Broekaert, *Spectrochim. Acta, Part B*, 2003, **58**, 1585–1596.
9. J. F. Camuna-Aguilar, R. Pereiro-Garcia, J. E. Sanchez-Uria and A. Sanz-Medel, *Spectrochim. Acta, Part B*, 1994, **49**, 545–554.
10. B. W. Pack, G. M. Hieftje and Q. Jin, *Anal. Chim. Acta*, 1999, **383**, 231–241.
11. W. Braun, N. C. Peterson, A. M. Bass and M. J. Kurylo, *J. Chromatogr.*, 1971, **55**, 237–248.
12. A. T. Zander and G. M. Hieftje, *Appl. Spectrosc.*, 1989, **35**, 357–371.
13. M. McKenna, I. L. Marr, M. S. Cresser and E. Lam, *Spectrochim. Acta, Part B*, 1986, **41**, 669–676.
14. K. B. Olsen, J. C. Evans, D. S. Sklarew, J. S. Fruchter, D. C. Girvin and C. L. Nelson, *Environ. Sci. Technol.*, 1990, **24**, 258–263.
15. F. A. Serravallo and T. H. Risby, *Anal. Chem.*, 1976, **48**, 673–676.
16. G. R. Ducatte and G. L. Long, *Appl. Spectrosc.*, 1994, **48**, 493–501.
17. J. F. Camuna-Aguilar, R. Pereiro-Garcia, J. E. Sanchez-Uria and A. Sanz-Medel, *Spectrochim. Acta, Part B*, 1994, **49**, 475–484.
18. W. Yang, H. Zhang, A. Yu and Q. Jin, *Microchem. J.*, 2000, **66**, 147–170.
19. A. Jackowska, PhD thesis, Warsaw University of Technology, 2006 (in Polish).
20. K. Wagatsuma, *Appl. Spectrosc. Rev.*, 2005, **40**, 229–243.
21. C. A. Junior, N. K. M. Galvao, A. Gregory, G. Henrion and T. Belmonte, *J. Anal. At. Spectrom.*, 2009, **24**, 1459–1461.
22. K. Jankowski, DSc thesis, Warsaw University of Technology, 2001 (in Polish).
23. B. D. Quimby and J. J. Sullivan, *Anal. Chem.*, 1990, **62**, 1027–1034.

24. K. Cammann, L. Lendero, H. Feuerbacher and K. Ballschmiter, *Fresenius' Z. Anal. Chem.*, 1983, **316**, 194–200.
25. Q. Jin, F. Wang, C. Zhu, D. M. Chambers and G. M. Hieftje, *J. Anal. At. Spectrom.*, 1990, **5**, 487–494.
26. A. Granier, E. Bloyet, P. Leprince and J. Marec, *Spectrochim. Acta, Part B*, 1988, **43**, 963–970.
27. E. Denkhaus, A. Golloch and H. M. Kuss, *Fresenius' J. Anal. Chem.*, 1995, **353**, 156–161.
28. U. Engel, A. M. Bilgic, O. Haase, E. Voges and J. A. C. Broekaert, *Anal. Chem.*, 2000, **72**, 193–197.
29. D. Boudreau, C. Laverdure and J. Hubert, *Appl. Spectrosc.*, 1989, **43**, 456–460.
30. H. E. Taylor, J. H. Gibson and R. K. Skogerboe, *Anal. Chem.*, 1970, **42**, 876–881.
31. S. Kirschner, A. Golloch and U. Telgheder, *J. Anal. At. Spectrom.*, 1994, **9**, 971–974.
32. H. E. Taylor, J. H. Gibson and R. K. Skogerboe, *Anal. Chem.*, 1970, **42**, 1569–1575.
33. C. Dietz, Y. Madrid, C. Camara and P. Quevauviller, *J. Anal. At. Spectrom.*, 1999, **14**, 1349–1355.
34. A. J. McCormack, S. C. Tong and W. D. Cooke, *Anal. Chem.*, 1965, **37**, 1470–1476.
35. W. Drews, G. Weber and G. Tölg, *Fresenius' Z. Anal. Chem.*, 1989, **332**, 862–865.
36. A. D. Campbell, *Pure Appl. Chem.*, 1992, **64**, 227.
37. Q. De-ren, *Trends Anal. Chem.*, 1995, **14**, 76–82.
38. P. Pohl, *Trends Anal. Chem.*, 2004, **23**, 87–101.
39. C. Schickling, J. Yang and J. A. C. Broekaert, *J. Anal. At. Spectrom.*, 1996, **11**, 739–745.
40. R. J. Watling, *Anal. Chim. Acta*, 1975, **75**, 281–288.
41. K. Tanabe, K. Chiba, H. Haraguchi and K. Fuwa, *Anal. Chem.*, 1981, **53**, 1450–1453.
42. Y. Nojiri, A. Otsuki and K. Fuwa, *Anal. Chem.*, 1986, **58**, 544–547.
43. T. Nakahara and T. Wasa, *Chem. Express*, 1990, **5**, 121.
44. T. Nakahara and T. Wasa, *Microchem. J.*, 1990, **41**, 148.
45. N. W. Barnett, L. S. Chen and G. F. Kirkbright, *Spectrochim. Acta, Part B*, 1984, **39**, 1141–1147.
46. H. Matusiewicz and M. œlachciński, *Microchem. J.*, 2006, **82**, 78–85.
47. P. Pohl, I. Jimenez Zapata, E. Voges, N. H. Bings and J. A. C. Broekaert, *Microchim. Acta*, 2008, **161**, 175–184.
48. H. Matusiewicz and R. E. Sturgeon, *Spectrochim. Acta, Part B*, 1996, **51**, 377–397.
49. E. Bulska, J. A. C. Broekaert, P. Tschöpel and G. Tölg, *Anal. Chim. Acta*, 1993, **276**, 377–384.
50. C. Dietz, Y. Madrid and C. Camara, *J. Anal. At. Spectrom.*, 2001, **16**, 1397–1402.

51. E. Bulska, E. Beirohr, P. Tschöpel, J. A. C. Broekaert and G. Tölg, *Chem. Anal. (Warsaw)*, 1996, **41**, 615–623.

52. J. P. Snell, W. Frech and Y. Thomassen, *Analyst*, 1996, **121**, 1055–1060.

53. J. Yang, C. Schickling, J. A. C. Broekaert, P. Tschöpel and G. Tölg, *Spectrochim. Acta, Part B*, 1995, **50**, 1351–1363.

54. H. Tao and A. Miyazaki, *Anal. Sci.*, 1991, **7**, 55–59.

55. Z. Gong, W. F. Chan, X. Wang and F. S. C. Lee, *Anal. Chim. Acta*, 2001, **450**, 207–214.

56. A. Matsumoto and T. Nakahara, *Tetsu Hagane*, 2003, **89**, 1–9.

57. P. Liang and A. Li, *Fresenius' Z. Anal. Chem.*, 2000, **368**, 418–420.

58. F. Lunzer, R. Pereiro-Garcia, N. Bordel-Garcia and A. Sanz-Medel, *J. Anal. At. Spectrom.*, 1995, **10**, 311–315.

59. J. M. Costa-Fernandez, R. Pereiro-Garcia, A. Sanz-Medel and N. Bordel-Garcia, *J. Anal. At. Spectrom.*, 1995, **10**, 649–653.

60. S. Schermer, L. Jurica, J. Paumard, E. Beinrohr, F. M. Matysik and J. A. C. Broekaert, *Fresenius' Z. Anal. Chem.*, 2001, **371**, 740–745.

61. R. K. Skogerboe, D. L. Dick, D. A. Pavlica and F. E. Lichte, *Anal. Chem.*, 1975, **47**, 568–570.

62. K. C. Tran, C. Lauzon, R. Sing and J. Hubert, *J. Anal. At. Spectrom.*, 1998, **13**, 507–513.

63. J. F. Camuna, M. Montes, R. Pereiro, A. Sanz-Medel, C. Katschthaler, R. Gross and G. Knapp, *Talanta*, 1997, **44**, 535–544.

64. N. W. Barnett, *J. Anal. At. Spectrom.*, 1988, **3**, 969–972.

65. T. Nakahara and T. Nishida, *Spectrochim. Acta, Part B*, 1992, **53**, 211–217.

66. T. Nakahara, S. Morimoto and T. Wasa, *J. Anal. At. Spectrom.*, 1998, **7**, 507–513.

67. M. D. Calzada, M. C. Quintero, A. Gamero and M. Gallego, *Anal. Chem.*, 1992, **64**, 1374–1378.

68. S. R. Koirtyohann, *Anal. Chem.*, 1983, **55**, 374–376.

69. T. Nakahara, T. Mori, S. Morimoto and H. Ishikawa, *Spectrochim. Acta, Part B*, 1995, **50**, 393–403.

70. K. A. Anderson and P. Markowski, *J. AOAC Int.*, 2000, **83**, 225–230.

71. Y. Duan, M. Wu, Q. Jin and G. M. Hieftje, *Spectrochim. Acta, Part B*, 1995, **50**, 1095–1108.

72. T. H. Risby and Y. Talmi, *Crit. Rev. Anal. Chem.*, 1983/84, **14**, 231–265.

73. P. C. Uden, *Trends Anal. Chem.*, 1987, **6**, 238–246.

74. E. Bulska, *J. Anal. At. Spectrom.*, 1992, **7**, 201–210.

75. S. R. Goode and C. Thomas, *Spectroscopy (Eugene)*, 1994, **9**, 14–22.

76. P. C. Uden, *J. Chromatogr. A*, 1995, **703**, 393–416.

77. R. Łobiński and F. C. Adams, *Spectrochim. Acta, Part B*, 1997, **52**, 1865–1903.

78. R. Łobiński, V. Sidelnikov, Y. Patrushev, I. Rodriguez and A. Wasik, *Trends Anal. Chem.*, 1999, **18**, 449–460.

79. L. L. P. van Stee and U. A. T. Brinkman, *J. Chromatogr. A.*, 2008, **1186**, 109–122.

80. *Selective Detectors*, ed. R. E. Sievers, Wiley, New York, 1995.
81. E. H. Evans, J. J. Giglio, T. M. Castillano and J. A. Caruso, *Inductively Coupled and Microwave Induced Plasma Sources for Mass Spectrometry*, RSC, Cambridge, 1995.
82. R. P. W. Scott, *Tandem Techniques*, Wiley, New York, 1997.
83. R. L. Grob and E. F. Barry, *Modern Practice of Gas Chromatography*, Wiley, Hoboken, NJ, 2004.
84. A. M. Carro, L. Neira, R. Rodil and R. A. Lorenzo, *Chromatography*, 2002, **56**, 733–738.
85. S. M. Lee and P. L. Wylie, *J. Agric. Food Chem.*, 1991, **39**, 2192–2199.
86. J. R. Baena, M. Gallego, M. Valcarcel, J. Leenaers and F. C. Adams, *Anal. Chem.*, 2001, **73**, 3927–3934.
87. S. Aguerre, G. Lespes, V. Desauziers and M. Potin-Gautier, *J. Anal. At. Spectrom.*, 2001, **16**, 263–269.
88. M. B. de la Calle-Guntinas, C. Brunori, R. Scerbo, S. Chiavarini, P. Quevauviller, F. C. Adams and R. Morabito, *J. Anal. At. Spectrom.*, 1997, **12**, 1041–1046.
89. B. D. Quimby, P. C. Uden and R. M. Barnes, *Anal. Chem.*, 1978, **50**, 2112–2118.
90. L. Zhang, J. W. Carnahan, R. E. Winans and P. H. Neill, *Anal. Chim. Acta*, 1990, **283**, 149–154.
91. K. M. L. Holden, K. D. Bartle and S. R. Hall, *J. High Resolut. Chromatogr.*, 1999, **22**, 159–163.
92. M. E. Birch, *Anal. Chim. Acta*, 1993, **282**, 451–458.
93. J. P. J. van Dalen, P. A. de Lezenne Coulander and L. De Galan, *Anal. Chim. Acta*, 1977, **94**, 1–19.
94. N. Campillo, P. Vinas, I. Lopez-Garcia, N. Aguinaga and M. Hernandez-Cordoba, *J. Chromatogr. A*, 2004, **1035**, 1–8.
95. E. Dimitrakakis, C. Haberhauer-Troyer, Y. Abe, M. Ochsenkühn-Petropoulou and E. Rosenberg, *Anal. Bioanal. Chem.*, 2004, **379**, 842–848.
96. I. Rodriguez Pereiro, V. O. Schmitt and R. Łobiński, *Anal. Chem.*, 1997, **69**, 4799–4807.
97. W. Frech, J. P. Snell and R. E. Sturgeon, *J. Anal. At. Spectrom.*, 1998, **13**, 1347–1353.
98. R. R. Freeman and W. E. Wentworth, *Anal. Chem.*, 1971, **43**, 1987–1991.
99. G. Gremaud, W. E. Wentworth, A. Zlatkis, R. Swatloski, E. C. M. Chen and S. D. Stearns, *J. Chromatogr. A.*, 1996, **724**, 235–250.
100. R. M. Alvarez Bolainez and C. B. Boss, *Anal. Chem.*, 1991, **63**, 159–163.
101. R. M. Alvarez Bolainez, M. P. Dziewiatkoski and C. B. Boss, *Anal. Chem.*, 1992, **64**, 541–544.
102. W. C. Story and J. A. Caruso, *J. Anal. At. Spectrom.*, 1993, **8**, 571–575.
103. L. K. Olson and J. A. Caruso, *Spectrochim. Acta, Part B*, 1994, **49**, 7–30.
104. A. M. Zapata, C. L. Bock and A. Robbat, Jr., *J. Anal. At. Spectrom.*, 1999, **14**, 1187–1192.
105. G. O'Connor, S. J. Rowland and E. H. Evans, *J. Sep. Sci.*, 2002, **25**, 839–846.

106. J. T. Andersson, *Anal. Bioanal. Chem.*, 2002, **373**, 344–355.
107. R. Łobiński and F. C. Adams, *J. Anal. At. Spectrom.*, 1992, **7**, 987–992.
108. A. Wasik, I. Rodriguez Pereiro and R. Łobiński, *Spectrochim. Acta, Part B*, 1998, **53**, 867–879.
109. S. Slaets, F. Laturnus and F. C. Adams, *Fresenius' Z. Anal. Chem.*, 1999, **364**, 133–140.
110. N. Campillo, P. Vinas, I. Lopez-Garcia, N. Aguinaga and M. Hernandez-Cordoba, *Talanta*, 2004, **64**, 584–589.
111. J. W. Carnahan, K. J. Mulligan and J. A. Caruso, *Anal. Chim. Acta*, 1981, **130**, 227–241.
112. R. Reuther, L. Jaeger and B. Allard, *Anal. Chim. Acta*, 1999, **394**, 259–269.
113. M. Zabaljauregui, A. Delgado, A. Usobiaga, O. Zuloaga, A. De Diego and J. M. Madariaga, *J. Chromatogr. A*, 2007, **1148**, 78–85.
114. M. Ceulemans and F. C. Adams, *J. Anal. At. Spectrom.*, 1996, **11**, 201–206.
115. M. Y. Khuhawar, A. Sarafraz-Yazdi, J. A. Seeley and P. C. Uden, *J. Chromatogr. A*, 1998, **824**, 223–229.
116. M. Y. Khuhawar, A. Sarafraz-Yazdi, J. A. Seeley and P. C. Uden, *Chromatography*, 2002, **56**, 729–732.
117. L. L. P. van Stee, J. Beens, R. J. J. Vreuls and U. A. T. Brinkman, *J. Chromatogr. A*, 2003, **1019**, 89–99.
118. L. L. P. van Stee, P. E. Leonards, W. M. van Loon, A. J. Hendriks, J. L. Maas, J. Struijs and U. A. T. Brinkman, *Water Res.*, 2002, **36**, 4455–4470.
119. K. L. Sobeih, M. Baron and J. Gonzalez-Rodriguez, *J. Chromatogr. A*, 2008, **1186**, 51–66.
120. A. B. Ross, S. Junyapoon, K. D. Bartle, J. M. Jones and A. Williams, *J. Anal. Appl. Pyrolysis*, 2001, **58/59**, 371–385.
121. H. Kipphardt, M. Czerwensky and R. Matschat, *J. Anal. At. Spectrom.*, 2008, **23**, 588–591.
122. J. A. Seeley, Y. Zeng, P. C. Uden, T. I. Eglinton and I. Ericson, *J. Anal. At. Spectrom.*, 1992, **7**, 979–985.
123. M. P. M. van Lieshout, H. G. Janssen, C. A. Cramers and G. A. van den Bos, *J. Chromatogr. A*, 1997, **764**, 73–84.
124. U. Fuchslueger, H. J. Grether and M. Grasserbauer, *Fresenius' Z. Anal. Chem.*, 1999, **364**, 133–140.
125. M. Brebu, T. Bhaskar, K. Murai, A. Muto, Y. Sakata and M. Uddin, *Chemosphere*, 2004, **56**, 433–440.
126. Z. Mester, R. Sturgeon and J. Pawliszyn, *Spectrochim. Acta, Part B*, 2001, **56**, 233–260.
127. V. Kaur, A. K. Malik and N. Verma, *J. Sep. Sci.*, 2006, **29**, 333–345.
128. S. Tutschku, M. M. Schantz and S. A. Wise, *Anal. Chem.*, 2002, **74**, 4694–4701.
129. F. Yang and Y. K. Chau, *Analyst*, 1999, **124**, 71–73.

130. M. Crnoja, C. Haberhauer-Troyer, E. Rosenberg and M. Grasserbauer, *J. Anal. At. Spectrom.*, 2001, **16**, 1160–1166.
131. N. Campillo, R. Penalver and M. Hernandez-Cordoba, *Food Addit. Contam.*, 2007, **24**, 777–783.
132. N. Campillo, R. Penalver, I. Lopez-Garcia and M. Hernandez-Cordoba, *J. Chromatogr. A*, 2009, **1216**, 6735–6740.
133. A. Chatterjee, Y. Shibata, M. Yoneda, R. Banerjee, M. Uchida, H. Kon and M. Morita, *Anal. Chem.*, 2001, **73**, 3181–3186.
134. A. Delgado, A. Usobiaga, A. Prieto, O. Zuloaga, A. de Diego and J. M. Maradiaga, *J. Sep. Sci.*, 2008, **31**, 768–774.
135. C. Dietz, J. Sanz Landaluze, P. Ximenez-Embun, Y. Madrid-Albarran and C. Camara, *J. Anal. At. Spectrom.*, 2004, **19**, 260–266.
136. C. Dietz, J. Sanz Landaluze, P. Ximenez-Embun, Y. Madrid-Albarran and C. Camara, *Anal. Chim. Acta*, 2004, **501**, 157–167.

CHAPTER 7

Solution and Slurry Nebulization Coupling with Microwave Plasmas

7.1 Nebulization Techniques Compatible with Microwave Plasmas

Nebulization of liquids is commonly applied for sample introduction in various spectroscopic techniques. In 1984, Browner and Boorn[1] described solution nebulization as the *Achilles' heel* of spectroscopic techniques. In comparison with an inductively coupled plasma (ICP) or flame, a microwave plasma (MWP) is more susceptible to changes in plasma composition resulting from sample introduction. This method of sample introduction is not well suited to the MWP method since the low gas temperature of the plasma does not promote efficient desolvation or atomization of the aerosol. It was even recognized that regardless of efforts to match closely the MWP and nebulizer characteristics, a pneumatic or ultrasonic nebulizer will never gain widespread acceptance for low-power MWP devices. Nevertheless, further investigations have shown some solutions to solve this problem.

In the case of MWPs, it is possible to limit the amount of water brought into the plasma by providing desolvation of wet aerosols before they enter the plasma. Various desolvation systems are widely used in cooperation with both pneumatic and ultrasonic nebulizers.[2–9] This process reduces the amount of solvent from the sample but has been found to suffer from memory effects. Another approach to solve the problem of low plasma tolerance to solvent loading is the use of moderate or high power levels of a MWP source. In systems operating at power levels exceeding 300 W, aerosol introduction without the use of a desolvation system is easily achieved.[8–17]

RSC Analytical Spectroscopy Monographs No. 12
Microwave Induced Plasma Analytical Spectrometry
By Krzysztof J. Jankowski and Edward Reszke
© The Royal Society of Chemistry 2011
Published by the Royal Society of Chemistry, www.rsc.org

A nebulizer should match to other elements of the excitation source as far as gas flow and aerosol density are concerned. The optimal gas flow at which a majority of conventional pneumatic nebulizers achieves maximum performance is *ca.* $1 \, L \, min^{-1}$. When MWPs have operated under such conditions, the concentric and cross-flow nebulizers were used.[10,12,13,15,18–21] In some cases, aerosols were fed directly into the plasma with no desolvation apparatus.

Numerous studies have shown that a high yield for converting a solution into an aerosol, characterized by a stable flow and a small diameter of the droplets, is expected from a good nebulizer.[22–24] If the sample is converted into an aerosol of very small diameter droplets, then plasma stability and the analytical performance achieved are good. However, this requires the use of specially designed nebulizers that can cooperate with low-power MWPs, in which usually relatively small gas (often below $0.5 \, L \, min^{-1}$) and analyte flows are used. The assuring of stable aerosol flows and reduction of memory effects are difficult to achieve under such conditions.

Three types of nebulizers have been designed and developed to operate with low-power microwave induced plasmas (MIPs).[22,23] The ultrasonic nebulizer operating at optimal conditions generates a stable flow of a fine aerosol (diameter of *ca.* $2 \, \mu m$), which allows the introduction of a larger sample to the plasma and assures good sensitivity and precision of measurement. Two pneumatic nebulizers, *i.e.* frit and multicapillary array nebulizers, can operate at nebulizer gas flow rates as low as $50 \, mL \, min^{-1}$. Producing a stable aerosol stream of satisfactory concentration and drop size distribution are major factors for the analytical performance of the whole spectrometric system.

In work with MWPs, micronebulizers can be used, as shown by a number of authors.[25–27] The devices performed better than a conventional pneumatic nebulizer. Also, direct aerosol injection into the MWP has been proved.[28–30]

Recently, it has been shown for some new MWP source designs that providing a highly symmetrical coupling of microwaves has allowed progress in MWP tolerance for aerosol loading.[31–33]

7.2   Plasma Tolerance to Solvent Loading

The analysis of aqueous and organic solutions by MWP spectrometry encounters difficulties connected with the introduction to the plasma of relatively large quantities of solvent, the evaporation of which consumes a considerable amount of energy. Hence, in order to avoid these problems, desolvation of the aerosol is often applied. There is a limited amount of reliable data concerning MWP tolerance to solvent loading. In their early studies on aqueous solution nebulization into argon MIPs, Skogerboe and Coleman[2] found a 16-fold improvement in detection limits of some elements when the sample was desolvated in comparison with wet aerosol introduction. When both wet and dry aerosols were delivered into a TM_{010} cavity, the use of a wet aerosol decreased the sensitivity by a factor of between 2 and 10, depending on the element involved.[18] Nevertheless, Haas *et al.*[34] were able to nebulize solutions directly into a moderate power argon MIP at a sample flow rate of $1.5\,mL\,min^{-1}$ using a pneumatic nebulizer without desolvation unit. The microwave plasma torch (MPT) offers the advantages of a high tolerance to the introduction of aqueous aerosols, with a central channel that exhibits enhanced energy coupling between the plasma and sample species. The amount of water that an argon MPT withstands before being extinguished is 10–100 times higher than those observed for a TM_{010} cavity or surfatron.[35] However, it was also reported that the presence of a little water in the plasma improved the stability of the capacitive microwave plasma (CMP),[36] the MPT[37] and the MIP discharges.[3]

Since helium has a low thermal capacity, the He-MWP is more easily extinguished by the introduction of foreign material. Additionally, most of the pneumatic nebulizers exhibit poor performance when operating with helium gas. Beenakker,[38] utilizing a special low-flow nebulizer, was able to maintain a helium MIP with a sample flow rate of $1.7\,mL\,min^{-1}$ aqueous sample without desolvation. Assuming 3% nebulization efficiency, one can obtain approximately $50\,mg\,min^{-1}$ of water aerosol loading. However, it was observed that the presence of larger droplets degraded the signal stability. Because of the relatively high tolerance of the CMP to water and organic aerosols, direct aerosol introduction into the helium CMP at a sample flow rate of $1.2\,mL\,min^{-1}$ was carried out with the use of a concentric nebulizer.[39] The filament helium MIP in the TE_{101} cavity can take up to $35\,mg\,min^{-1}$ of water aerosol before being extinguished, while the annular-shaped helium plasma in the TEM even manages $80\,mg\,min^{-1}$.[32] The three-phase rotating field helium plasma can take more than $100\,mg\,min^{-1}$ of water without being extinguished.[33]

Again, early reports on MWP tolerance to organic liquids argued that organics rapidly extinguish the plasma. Systematic studies have changed this meaning due to some essential improvements in nebulizers and MWP source operations. The MPT can handle large amounts of organics without being extinguished and without a substantial deterioration in its performance. Both ethanol and methanol aerosols can be loaded into an Ar-MPT at liquid flow rates up to $2.0\,mL\,min^{-1}$ and $2.5\,mL\,min^{-1}$ by means of ultrasonic nebulization with aerosol desolvation[40] and pneumatic nebulization without aerosol desolvation,[41] respectively. The coupling of a thermospray nebulizer with an Ar-MPT for introduction of 20% methanol showed no effect on the stability of the MPT discharge.[7]

Ng and Culp[28] have developed a direct injection nebulizer for liquid handling into argon and helium plasmas sustained in a modified Beenakker cavity. They reported the possibility of the continuous introduction of methanol, ethanol and acetonitrile up to $15\,\mu L\,min^{-1}$. In other studies, $5\,\mu L\,min^{-1}$ of carbon tetrachloride has been introduced into a helium MIP.[4] Kerosene has been introduced through a pneumatic nebulizer at a flow rate of $1.2\,mL\,min^{-1}$ into a He-CMP when determining organic fluorine and chlorine.[39]

The use of MWPs containing oxygen has led to a substantial improvement in direct organic solvent introduction due to the combustion of organic vapours. The air-MIP, air-MPT and air-CMP have been investigated for organic sample aerosol introduction with the use of a special V-grove pneumatic nebulizer in several papers.[42–44] The estimated aerosol loading of acetonitrile and cyclohexane was as large as 4.4 mg min^{-1}.

The nitrogen/oxygen mixed-gas MIP with an Okamoto cavity has been presented as highly robust to the loading of organic solvents.[45,46] However, reports on the introduction of 80% v/v ethanol and 4-methylpentan-2-one into the plasma did not contain any detail concerning the amount of sample load. The direct aspiration of organic solvents into the plasma for emission analysis can be useful for liquid chromatography applications. The pneumatic nebulization of an HPLC eluent containing 90% v/v methanol at liquid flow rates up to 5 mL min^{-1} is possible into an oxygen/argon MIP.[47]

7.3 Nebulizer Designs

Continuous nebulization is the most common method for liquid aerosol generation. It basically comprises a nebulizer, which converts the liquid into an aerosol, and a spray chamber, which modifies both the characteristics and transport efficiency of the aerosol into the plasma. Sometimes, a desolvation system is employed to remove the solvent from the sample prior to its entering the plasma. The use of nebulization techniques for MWP-OES is restricted by the relatively low flow rates ordinarily employed with a MIP and because of the limited tolerance of a MWP to solvent loading.

Pneumatic nebulizers (PNs) are readily applied due to the simplicity in operation and low cost. However, conventional nebulizers suffer from very low nebulization efficiency. Both pneumatic micronebulizers and ultrasonic nebulizers (USNs) have been shown to be more efficient and can produce aerosols with smaller and more uniform droplet size. The other type of nebulizer, causing much interest in liquid chromatography techniques coupled with plasma OES, is the thermospray (TSP) nebulizer.

A variety of nebulizers have been applied to MWP-OES and MWP-MS, including concentric nebulizers,[10–13,19,21,24,34,45,46] cross-flow nebulizers,[15,18] V-groove Babington-type nebulizers,[48,49] V-groove Legere-type nebulizers,[42–44] Hildebrand grid nebulizers,[50] microconcentric nebulizers,[25] direct injection nebulizers (DINs),[28–30] sonic nebulizers,[24,26] multi-microspray nebulizers,[27] flow focusing nebulizers (FFPNs),[25] frit nebulizers,[4,5,22,51] microcapillary array nebulizers (MCAs),[22,25,52] USNs,[6,8,9,13,14,16,23] hydraulic high-pressure nebulizers[17,53] and TSP nebulizers.[7]

7.3.1 Pneumatic Nebulizers

In pneumatic nebulization, the gas pressure gradient is the driving force of the aerosol generation process, and its efficiency is directly dependent on the gas flow. The fundamentals of the design and operation of these devices have been described in some reviews.[54–57] For conventional pneumatic nebulizers the nebulization efficiency does not exceed a few per cent. To overcome this drawback, different nebulizer designs have been developed. A considerable improvement in the efficiency has been achieved in the design called the frit nebulizer. Some reports record that its nebulization efficiency can reach even 95%.[58] The other problem with PNs is that they do not work well at relatively low gas flow rates often used in MWPs. This degrades the nebulizer performance, including the nebulization efficiency and the droplet size distribution. Fortunately, micronebulizers offer much better characteristics.[24–27] The other shortcoming of PNs is that they do not operate well with helium. To overcome this, specially designed "helium" nebulizers have been developed.[38,39]

A majority of the earlier nebulizer designs can be expressed as conventional nebulizers with a solution consumption of about $1\,\mathrm{mL\,min^{-1}}$. In recent years, micronebulizers have been developed which are able to generate a stable aerosol at liquid flow rates of the order of several tens of $\mu\mathrm{L\,min^{-1}}$.[57] To show one example, Tarr *et al.*[59] developed a micro-flow ultrasonic nebulizer, in which the sample consumption was $5\,\mu\mathrm{L\,min^{-1}}$. This results from the specific requirements of the new applications of plasma techniques, which can be coupled with liquid chromatography or serve as an ion source for MS. Moreover, in many fields the sample volume is the limiting factor for the analysis. This is coincident with MWP requirements, where a limited amount of a sample can be introduced into the plasma.

In the case of pneumatic microconcentric nebulizers, the inner diameter and wall thickness of the liquid capillary are reduced with respect to conventional ones. As a result of these modifications, finer primary aerosols are generated with high efficiency that promote the analyte transport to the plasma, leading to higher sensitivities and lower limits of detection.

The DIN shows promise as a liquid sample introduction device for MWPs.[28–30] The advantages of DIN over conventional nebulizer-spray chamber systems for MWP-OES/MS are higher sensitivities, better signal stability, lower limits of detection and shorter wash-out times. A short-term stability test with a DIN was found to give less than 0.5% relative standard deviation for 20 min. The drawbacks of the DIN are that it produces relatively high matrix effects and it is prone to tip blocking because of the narrow capillary used. In general, the matrix effects are found to be more pronounced when working at low liquid flow rates that at conventional rates.

In the SSN a silica capillary supported by a stainless steel tube is fixed in the middle of an orifice formed in a polyamide material.[26] A multimicrospray nebulizer with three orifices was found to be suitable for MIP-MS. It can operate within a wide range of sample flow rates between 5 and $250\,\mu\mathrm{L\,min^{-1}}$ and provide high nebulization efficiency.[27]

Matusiewicz *et al.*[25] compared several micronebulizers coupled to a MIP-OES and found that the analytical performance of FFPNs was superior to those obtained with the MCA nebulizer and microconcentric nebulizers. Nevertheless, all the micronebulizers studied were successfully applied to several analytical applications of the MIP-OES technique.

The nebulizer based on a glass frit has gas flow requirements similar to those of a MWP and is readily coupled with a low-power helium MIP at atmospheric pressure. Its low sample consumption rate is useful where sequential

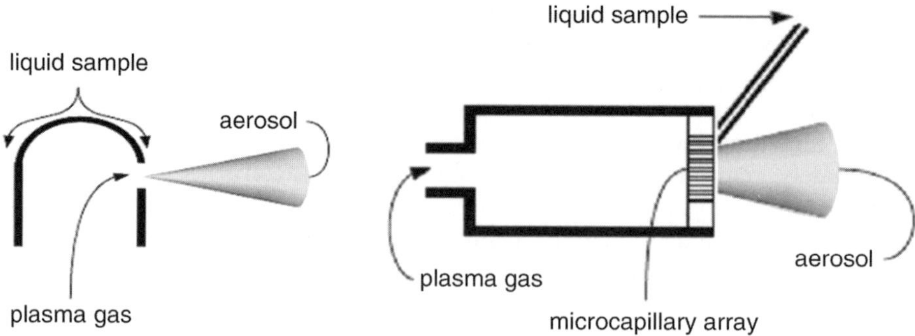

Figure 7.1   The principle of Babington and MCA nebulizers.

multi-element analysis is required, or where limited sample volume is available. However, a desolvation system must be used to obtain better analytical results due to the high aerosol density, leading to a high water loading to the plasma.[4,5] The performance of the frit nebulizer can be improved by applying ceramic material to obtain the frit structure.[22] An important disadvantage of the frit nebulizer is the considerable memory effects, causing the necessity of washing for several minutes before introduction of the next sample, which is not acceptable in routine analysis.

In contrast, the V-groove Babington-type nebulizer exhibits a relatively low nebulization efficiency, but is easy to wash. The concept of a multispray micronebulizer is a combination of these two concepts. The nebulizer gas is divided into hundreds of orifices that lead to the increase of aerosol generation efficiency and a decrease of sample consumption.[22,25] The microcapillary array (MCA) nebulizer is a flexible solution, the design details of which can, in a simple way, be adapted to various conditions of the nebulizer operation dependent on the parameters of the excitation source with which it coop- erates.[22,52] Both the sample and nebulizer gas flow rate should be matched with the frit or MCA parameters to prevent flooding (Figure 7.1).

7.3.2   Ultrasonic Nebulizers

Ultrasonic nebulizers reach a high efficiency of aerosol formation, but they are more complicated in operation and relatively expensive. Dispersion of the liquid occurs directly on the specially coated piezoelectric plate or on the additional chemically resistant vibrating plate to avoid fast degradation of the transducer. Since a considerable part of the acoustic energy undergoes conversion into heat, the converter requires cooling with air or water. In most USNs, desolvation is usually performed after aerosol generation to introduce into the plasma the so-called dry aerosol for improving vaporization and atomization of the sample. An advantage of USNs is the independence of the aerosol generation process from the nebulizer gas flow, which allows achieving high nebulization efficiency even at small gas flows. This is especially favourable when coupling a USN with MWPs. The small droplet size and narrow droplet size distribution are additional advantages of these sample introduction devices. A drawback of a USN is a relatively serious memory effect; however, a desolvation system is a primary source of this effect. The cost of a commercial USN system is high, compared to a traditional PN. The application of USNs in spectrochemical analysis has allowed considerably lower detection limits for many elements and improved the precision of the results obtained.[13,60]

Some designs of USN coupled with MWPs are described in the literature.[23,59] Continuous-type and geyser-type devices can be mentioned when taking into account the nebulizer operating mode. Water-cooled USNs for MWPs have been developed by Jankowski *et al.*[23] (Figure 7.2). When the temperature of the cooling water and the sample solution volume are controlled, the nebulizaton process is stable over a relatively long period of time. The effectiveness of generating a primary aerosol exceeds 90% over a broad range of analyte flows. Jin *et al.*[60–62] have developed a low-cost semi-continuous geyser-type USN device, based on a household humidifier.

Thermospray is a relatively new liquid sample introduction approach.[63] The generation of a fine aerosol results from partial vaporization of the liquid, when forced through a heated capillary. It offers relatively high aerosol generation efficiency (40–50%) and aerosol transport efficiency. This is because the size of the aerosol droplets produced are relatively small (mean droplet diameter *ca.* 2 μm). As for a USN, the nebulization efficiency of a TSP nebulizer is independent of the type and flow rate of the nebulizer gas. However, for a conventional TSP system a high-pressure pump has to be used to transmit the sample solution, owing to the high pressure built up inside the working capillary. A modified TSP device using a common peristaltic pump for MPT-OES has been evaluated.[7] The experimental results indicated that the detection limits

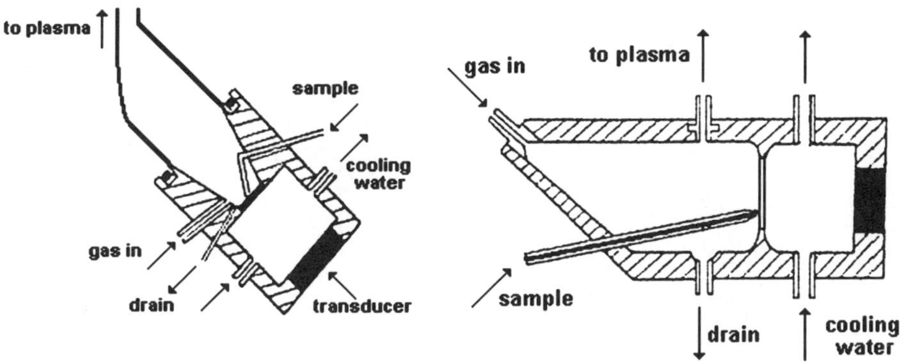

Figure 7.2 Water-cooled ultrasonic nebulizers for MIP-OES: with single-pass spray chamber (left); compact HF-resistant design (right).

of some elements obtained were improved in comparison with those obtained with PN by a factor of 5–10, but the precision was poor.

The limitation of the TSP system is that overheating may cause clogging of the outlet of the TSP capillary. To solve this problem, Bordera *et al.*[64] proposed a TSP system based on sample heating by means of microwave radiation. Ding *et al.*[65] developed a microwave TSP device that could work at low microwave power levels based on a TM_{010} cavity. A Teflon capillary coil was utilized as the TSP capillary, minimizing risk of contamination. It was shown that the performance of this new TSP device was satisfactory.

7.3.3   Spray Chambers and Desolvation Systems

Once the sample aerosol is generated by the nebulizer it must be transported to the plasma, usually passing through the spray chamber (SC). The primary function of the SC is to allow droplets with diameters of about 10 µm or smaller to pass to the plasma, because only a fine aerosol is suitable for MWPs. A secondary purpose of the SC is to smooth out pulses that occur during nebulization, often due to pumping of the solution. The material from which a spray chamber is constructed can be an important characteristic of a SC. Glass spray chambers are usually used; however, SCs made from corrosion-resistant materials are available for the introduction of samples containing hydrofluoric acid.[23] Some typical SC designs are shown in Figure 7.3.

In conventional PNs the aerosol stream is highly turbulent ($Re = ca.$ 15 000), but already at a small distance from the capillary outlet it becomes laminar. The most important aerosol losses come from inertial impact against the frontal wall of the SC and these types of losses are drop size and velocity dependent.[55,56]

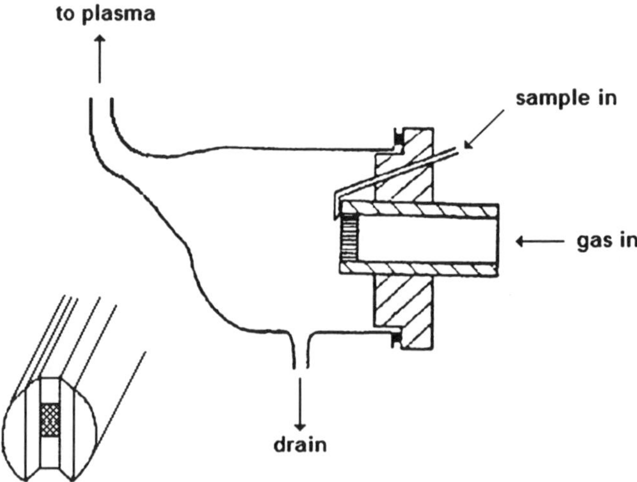

Figure 7.3   Multicapillary array nebulizer with single-pass spray chamber (top right), cyclonic spray chamber (bottom left) and double-pass spray chamber (bottom right).

Since the gas and particle velocities in MWPs are usually relatively low, three types of loss process have been observed. The use of small gas flows leads to an increase in gravitational separation and an increase in memory effects. To solve this problem, the volume of the SC can be minimized and the analyte route to the plasma can be shortened, providing that the aerosol obtained is sufficiently fine. For a single-pass SC at argon flow rates up to $150\,mL\,min^{-1}$ the aerosol droplets are separated primarily by gravitational settling. For higher flows the turbulent or impact losses increase.[22]

The use of nebulization techniques for MWP-OES is often restricted by the relatively large dead volumes that are not easily renewed by the nebulizer gas, and relatively long washout times of the nebulizer–spray chamber system (*i.e.* the time required by the system to achieve 1% of the steady-state signal). Low-volume SCs coupled with MWPs provide shorter washout times than conventional ones.[22,23,25,41]

Three types of SCs are used for MWP applications. A single-pass SC design is used with systems that do not require a strong filtering action of the aerosol.[11,19,22,23,50] The aerosol path is very simple, with its main advantages being the increase in analyte-transport efficiency and the shortening in wash-out times, especially for low-volume designs. In a cyclone spray chamber a swirling action serves to remove large droplets. A cyclone spray chamber exhibits low memory effects, high transport efficiency and minimal back-pressure fluctuation and can be applied to MWP-OES.[17,25,41] In a double-pass spray chamber the aerosol goes through a tube and then is forced to modify its trajectory by 180°. When working at low liquid flow rates with this chamber, the wash-out times are long with a subsequent drop in analysis throughput. A long aerosol chamber, a Scott chamber for instance, becomes detrimental to the operation of the whole system, since wash-out times may reach several minutes at low gas flow rates.[10,26,27,45,46] The use of low-volume designs may partially solve this limitation. The wash-out times are lower for the cyclone SC than for the double-pass SC. However, in general terms the double-pass SC produces finer tertiary aerosols than the single-pass and cyclone SCs.

The SC operation was found to be an important source of matrix effects in MWP spectrometric techniques. The so-called steady-state interferences caused by inorganic acids and dissolved salts lead to a change of the analytical signal level with respect to a plain water solution. The so-called transient effects caused by inorganic matrices consist of an increase in the time required for signal stabilization when analyzing samples having different matrix compositions. It has been found that the single-pass or cyclonic SCs significantly improve the signal stabilization times in comparison with a double-pass SC.

The main goal of a desolvation system is to decrease the load of solvent in the plasma in order to prevent low-power MWPs from being extinguished by the solvent. Therefore, to take full advantage of the MWP as an excitation source for OES, many researchers adopt the desolvation stage. The most common desolvation device in atomic spectrometry is a combination of a heating tube and a water condenser. Such a device can remove more than 80% of the water vapour. Sometimes, a heated single-pass SC can be used instead of a heating tube. Jin *et al.*[60] found that by using a desiccator with sulfuric acid for secondary desolvation, the overall desolvation efficiency can be greater than 99%. The detection limits obtained with the MPT-OES system without deso-lvation were higher than those obtained with desolvation by a factor of 10–100.

Desolvation systems equipped with a membrane have been used to reduce solvent loading into MWPs. Akinbo and Carnahan[66,67] used a flat-sheet membrane desolvator for both aqueous and organic desolvation. Jin *et al.*[68] examined a Nafion-dryer device for MPT-OES. Interestingly, desolvators uti-lizing microwave radiation have been extensively investigated for both OES and MS.[69–72]

The overall nebulization efficiency results from both aerosol generation and transport efficiencies. The aerosol generation efficiency is usually high and for the majority of recent nebulizers is close to 100%. However, the aerosol transport efficiency depends on the characteristics of the nebulizer–spray chamber system used.

The nebulization efficiency (E_n) can be calculated from experimental data using the following equation:

$$E_n = \frac{Q_{total} - Q_{waste}}{Q_{total}} \times 100\%$$

where Q_{total} is the mass of the sample solution delivered to the nebulizer in a period of time and Q_{waste} is the mass of waste solution produced in the same period of time.

The nebulization efficiency depends on several operation parameters, such as gas and analyte flow, the kind of gas, physicochemical properties of the nebulizer gas, the shape of the SC and others. These parameters should be optimized, since every nebulizer shows specific best operating conditions. Optimal values of these conditions are analyte- and matrix-dependent. Com-paring the efficiency of nebulizers, one should have in mind that, if possible, each of them should operate at its optimal conditions. High nebulization effi-ciency allows decreasing in the consumption of the sample solution.[22,25,55,56]

The DIN generates the aerosol at the base of the plasma. As a result, the analyte-transport efficiency is close to 100%. For a pneumatic micronebulizer coupled to a double-pass SC, it has been found that this parameter goes from 10% to 60% as the liquid flow rate decreases from 100 to 10 µL min^{-1}. The TSP nebulizer efficiency is 38–40% with a sample uptake rate of 1.5 mL min^{-1}.

It is known that the size of the aerosol droplets influences both the aerosol transport efficiency and the analytical performance of the plasma source. Jankowski *et al.*[22] estimated that, for a low-power MIP, droplets injected into the plasma should be smaller than 15 µm. A small diameter of the aerosol droplets leads to rapid and efficient evaporation and induction of the sample in the plasma region. This is especially favourable in the case of low-power MIPs, which operate at a limited level of energy in comparison to other plasma sources. Matusiewicz *et al.*[25] stated that, for all the nebulizers tested, the majority of the tertiary aerosol is contained in droplets smaller than 20 µm.

Under optimized working conditions for each nebulizer, the number percentages of the tertiary aerosol contained in droplets smaller than 8 μm are 98% for FFPNs, 86% for NAR-1 and near 100% for the concentric micronebulizers.

7.3.4 Flow Injection Analysis

In a flow injection (FI) system, a low sample volume is injected into a liquid-carrier stream. Then, the sample plug is conducted towards the nebulizer. A transient signal is produced despite continuous operation of the nebulizer. Flow injection analysis (FIA) is found to offer several advantages over continuous sample introduction, including rapid sample throughput, low sample consumption and a reduction of memory effects without a loss in sensitivity or precision. Gehlhausen and Carnahan determined aqueous fluorine by coupling FI system with a He-MIP using an ultrasonic nebulizer.[73]

The basic design criteria for a FI-MWP-OES interface are to reduce dead volumes and to increase the nebulization efficiency regarding low sample volume. It is observed that plasma tolerance to matrices, such as high salt content solutions and organic solvents, increases and the possibility of clogging the tip of the nebulizer decreases due to the continuous washing typical for FI systems.

Organic solvents were explored as the carrier for FIA by Madrid *et al.*[41] Under optimized conditions the sensitivity decreased less than half for ethanol and methanol, compared with performance with water as the carrier.

Procedures involving on-line preconcentration and matrix separation can be easily applied to the FIA-MWP technique. The combination of FI with on-line preconcentration was shown to improve the overall analytical performance of the system due to the higher sensitivity, better plasma stability and elimination of interferences caused by easily ionized elements.[74] A flow-injection on-line preconcentration system was employed to determine heavy metals,[75,76] noble metals[77] and rare earth metals[78] by MPT-OES.

7.4 Nebulization Methods Appropriate for Different Sample Classes

Various parameters are proposed for the evaluation of nebulizer performance. Physical parameters such as aerosol generation efficiency, aerosol transport efficiency and aerosol density or drop size distribution are connected precisely with nebulizer operation. These parameters characterize the nebulizer itself, but do not really correspond to its analytical utilization. Numerous studies have shown that knowledge of the efficiency of various types of nebulizers is not sufficient for both nebulizer comparison and prediction of the sensitivity of the entire measurement system.[68–70] Spectroscopic parameters, such as intensity of the signal measured, signal stability and detection limit, are more of advantage for an analyst, but are dependent on the operation of other parts of the spectrometer (torch, power generator, analyte pump, monochromator). These parameters characterize the overall performance of the entire spectrometer. Additional characteristics include wash-out time, resistance to clogging, sample and gas consumption rates, the simplicity of use and cost. An ideal liquid sample-introduction system must fulfil the following requirements: high analyte-transport efficiency, low solvent-transport efficiency, good reproducibility, low memory effects and robustness, *i.e.* stability of the system against changes in the sample matrix. The key factor in the success of analysis is the production of an aerosol as fine and monodisperse as possible in order to improve aerosol transport to the plasma. An overall comparison of various nebulizer classes utilized for the SN-MWP-OES technique is presented in Table 7.1.

Despite the fact that the comparison is somewhat subjective, one can conclude that none of the nebulizers fulfils the above requirements completely.

Evaluating three nebulizers designed for MIP-OES applications, Jankowski *et al.*[23] showed that the aerosol density can be a more precise parameter for evaluation of a nebulizer than the nebulization efficiency. Further, they found that for an argon MIP sustained in a TM_{010} cavity the optimal aerosol density (expressed as the volume or mass of the aerosol per 1 L of nebulizer gas) at which maximum sensitivity of the emission measurement is achieved lies in the $0.1–0.15\,mL\,L^{-1}$ range. Moreover, there exists a certain critical density (*ca.* $0.5\,mL\,L^{-1}$) at which the plasma discharge stability starts to worsen. In conclusion, they stated that a nebulizer appropriate for cooperation with low-power MWPs should fulfil the following requirements: optimal gas flow in the range $50–1000\,mL\,min^{-1}$, optimal sample flow in the range $10–100\,\mu L\,min^{-1}$, optimal aerosol density of *ca.* $0.15\,mL\,L^{-1}$ and average diameter of drops below 15 μm.

Browner *et al.*[79] considered the concept of the total aerosol mass reaching the excitation region during 1 s (W_{tot}). They showed that such a physical parameter is well correlated with the emission intensity at various analyte flows. An extension of this conception was the defining of the useful mass transport rate and the net analyte mass reaching the plasma in a time unit, and the total solvent loading.[59] This corresponds to the above-mentioned evaluation of aerosol introduction into the MIP. Attention should be drawn to the fact that the W_{tot} parameter does not depend directly on the gas flow, and in consequence on the analyte residence time in the plasma region. However, it was shown that in the case of low-power MIPs[22,23] the residence time has a great effect on the value of the signal measured. This may justify the better correlation between the aerosol density and signal-to-background ratio for MIPs, since the aerosol density depends on the nebulizer gas flow. By multiplying the aerosol density (C_a) by the plasma volume (V_p), the total mass of aerosol which can undergo vaporization, atomization and excitation in a given moment ($M_{tot} = C_a \times V_p$) is obtained. The same result is obtained when multiplying W_{tot} by the aerosol residence time in the plasma region. In the case of MIPs, where the plasma discharge occurs inside the discharge tube, it is easy to estimate the plasma volume as part of the discharge tube internal space present in the

Table 7.1 Comparison of the analytical performance of various nebulizers.

	Concentric	Cross-flow	Babington	MCA	Frit	USN	TSP
Aerosol efficiency	+	+	+	+ + +	+ + +	+ + +	+ +
Fineness of aerosol	+ +	+ +	+	+ + +	+ +	+ + +	+ + +
Detection limit	+	+	+	+ +	+ +	+ + +	+ +
Precision	+ +	+ +	+	+ +	+ +	+ + +	+ +
Sample	+	+	+	+ + +	+ +	+ + +	+ +
Gas consumption	+	+	+	+ +	+ +	+ + +	+ + +
Memory effects	+ + +	+ + +	+ + +	+ +	+	+ +	+ +
Salt tolerance	+ +	+ +	+ + +	+ +	+	+ +	+
Particle tolerance	+ +	+ +	+ + +	+ +	+	+ +	+
Organic solvents	+ +	+ +	+ +	+ + +	+ + +	+ +	+ + +
Cost	+ +	+ + +	+ + +	+ +	+ +	+	+
Summary	19	20	20	26	21	27	22

resonator. Therefore, M_{tot} may be a parameter useful for predicting the signal level, depending on the plasma dimensions.

Several factors must be taken into account in making the selection of the most appropriate liquid sample introduction system for a given MWP spectrometry application. The most important include the required analytical figures of merit, the composition of the sample matrix and the available volume of the sample. Also the performance of the system is decisive for elimination or at least minimization of matrix interferences. The primary requirement is the use of a nebulizer which produces droplets in the 1–10 μm mean size range.

Taking into account the above-mentioned criteria and characteristics of nebulizers for MWPs, we can suggest some recommendations. For routine analysis of aqueous solutions, micronebulizers with single-pass SC are suitable, offering relatively high sample throughput. When high-efficiency nebulizers are used, a single-pass SC may be replaced by a cyclone SC. A double-pass SC is not recommended owing to the long wash-out times, unless the nebulizer gas flow rate is about $1\,L\,min^{-1}$. A V-groove Legere-type nebulizer coupled to a double-pass SC is adequate for the nebulization of organic liquids. For high-power MWPs a conventional concentric nebulizer may be used as well. For MWP-MS applications, a micronebulizer–double-pass SC system shows considerable promise for obtaining a highly stable aerosol introduction of relatively low density. On the basis of the literature data, the USN with a desolvation unit seems to be the general use liquid-sample introduction system, including the determination of low analyte concentrations close to the quantitation limit.

7.5   Microsampling Techniques for Liquids

In the discrete sample injection mode, a microvolume of sample is injected into a gas stream. The sample is pumped into the nebulizer, rather than being aspirated. Since the sample introduction system remains dry between injections, high analytical signals may be achieved due to better evaporation. If the sample is introduced using a discrete sampling device, transient signals are achieved. Matrix effects can be more severe than when steady-state measurements are made. Problems associated with the background correction procedure are also found. Among the drawbacks of this methodology are that the signal noise can be very high and the time between two consecutive sample injections is very long because a washing cycle is required.

Matusiewicz[30] developed a discrete sample introduction system for simultaneous multi-element analysis by MIP-OES with the aid of a microconcentric nebulizer. A SC was eliminated from the system, thus obtaining a direct injection nebulizer. Reliable analytical results were obtained when operating at $250\,W$ in a TE_{101} resonator, with a sample volume of $20\,\mu L$ into the helium MIP.

Recently, jet systems delivering the analyte to the plasma have been proposed in order to increase the sample introduction efficiency and reproducibility without affecting the plasma stability. Aqueous solutions are not nebulized but directly injected into the plasma as a single droplet or a series of small droplets. Elemental analyses of extremely small samples, such as single cells and nano-particles, are predicted. The piezoelectric droplet generator produces mono-disperse droplets of controllable size and reproducibly and introduces the droplets into a plasma with 100% efficiency, allowing a very small and accurate sample introduction to be achieved.[80,81] A design for microelectromechanical systems has been proposed.[82] The possibility of changing the droplet size on demand and high reproducibility of the generated droplet volume gives new possibilities for calibration procedures. The absolute detection limits for some elements obtained by ICP-OES are on the sub-picogram level.[83,84] Not yet applied to MWPs, the technique seems to be promising and may open a new field of analytical applications of MWPs.

Other developments involving MIPs have also relied on microsampling of liquids. They consist of thermal vaporization with the use of various devices. The solvent is usually removed before vaporization of the analyte-containing residue. In most cases the same devices may be alternatively used for vaporization of solid microsamples. Consequently, these techniques will be discussed in the next chapter.

7.6 Dual-flow Nebulization Techniques

In recent years, a dual-flow nebulization technique has been explored. The system consists of two pneumatic nebulizers (or two pump tubings delivering two solutions simultaneously connected to a Y-adapter before entering the nebulizer) connected to one spray chamber. The use of nebulizers with dual solution loading enables on-line internal standardization and calibration with standard addition. The so-called tandem calibration methodology achieves the accuracy of standard addition while maintaining the simplicity of external calibration.[85–87] Significant progress has been made in terms of accuracy compared with conventional calibration with synthetic solutions without matrix matching; however, the precision and detection limits are worse. This methodology could be applicable to a variety of spectrometric techniques, whether continuous or discrete sample introduction systems are used.

The other application of dual-flow design consists in mixing of aerosols of non-miscible solutions for elemental determinations in organic solutions.[88] This offers both the use of aqueous standards for calibration and improved plasma tolerance to loading of organic liquids.

A dual-sample introduction system that combines the benefits of nebulization and vapour generation in a single device has been described.[89] Better sensitivities and similar precision can be obtained compared with those obtained with the single-mode systems. Moreover, hydride and non-hydride forming elements can be determined simultaneously. Matusiewicz and Ślachciński applied this methodology for the multi-element determinations by

MIP-OES combining aqueous nebulization and HG[90] or slurry nebulization and HG.[91]

Duan *et al.*[92] explored on-line vapour generation nebulization for the determination of non-metals by means of a dual-flow concentric nebulizer and MPT-MS. Higher sampling efficiency and improved sensitivity were attained in comparison with conventional nebulization because the analytes were introduced to the plasma in the form of volatile compounds.

7.7 Slurry Nebulization

Slurry nebulization is an alternative technique to solution nebulization. The technique offers the elimination of complex digestion procedures and the avoidance of matrix effects from inorganic acids. In addition, the technique requires little instrument modification, including selection of the nebulizer. The most significant advantage of slurry nebulization is that it can be calibrated using aqueous solutions, while other solid sampling techniques are plagued by problematic calibration procedures. It is postulated that a narrow particle size distribution with a mean particle size of less than $2\,\mu m$ will ensure sample transport and recoveries comparable with those for the equivalent solutions for most samples. If this is achieved, then simple aqueous calibration may be used successfully. To achieve particle size distribution, a wide range of grinding techniques have been employed. Slurries have been homogenized using ultrasonic agitation. The finest particle size slurry gave near 100% recovery, indicating that the transport and atomization efficiencies of the slurry are close to those of a solution.

The stability and homogeneity of a slurry is the second crucial prerequisite for efficient and reproducible nebulization. Preparing slurries in aqueous solution alone is unsuitable for the majority of samples owing to flocculation effects which result in rapid sedimentation of the powdered material. This is achieved by employing dispersants or even by adjusting a suitable pH value of the solution. The particle size is the limiting parameter for efficient slurry nebulization and transport to the plasma. Also, particle size distribution affects the atomization efficiency of elements in slurries using plasma spectrometry. It was observed that suspensions prepared from different particle size ranges produced different emission intensities for the elements determined.

MWPs can handle slurries. Since in low-power MWP atomization the efficiency of solid samples is limited, a particle size of less than $5\,\mu m$ is optimum. The nebulization of slurries into the plasma requires a nebulizer that is unlikely to clog when solid particles are passed through it. A V-groove Babington nebulizer has been used successfully with MIPs. This design of nebulizer contains a small gas orifice and an impact ball to break up large droplets. The high pressure of the nebulizer gas and easy passage of the sample ensures efficient nebulization and prevents clogging. An in-house SC was designed to complement a V-groove nebulizer for MIP spectrometry. Biological and environmental samples were successfully analyzed, together with certified materials by

Matusiewicz and Sturgeon[48] and by Matusiewicz and Golik[49] using MIP-OES. Fine powders were suspended in a 10% solution of nitric acid to make a 1% m/v slurry. The suspensions were pre-treated by sonication prior to analysis for 5 min. However, as 10% nitric acid was used to prepare the slurries, solid slurry particles were partially dissolved and elements of interest were extracted into the aqueous phase.

7.8 Separation/Preconcentration Techniques and Solution Nebulization

Liquid chromatography techniques have been investigated for separation of sample components prior to excitation in a MWP. Although generally not as popular as GC, high-performance chromatography (HPLC) and replacement ion chromatography (RIC), as well as capillary zone electrophoresis (CZE) separation techniques, have been successfully applied using the MWP-OES/MS detection systems. In general, problems with chromatographic column and MWP interfacing are very severe since with normal operation the large amounts of mobile phase continuously introduced into the discharge will extinguish it. The high-power MWPs accommodate the continuous solvent flow, as long as the solvent is efficiently nebulized with the use of selected liquid introduction systems usually equipped with a desolvator.[53,93–96] The other strategy consists in converting the liquid eluent into the gaseous phase before entering the plasma using hydride generation,[97] thermal vaporization[98] or GC systems.[99]

A unique system which couples thin layer chromatography (TLC) and the MIP has been developed by Jansen and co-workers.[100] This system creates the plasma directly on the TLC plate by connecting the inner conductor of the coaxial cable to a stainless steel capillary tube through which the helium support gas flows. The impedance of this plasma electrode is matched simply by controlling the distance between the capillary tubing and the grounded metal plate below the TLC silica support. However, irreproducibility in the signal and background were observed because of irregular experimental conditions.

With on-line preseparation, the matrix effect can be eliminated or reduced, while with on-line preconcentration the analyte can be selectively concentrated before being determined. The preseparation for eliminating the matrix effect and the preconcentration for enhancing the signal of the analyte can be implemented using solid-phase extraction, size exclusion chromatography or ion exchange columns. For the determination of selenite and selenate in drinking water, anionic exchange column chromatography coupled to a MIP-MS system using a nitrogen plasma can be used. Isotope dilution was applied for calibration.[101]

The MIPs have found much greater use in GC than in HPLC interfacing, although the application of the high-power MWP sources as well as MWP sources of high immunity towards solvent loading may increase its analytical potential.[94,102] Several workers directly combined the MPT-OES system with a liquid chromatographic system without using any special interface.[44] The separation and determination of underivatized amino acids with the use of a HPLC-MIP-MS system equipped with a dual oscillating nebulizer was presented by Kwon and Moini.[103]

Liu and Lopez-Avila[104] investigated the possibility of coupling CZE and MWP-OES. The CZE capillary was interfaced to the MIP through an ion exchange interfacing capillary, which was also used as the electrical junction connector to complete the CZE circuit outside the detector. Only preliminary work was carried out, which showed some limitations of this approach.

References

1. R. F. Browner and A. W. Boorn, *Anal. Chem.*, 1984, **56**, 786A–798A.
2. R. K. Skogerboe and G. N. Coleman, *Appl. Spectrosc.*, 1976, **30**, 504–507.
3. K. Fallgatter, V. Svoboda and J. D. Winefordner, *Appl. Spectrosc.*, 1971, **25**, 347–352.
4. R. G. Stahl and K. J. Timmins, *J. Anal. At. Spectrom.*, 1987, **2**, 557–559.
5. L. J. Galante, M. Selby and G. M. Hieftje, *Appl. Spectrosc.*, 1988, **42**, 559–567.
6. Y. Su, Z. Jin, Y. Duan, M. Koby, V. Majidi, J. A. Olivares and S. P. Abeln, *Anal. Chim. Acta*, 2000, **422**, 209–216.
7. C. Yang, Z. Zhuang, Y. Tu, P. Yang and X. Wang, *Spectrochim. Acta, Part B*, 1998, **53**, 1427–1435.
8. M. Wu and J. W. Carnahan, *Appl. Spectrosc.*, 1992, **46**, 163–168.
9. T. Okamoto and Y. Okamoto, *IEEJ Trans. FM*, 2007, **127**, 272–276.
10. F. Leis and J. A. C. Broekaert, *Spectrochim. Acta, Part B*, 1984, **39**, 1459–1463.
11. D. L. Haas and J. A. Caruso, *Anal. Chem.*, 1984, **56**, 2014–2019.
12. K. G. Michlewicz, J. J. Urh and J. W. Carnahan, *Spectrochim. Acta, Part B*, 1985, **40**, 493–499.
13. K. G. Michlewicz and J. W. Carnahan, *Anal. Chem.*, 1986, **58**, 3122–3125.
14. M. Wu and J. W. Carnahan, *J. Anal. At. Spectrom.*, 1992, **11**, 1249–1252.
15. M. M. Mohamed and Z. F. Ghatass, *Fresenius' J. Anal. Chem.*, 2000, **368**, 449–455.
16. H. Yamada and Y. Okamoto, *Appl. Spectrosc.*, 2001, **55**, 114–119.
17. A. Geiger, S. Kirschner, B. Ramacher and U. Telgheder, *J. Anal. At. Spectrom.*, 1997, **12**, 1087–1090.
18. C. I. M. Beenakker, B. Bosman and P. W. J. M. Boumans, *Spectrochim. Acta, Part B*, 1978, **33**, 373–381.
19. Y. K. Zhang, S. Hanamura and J. D. Winefordner, *Appl. Spectrosc.*, 1985, **39**, 226–230.
20. H. Matusiewicz, B. Golik and A. Suszka, *Chem. Anal. (Warsaw)*, 1999, **44**, 559–566.
21. G. L. Long and L. D. Perkins, *Appl. Spectrosc.*, 1987, **41**, 980–985.

22. K. Jankowski, D. Karmasz, L. Starski, A. Ramsza and A. Waszkiewicz, *Spectrochim. Acta, Part B*, 1997, **52**, 1801–1812.

23. K. Jankowski, D. Karmasz, A. Ramsza and E. Reszke, *Spectrochim. Acta, Part B*, 1997, **52**, 1813–1823.

24. M. Huang, H. Kojima, T. Shirasaki, A. Hirabayashi and H. Koizumi, *Anal. Chim. Acta*, 2000, **413**, 217–222.

25. H. Matusiewicz, M. Ślachciński, M. Hidalgo and A. Canals, *J. Anal. At. Spectrom.*, 2007, **22**, 1174–1178.

26. M. Huang, H. Kojima, T. Shirasaki, A. Hirabayashi and H. Koizumi, *Anal. Chem.*, 1999, **71**, 427–432.

27. M. Huang, H. Kojima, T. Shirasaki, A. Hirabayashi and H. Koizumi, *Anal. Chem.*, 2000, **72**, 2463–2467.

28. K. C. Ng and R. C. Culp, *Appl. Spectrosc.*, 1997, **51**, 1447–1452.

29. J. J. Giglio, J. Wang and J. A. Caruso, *Appl. Spectrosc.*, 1995, **49**, 314–319.

30. H. Matusiewicz, *Spectrochim. Acta, Part B*, 2002, **57**, 485–494.

31. K. Wagatsuma, *Appl. Spectrosc. Rev.*, 2005, **40**, 229–243.

32. K. Jankowski, A. Jackowska, A. P. Ramsza and E. Reszke, *J. Anal. At. Spectrom.*, 2008, **23**, 1234–1238.

33. K. Jankowski, A. P. Ramsza, E. Reszke and M. Strzelec, *J. Anal. At. Spectrom.*, 2010, **25**, 44–47.

34. D. L. Haas, J. W. Carnahan and J. A. Caruso, *Appl. Spectrosc.*, 1983, **37**, 82–85.

35. J. F. Camuna-Aguilar, R. Pereiro-Garcia, J. E. Sanchez-Uria and A. Sanz-Medel, *Spectrochim. Acta, Part B*, 1994, **49**, 475–484.

36. B. Kirsch, S. Hanamura and J. D. Winefordner, *Spectrochim. Acta, Part B*, 1984, **39**, 955–963.

37. Y. Madrid, M. W. Borer, C. Zhu, Q. Jin and G. M. Hieftje, *Appl. Spectrosc.*, 1994, **48**, 994–1002.

38. C. I. M. Beenakker, *Spectrochim. Acta, Part B*, 1976, **31**, 483–486.

39. B. M. Spencer, A. R. Raghani and J. D. Winefordner, *Appl. Spectrosc.*, 1994, **48**, 643–646.

40. Y. Madrid, M. Wu, Q. Jin and G. M. Hieftje, *Anal. Chim. Acta*, 1993, **277**, 1–8.

41. M. Wu, Y. Madrid, J. A. Auxier and G. M. Hieftje, *Anal. Chim. Acta*, 1994, **286**, 155–167.

42. M. Seeling, N. H. Bings and J. A. C. Broekaert, *Fresenius' J. Anal. Chem.*, 1998, **360**, 161–166.

43. C. Prokisch and J. A. C. Broekaert, *Spectrochim. Acta, Part B*, 1998, **53**, 1109–1119.

44. J. A. C. Broekaert, N. H. Bings, C. Prokisch and M. Seeling, *Spectrochim. Acta, Part B*, 1998, **53**, 331–338.

45. T. Maeda, K. Wagatsuma and Y. Okamoto, *Anal. Bioanal. Chem.*, 2005, **382**, 1152–1158.

46. T. Maeda and K. Wagatsuma, *Spectrochim. Acta, Part B*, 2005, **60**, 81–87.

47. D. Kollotzek, D. Oechsle, G. Kaiser, P. Tschöpel and G. Tölg, *Fresenius' Z. Anal. Chem.*, 1984, **318**, 485–489.
48. H. Matusiewicz and R. E. Sturgeon, *Spectrochim. Acta, Part B*, 1993, **48**, 723–727.
49. H. Matusiewicz and B. Golik, *Spectrochim. Acta, Part B*, 2004, **59**, 749–754.
50. H. Matusiewicz, *J. Anal. At. Spectrom.*, 1993, **8**, 961–964.
51. Y. N. Pak and S. R. Koirtyohann, *Appl. Spectrosc.*, 1991, **45**, 1132–1142.
52. K. Jankowski, A. Karaś, D. Pysz, A. P. Ramsza and W. Sokołowska, *J. Anal. At. Spectrom.*, 2008, **23**, 1290–1293.
53. G. Heltai, T. Jozsa, K. Percsich, I. Fekete and Z. Tarr, *Fresenius' Z. Anal. Chem.*, 1999, **363**, 487–490.
54. B. L. Sharp, *J. Anal. At. Spectrom.*, 1988, **3**, 613–652.
55. B. L. Sharp, *J. Anal. At. Spectrom.*, 1988, **3**, 939–963.
56. J. Mora, S. Maestre, V. Hernandis and J. L. Todoli, *Trends Anal. Chem.*, 2003, **22**, 123–132.
57. J. L. Todoli and M. Mermet, *Trends Anal. Chem.*, 2005, **24**, 107–116.
58. L. R. Layman and F. E. Lichte, *Anal. Chem.*, 1982, **54**, 638–642.
59. M. A. Tarr, G. Zhu and R. F. Browner, *Anal. Chem.*, 1993, **65**, 1689–1695.
60. Q. Jin, H. Zhang, Y. Wang, X. Yuan and W. Yang, *J. Anal. At. Spectrom.*, 1994, **9**, 851–856.
61. Q. Jin, F. Liang, Y. Huan, Y. Cao, J. Zhou, H. Zhang and W. Yang, *Lab. Robotics Autom.*, 2000, **12**, 76–80.
62. Q. Jin, H. Zhang, D. Ye and J. Zhang, *Microchem. J.*, 1993, **47**, 278–286.
63. J. A. Koropchak and M. Veber, *CRC Crit. Rev. Anal. Chem.*, 1992, **23**, 113–141.
64. L. Bordera, J. L. Todoli, J. Mora, A. Canals and V. Hernandis, *Anal. Chem.*, 1997, **69**, 3578–3586.
65. L. Ding, F. Liang, Y. Huan, Y. Cao, H. Zhang and Q. Jin, *J. Anal. At. Spectrom.*, 2000, **15**, 293–296.
66. O. T. Akinbo and J. W. Carnahan, *Appl. Spectrosc.*, 1998, **52**, 1079–1085.
67. O. T. Akinbo and J. W. Carnahan, *Anal. Chim. Acta*, 1999, **390**, 217–226.
68. Y. Huan, J. Zhou, Z. Peng, Y. Cao, A. Yu, H. Zhang and Q. Jin, *J. Anal. At. Spectrom.*, 2000, **15**, 1409–1411.
69. K. O. Douglass, N. Fitzgerald, B. J. Ingebrethsen and J. F. Tyson, *Spectrochim. Acta, Part B*, 2004, **59**, 261–270.
70. L. Gras, J. Mora, J. L. Todolí, V. Hernandis and A. Canals, *Spectrochim. Acta, Part B*, 1997, **52**, 1201–1213.
71. J. Mora, A. Canals, V. Hernandis, E. H. van Veen and M. T. C. de Loos-Vollebregt, *J. Anal. At. Spectrom.*, 1998, **13**, 175–181.
72. G. Grindlay, S. Maestre, J. Mora, V. Hernandis and L. Gras, *J. Anal. At. Spectrom.*, 2005, **20**, 455–461.
73. J. M. Gehlhausen and J. W. Carnahan, *Anal. Chem.*, 1989, **61**, 674–677.
74. H. Zhang, D. Ye, J. Zhao, J. Yu, R. Men, Q. Jin and D. Dong, *Microchem. J*, 1996, **53**, 69–78.
75. D. Ye, H. Zhang and Q. Jin, *Talanta*, 1996, **43**, 535–544.

76. H. Zhang, X. Yuan, X. Zhao and Q. Jin, *Talanta*, 1997, **44**, 1615–1623.
77. F. Liang, D. Zhang, Y. Lei, H. Zhang and Q. Jin, *Microchem. J.*, 1995, **52**, 181–187.
78. Q. Jia, X. Kong, W. Zhou and L. Bi, *Microchem. J.*, 2008, **89**, 82–87.
79. R. F. Browner, A. W. Boorn and D. D. Smith, *Anal. Chem.*, 1982, **54**, 1411–1419.
80. J. B. Frech, J. B. Etkin and R. Jong, *Anal. Chem.*, 1994, **66**, 685–691.
81. S. Groh, C. C. Garcia, A. Murtazin, V. Horvatic and K. Niemax, *Spectrochim. Acta, Part B*, 2009, **64**, 247–254.
82. S. Yuan, Z. Zhou, G. Wang and C. Liu, *Microelectron. Eng.*, 2003, **66**, 767–772.
83. A. C. Lazar and P. B. Farnsworth, *Anal. Chem.*, 1997, **69**, 3921–3929.
84. A. C. Lazar and P. B. Farnsworth, *Appl. Spectrosc.*, 1997, **51**, 617–624.
85. J. Hamier and E. D. Salin, *J. Anal. At. Spectrom.*, 1998, **13**, 497–505.
86. V. Huxter, J. Hamier and E. D. Salin, *J. Anal. At. Spectrom.*, 2003, **18**, 71–75.
87. M. Bauer and J. A. C. Broekaert, *Spectrochim. Acta, Part B*, 2007, **62**, 145–154.
88. M. Bauer and J. A. C. Broekaert, *J. Anal. At. Spectrom.*, 2008, **23**, 479–486.
89. Z. Benzo, D. Maldonado, J. Chirinos, E. Marcano, C. Gomez, M. Quintal and J. Salas, *Instrum. Sci. Technol.*, 2008, **36**, 598–610.
90. H. Matusiewicz and M. Ślachciński, *Microchem. J.*, 2010, **95**, 213–221.
91. H. Matusiewicz and M. Ślachciński, *Microchem. J.*, 2007, **86**, 102–111.
92. Y. Duan, M. Wu, Q. Jin and G. M. Hieftje, *Spectrochim. Acta, Part B*, 1995, **50**, 971–974.
93. G. Heltai, B. Feher and M. Horvath, *Chem. Pap.*, 2007, **61**, 438–445.
94. A. Chatterjee, Y. Shibata, J. Yoshinaga and M. Morita, *Anal. Chem.*, 2000, **72**, 4402–4412.
95. D. Das and J. W. Carnahan, *Anal. Chim. Acta*, 2001, **444**, 229–240.
96. L. J. Galante, D. A. Wilson and G. M. Hieftje, *Anal. Chim. Acta*, 1988, **215**, 99–109.
97. J. M. Costa-Fernández, F. Lunzer, R. Pereiro-García, A. Sanz-Medel and N. Bordel-García, *J. Anal. At. Spectrom.*, 1995, **10**, 1019–1025.
98. H. A. H. Billiet, J. P. J. van Dalen, P. J. Schoenmakers and L. De Galan, *Anal. Chem.*, 1983, **55**, 847–851.
99. C. Jia, X. Wang and Q. Ou, *Anal. Commun.*, 1997, **34**, 53–56.
100. G. W. Jansen, F. A. Huf and H. J. de Jong, *Spectrochim. Acta, Part B*, 1985, **40**, 307.
101. H. Minami, W. Cai, T. Kusumoto, K. Nishikawa, Q. Zhang, S. Inoue and I. Atsuya, *Anal. Sci.*, 2003, **19**, 1359–1363.
102. A. Chatterjee, Y. Shibata, H. Tao, A. Tanaka and M. Morita, *J. Chromatogr. A*, 2004, **1042**, 99–106.
103. J. Y. Kwon and M. Moini, *J. Am. Soc. Mass Spectrom.*, 2001, **12**, 117–122.
104. Y. Liu and V. Lopez-Avila, *J. High Resolut. Chromatogr.*, 1993, **16**, 717–720.

CHAPTER 8

Solid Sampling Techniques for Microwave Plasmas

8.1 Introduction

Direct solid sampling is desirable to simplify the sample preparation stage and avoid contamination or dilution of the analytes with reagents. A robust solid sampling technique should be applicable to a wide range of sample matrices and capable of reproducible sampling of large sample amounts to minimize the sampling error. The other advantages of direct solid sampling are well known and include reduced analysis time; thus, an inexpensive and quick determination of the major and minor components in solid samples becomes possible, as is required in quality control. Simple operation of a sampling system and simple changing of samples are also valuable. Potential problems with solid sampling in plasma spectrometry include incomplete vaporization and atomization of particles and a high background continuum. Moreover, calibration is difficult to achieve owing to matrix effects. Most solid sampling techniques use microgram to milligram amounts of sample that may be unrepresentative of the bulk composition of the sample. The basic requirement of solid sampling analysis is conversion into the vapour state. Various solid sampling procedures have been developed for atomic emission or mass spectrometry during recent years, including electrothermal vaporization (ETV), arc, spark or laser ablation and direct sample insertion (DSI) and slurry nebulization. The slurry nebulization technique for MWP-OES has been discussed in the previous chapter. A short description of various sample introduction techniques used in spectroscopy can be found in the book by Sneddon,[1] in some reviews[2–5] and, in regard exclusively to MIPs, in the reviews of Matousek *et al.*[6] and Matusiewicz.[7]

RSC Analytical Spectroscopy Monographs No. 12
Microwave Induced Plasma Analytical Spectrometry
By Krzysztof J. Jankowski and Edward Reszke
© The Royal Society of Chemistry 2011
Published by the Royal Society of Chemistry, www.rsc.org

8.2 Methods that Convert Solid Samples into an Aerosol or Vapour

8.2.1 Spark and Arc Ablation

For compact solid samples, arc or spark ablation in the case of electrically conducting samples and, in general, laser ablation (LA) have been reported. The ablation techniques (and also ETV) use external generation of sample vapours and aerosols to improve sample atomization in the plasma. The temporal separation of vaporization, atomization and excitation of the sample proves to be beneficial for analytical performance. However, these systems involve more instrumentation and, therefore, are more difficult to operate and optimize.

Contrary to the use of spark OES, the use of spark ablation (SA) for solid sampling causes separation of the vaporization and excitation of the sample, which leads to lower matrix effects and less complex emission spectra. The coupling of SA to MWP-OES is attractive for analysis of metal alloys containing Fe, Ti, Co or Cr, because the MWP produces relatively simple spectra. Self-reversal, which could be a problem in spark OES, in addition could be avoided as a result of the lower analyte number densities in the plasma, which favours the linearity of calibration curves.[8]

Both high- and medium-voltage sparks are used for sampling. Aerosol from a solid sample is not transported directly from the SA cell into the MWP, but is passed through the mixing chamber for homogenization. A low-power MPT and MIP,[8,9] a moderate-power MIP[10] or a pulsed MIP[11] may be used for excitation of spark-originated aerosols.

SA has been combined with MPT-OES for the direct analysis of compact metallic samples. The material is ablated by a medium-voltage spark in a point-to-plane configuration and delivered into the argon MPT.[8] The detection limits obtained by SA in the case of the MPT are in the $\mu g\,g^{-1}$ range and are about a factor of up to 20 higher than those obtained with spark ablation and ICP-OES. However, this is rather due to substantial differences between the optical systems used for MPT and ICP (the performance of that used for ICP was better) and only partly due to differences between both plasmas. The SA-MPT-OES system has been successfully used for the determination of Fe, Ni, Pb, and Sn in brass, Cr, Cu, Ni, Mn, Mo, Si and V in low-alloy steel and Cu, Fe, Mg, Mn, Si and Zn in the microgram to gram range. The stabilities of the emission signals depend on the element and relative standard deviations ranged from 0.5% to 3.5%. In the case of low-alloy steels, the linearity and the precision of the calibration can be improved by internal standardization.

Layman and Hieftje[12,13] have developed a dc microarc system for aqueous sample vaporization. The argon gas flows over the microarc electrodes after sample vaporization and the sample vapour is introduced to the MIP as shown in Figure 8.1. The microarc MIP combination exhibits detection limits and calibration curves comparable over 2–5 orders of magnitude of concentration, depending on the element determined. Matrix and ionization interferences

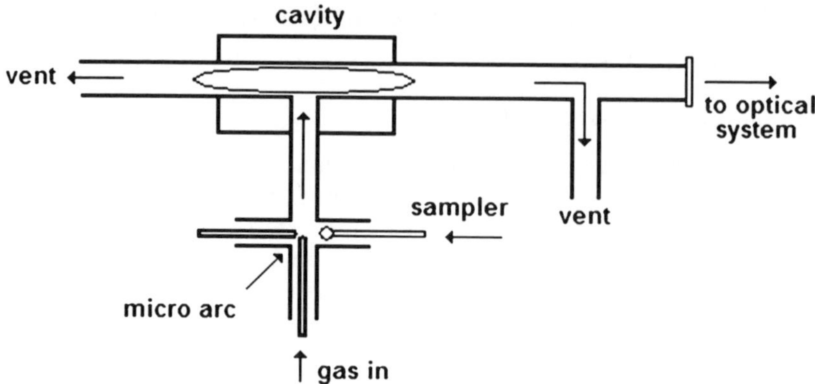

Figure 8.1 Micro arc sampling device for MWP-OES.

effects may be overcome easily. Precision, however, may be affected by the sample electrode material.

8.2.2 Laser Ablation

Laser ablation is increasing in popularity as a method of direct solid sample introduction. The power from a focused laser is used to vaporize an area of material directly from a solid surface. The current availability of a wide range of lasers, in particular Nd/YAG lasers and lasers of higher power, has made LA a popular technique for use with ICP-MS. Minimal loss of sample occurs during the vaporization stage, which is important, for example, in the analysis of materials where complete sample destruction is a disadvantage. Moreover, it is an advantage when spatial analysis is required. The basic conditions for accurate elemental analysis by LA are that the ablated material must have the stoichiometric composition of the solid sample irradiated by a focused laser beam and the material has to be atomized completely. The first condition can be fulfilled by the application of lasers with short pulses (nanoseconds).

The LA-MWP-OES technique was introduced by Leis and Laqua[14,15] and Ishizuka and Uwamino.[16] They used an open design LA cell. Subsequently, a compact LA-MIP design operating at low pressure was explored.[17–21] The sample is ablated inside the MIP cavity in the argon gas atmosphere and the discharge is formed just above the laser irradiated region of the sample surface, as shown in Figure 8.2. This approach has the advantage of more easily controlling material input rates to the MIP without extensive valving. As LA-MIP-OES offers very low detection limits, simultaneous measurements of up to 10 elements with single laser shots are possible. A Nd/YAG laser, operated in its fundamental mode at 1064 nm, is used for sampling and a 0.5-m echelle spectrometer, equipped with an intensified charge coupled device (ICCD) detector, is used for recording the spectra. However, the precision is relatively low and

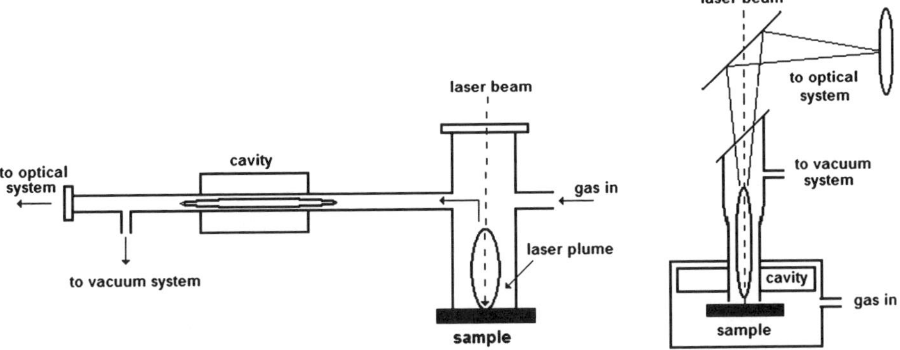

Figure 8.2 Possible arrangements of LA-MWP-OES: open (left) and compact (right) design of LA cell.

the calibration procedure is not straightforward. Measurements of trace elements in steel and aluminum and of sodium and lithium in high-purity quartz are given as examples. Further, Leis *et al.*[21] have investigated the potential of LA-MIP-OES for the analysis of polymer materials.

More recently, pulsed excimer LA instrumentation[22] was coupled with a low-pressure argon MIP by using a similar design of the system. The experiments confirmed the better efficiency of UV laser ablation over infrared LA. Further, a resonant LA has been presented as a method of selective vaporization.[23]

8.2.3 Electrothermal Vaporization

Electrothermal vaporization was used extensively for sample introduction into MWPs in the 1970s and 1980s.[4,6,7] Electrothermal atomization was recognized to be a nearly optimal technique for introduction of solution or solid samples into a MIP. Because the solvent can be vaporized at a lower temperature, and the analyte is then vaporized and partly atomized at a higher temperature, then the vaporized solvent or matrix does not enter into the plasma at the same time as the analyte. Hence, the matrix effects are reduced. Other advantages of ETV include high sample transport efficiency (almost 100%) and thus small sample consumption. Most ETV introduction systems operate in a discrete mode.

In general, such a system consists of the electrothermal vaporizer, a vapour transport interface, and a MWP excitation source. The vapour should be introduced directly into the plasma without any interfaces. The separative column atomizer is designed to fulfil this requirement.[24] In this way, analyte loss, peak broadening and tailing are minimized, as the analytes are detected immediately upon leaving the ETV chamber. In practice, however, this is not usually feasible, and the chamber is connected to the plasma by means of a short length of tubing. The operational vaporization devices include resistively heated graphite cups and rods,[25–27] tantalum and platinum boats,[28,29] a tantalum strip,[30] platinum or tungsten wires or loops[31] and metal filaments.[32–35] All versions of electrothermal atomizers can operate in the inert gas environment and can be readily inserted into the MIP support-gas supply system. These microsample vaporization devices are especially well-suited to use with non-thermal excitation sources such as the MIP and routinely operate efficiently with sample volumes compatible with the maximum sample loadings of MIPs.

In a more advanced ETV technique, a graphite furnace[36–42] with programmed temperature is used for selective vaporization of the sample components that differ with respect to volatility and their transportation to the MWP. Because of this, several substances can be determined simultaneously, including low-volatile substances.

In an early investigation, Runnels and Gibson[43] vaporized volatile metal acetylacetonates from a platinum filament into the gas stream of an Ar-MIP, while Volland *et al.*[27] vaporized metals from a graphite tube after electrolytic preconcentration of the analytes. Very low detection limits were obtained (10^{-10}–10^{-12} g). A series of studies has been carried out by Atsuya, Kawaguchi and co-workers,[31,32] who utilized a low-pressure He-MIP in conjunction with a tantalum filament vaporization system. They observed a substantial improvement of analytical performance for determination of metals in the presence of chloride and thus microsampling capability has been used to advantage in the determination of metals in biological materials, including the analysis of zinc metalloenzymes.

Although interfacing an ETV to a MWP is mechanically straightforward, tandem operation of the two components has required careful manipulation of the instrumental variables. When analyzing liquid samples, the rate of solvent evaporation must be carefully controlled to avoid plasma distortion or even extinguishing. Valve systems can be used for solvent venting.

Taking advantage of the high sensitivity of an He-MWP towards the determination of non-metals, the analytical performance of the ETV-MWP-OES system was examined. Microlitre liquid samples of chlorides, bromides, iodides, sulfides and ammonium nitrogen were introduced into a grooved graphite rod, where they were dried, ashed and vaporized into the plasma. Absolute detection limits for Cl, Br, I, S and N were on the lower nanogram level.[25] Hanamura *et al.*[44] applied thermal vaporization in conjunction with a CMP for speciation of some inorganic and organometallic compounds in biological matrices. A commercial instrument for thermal desorption and combustion coupled to a MIP was used for the simultaneous determination of adsorbable organically bound halides.[45]

Evans *et al.*[46] utilized a tantalum-tip electrothermal vaporizer combined with MPT-OES/MS for the determination of several trace elements. The method of standard addition could be used to determine trace elements in samples. An advanced analytical procedure with the use of slurry sampling, ETV and *in-situ* fusion and subsequent determination by isotope dilution/nitrogen MIP-MS has been employed for the determination of selenium in biological samples.[47]

8.3 Discrete Powder Introduction

The use of direct sample introduction (DSI) for solid sampling is attractive because this technique has the highest analyte transport efficiency.[2] A graphite rod is usually used as a sample elevator to introduce the solid through the discharge tube below the plasma region. Direct inductive heating of the carbon occurs and the sample vaporizes directly into the plasma. This direct insertion method may give low limits of detection and a wide dynamic range, but suffers from matrix effects and the need for closely matching standards.

A CMP operating in the range between 500 and 700 W was used for the analysis of solid samples deposited in the graphite electrode-cup. A number of elements were determined in coal and tomato leaves by the group led by Winefordner.[48,49] However, the DSI-CMP suffers from poor precision, possible contamination from the electrode materials and changes in background level, and it is necessary to reignite the plasma between samples, which may contribute to the poor precision. Further studies of this group have led to improvements in the design of the DSI device and various systems have been explored, including a tungsten cup and filament or direct insertion of the electrode made from metallic sample material.[50–55]

Gehlhausen and Carnahan[56] introduced milligram amounts of finely ground coal samples by means of discrete powder injection (DPI) into a 500-W helium MIP. The injection device is shown in Figure 8.3. A Teflon stopcock was used to control the injections. With the stopcock closed, 1-mg samples were loaded into the upper chamber and after opening the stopcock the injector gas-flow rate was maintained at about $0.5\,L\,min^{-1}$. This technique was used for the simultaneous determination of elemental ratios in coal with a variance above 10%.

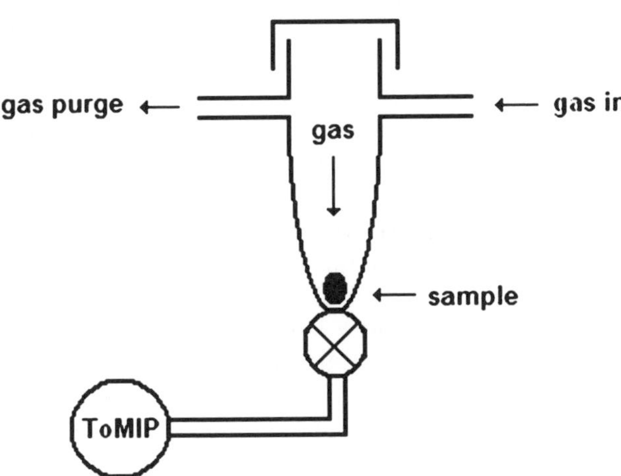

Figure 8.3   Discrete powder injection device.

8.4 Continuous Powder Introduction

The continuous powder introduction (CPI) of solid samples into the plasma may be implemented using a number of techniques.[1] A swirl cup device directly delivers powders into the plasma.[57] A carrier gas is introduced downwards into a mechanically agitated cup containing the powdered sample. A cloud of powder is formed by the gas displacement and is then led to an outlet and directed to the plasma. Once the sample loading is sufficiently great, the powder drops back into the base of the cup, into the path of the gas jet. Another technique for CPI employs a fluidized bed chamber. Argon gas flows through a sintered glass disc on to which the powdered sample is deposited. Mechanical vibration again enables an efficient and homogenous cloud of sample to be formed.[58,59] For CPI-ICP-OES, poor signal stability and severe matrix effects are observed. The continuous introduction of fine powders into the CMP was first reported by Kessler[60,61] and into the moderate-power MIP by Gehlhausen and Carnahan.[56] The CPI technique in conjunction with a low-power MIP has been extensively explored by Jankowski *et al.*[62–66]

For most solid sampling techniques, including ETV, LA and DPI, the emission signal generated with the sample introduction system is a transient peak, requiring the use of a simultaneous multichannel reading spectrometer for multi-element determinations. The CPI approach permits the generation of a steady-state signal and therefore a sequential spectrometer may be used for monitoring analytes and background emission.

The CPI system, based on a fluidized-bed concept, consists of a sample chamber in the form of a fluidized column, a tube adapter and a sample injector device directly attached to the MIP discharge tube. A diagram of the device is shown in Figure 8.4. The powdered sample is supported on a fritted disk of defined porosity, with the gas flow passing through the disk. A continuous gas flow goes up the outer tube of the sample injector to maintain the plasma. After the plasma is ignited, and the operation conditions adjusted, the sample aerosol is pneumatically inserted with the sampling gas through the inner tube of the sample injector device to the MIP discharge tube. Mechanical vibration is used to stabilize the formation of the fluidized bed. The handling of powders with the combination of mechanical agitation and gas flow allow the uniform delivery of particles over a wide range of diameters. Since a low injection velocity of the particles is desirable for optimal energy transfer from plasma to powder, the injection gas flow rate must be kept low. The vertical position of the CPI-MIP-OES system results in effective mixing of the powder and a relatively low gas flow rate assures a laminar passage of the sample particles through the plasma. The microwave arrangement of the system consists of a TE_{101} integrated microwave resonator with a power supply and a vertically positioned aerosol-cooled plasma discharge tube.

Each time the sample is removed from the source, the source is open to air. The discharge cavity will therefore contain some air every time the sample is changed, even if the source is flushed with argon during the positioning process. Two different types of pollution occur when opening and closing the sample

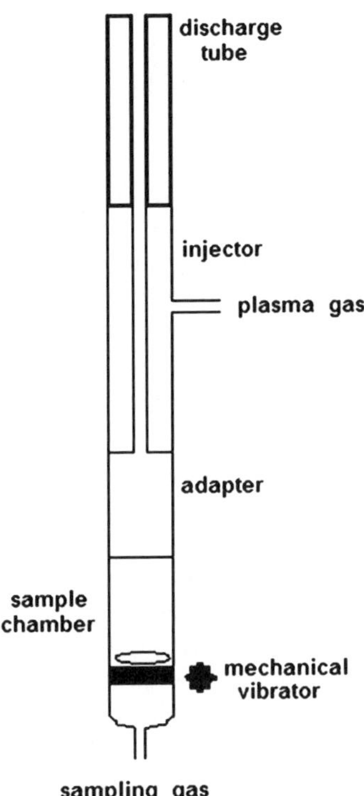

Figure 8.4 Schematic diagram of the CPI system.

chamber: residual gases, including N_2, O_2, CO_2 and H_2O from air, and deposition of H_2O vapour onto the CPI system walls. Gases in the argon atmosphere are removed before delivering the sample into the plasma. The effect of flushing the whole system prior to analysis is to reduce the time for the plasma to become stable during the analysis.

It is obviously difficult to define an optimal flush time for all applications and all needs. However, in our experience a flush time of 1 minute is sufficient for most analyses. Only when it is necessary to determine at trace levels (below 100 ppm) for elements such as N, O, H and C would it be necessary to increase the flush time to 2 or 5 minutes.

The CPI system has been designed to achieve a uniform delivery of particles lasting for at least several minutes. The essential requirements for a stable operation of the system are: vertical positioning of the CPI device and the discharge tube, the controlled and limited plasma loading and a relatively narrow particle size range. Additionally, for effective vaporization, atomization and excitation of the analytes, a relatively long residence time should be assured. Both the design of the CPI device and the discharge tube can fulfil these requirements. However, the physical characteristics of the sample should also be considered since the specific gravity and particle size distribution of the sample and the gas flow rate used are interdependent. Silica or carbon are low-density materials and can be efficiently converted into an aerosol by the CPI-MIP-OES system operating at relatively low gas flow rates. However, various industrial materials such as ceramics or glass powders could be directly introduced by the CPI system with a relatively low sample feed rate.

The CPI approach (as shown in Figure 8.4) is found useful for continuous and uniform introduction of powders with particle sizes between 20 and 80 µm. A suitable powder loading is about 1 g, depending on the specific gravity of the sample. The powder feed rate typically ranges between 5 and 20 mg min^{-1} and is easily controlled by the sampling gas flow rate ranging from 0.1 to 0.5 L min^{-1}. All powdered substances should be kept in a dry environment to avoid problems related to particle adsorption on the internal surfaces of the CPI system.

The continuous coal sampling device used by Gehlhausen and Carnahan[56] is shown in Figure 8.5. A powdered coal sample of about 2 g (100 mesh) was deposited onto the fritted glass disc. Powders of different particle size ranges were delivered for periods of time up to 20 min. The rate of coal introduction was about 1 mg s^{-1} and decreased somewhat with time. The authors state that the system was not useful for quantitative determinations. Taking into account the above-mentioned requirements for the CPI, one can conclude that the system has some disadvantages. A horizontal position of the plasma and a curved trajectory of the powder aerosol delivery do not serve to obtain a uniform aerosol delivery, owing to the impact losses and gravitational settling. Moreover, the plasma seems to be slightly overloaded even if a moderate power is applied.

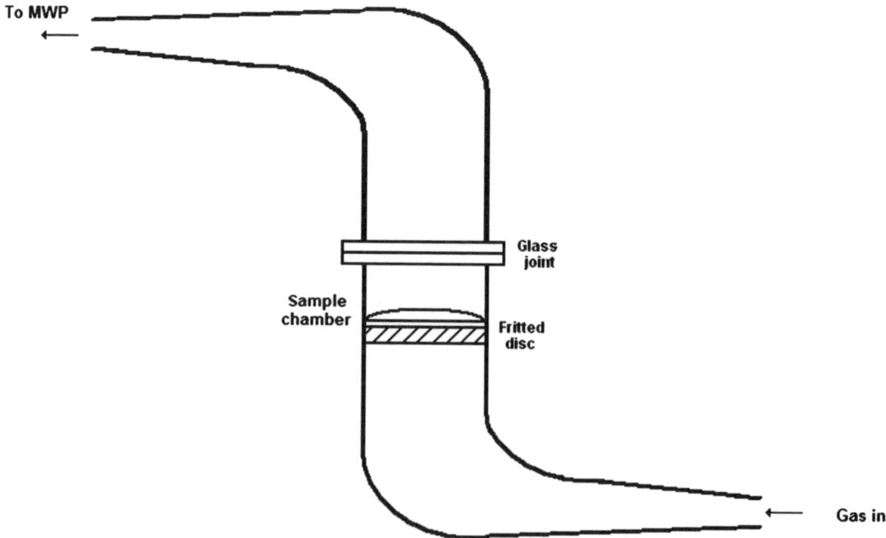

Figure 8.5   Continuous coal sampling device by Gehlhausen and Carnahan.

The CPI-MIP-OES system based on the fluidized-bed approach was successfully applied for the determination of trace elements in a number of geological and industrial materials.[62,65] The single calibration graph approach based on powdered reference materials was adequate for the determination of Cu, Mg, Si and Zn (detection limits around $100\,ng\,g^{-1}$) in Al_2O_3, SnO_2, La_2O_3 and MgO of spectrographic grade. A sensitive method for the determination of total fluorine content in soil, fly ash and coal has been developed and the experimental detection limit is in the range $3–6\,\mu g\,g^{-1}$. Another interesting approach of the CPI technique is to carry out direct multi-element analysis of elements preconcentrated on a suitable sorbent using solid-phase extraction.[63–64,67]

Moreover, the CPI-MWP-OES system can be applied to the determination of the chemical composition and size of individual particles,[68,69] as well as real-time air particulate monitoring.[70] This will be discussed in Chapter 12.

8.5 Separation Methods Coupled to Continuous Powder Introduction

Sorption techniques such as solid-phase extraction are commonly used for preconcentration and separation of analytes before determination. However, elution of analytes can be sometimes troublesome owing to irreversible binding of some elements. Additionally, the elution leads to sample dilution decreasing the analyte enrichment factor. Sometimes, matrix effects caused by the eluent present in the sample solution are observed in the determination step. These problems may be omitted by analyzing the analyte-on-sorbent particles directly by solid sample introduction into the plasma.

The coupling of analyte preconcentration by sorption on a suitable support material, with subsequent direct powder introduction to the plasma, has some advantages. By application of the analyte separation step, problems connected with possible matrix effects for individual sample materials as well as with particle size matching and plasma condition optimization can be avoided. Both the sensitivity of spectrometric measurements and the reproducibility of the results are good, because the analytes are mostly vaporized from the surface of the support particles. The presence of carbon in the plasma is beneficial for plasma stability and analyte excitation owing to its conductivity.

Activated carbon seems to be a suitable collector for direct introduction to plasmas because it is relatively easily decomposed by the plasma compared with chelating resins and gives relatively low spectral interferences, particularly in the UV region. In addition, it is available in analytical grade purity. Finally, activated carbon can be recognized as a universal carrier owing to its widespread sorption performance. Silica is the other sorbent of widespread use that is recommended for CPI applications. In general, it produces lower background than carbon; however, SiO bands are observed in the UV region. Powdered silica is also suitable as a carrier for immobilization of bacteria or yeasts when biosorption processes are applied. As the sorbent is decomposed under the measurement conditions, low-cost materials are preferred.

The CPI-MIP-OES system has been developed for multi-element determination of heavy metals in ground and tap water samples after preconcentration on activated carbon.[63] An enrichment factor of about 1000-fold for a sample volume of 1 L was obtained. The detection limits for the proposed method were $17–250\,ng\,L^{-1}$.

A second example of application of the method coupled with analyte preconcentration is the determination of precious metals in the platinum ore SARM-7 and in the Chinese soil CRMs.[64] The aliquots of the geological CRMs were digested and the analytes were separated on activated carbon, which was dried and introduced to the CPI-MIP-OES system. The possible matrix effects caused by alkali and alkaline earths elements are eliminated at the preconcentration stage as these metals are not retained on the activated carbon. The analyte recovery could be sometimes tedious and time consuming. The sorption studies have revealed that some noble metals are reduced to their elemental state when sorbed on activated carbon. This has made elution a critical step of the analytical procedure. A strong acid is usually required for elution, which may cause severe interference or lead to substantial sample dilution. The procedure of separation/preconcentration of trace amounts of precious metals with the aid of sorption on activated carbon, and subsequent CPI of the sorbent particles with the analytes collected into the plasma as a dry, particulate aerosol, substantially improves the analytical performance of the determination by OES.

Sorption on zirconium-modified activated carbon or calcium hydrogen phosphate has been used for the separation of aqueous fluoride.[67] The analyte was subsequently determined using CPI-MIP-OES. An absolute detection limit of $4.7\,\mu g$ was achieved. It gives the possibility of determining aqueous fluoride at the nanogram per millilitre level, including analyte preconcentration by a factor of 1000.

8.6 Analysis of Powdered Samples by CPI-MWP-OES

Any material to be analyzed by the CPI-MIP-OES system should be ground into a fine particle size, usually below $80\,\mu m$, and then classified into suitable size fractions. For a stable delivery rate and efficient vaporization of the analytes the particle size range is critical. When a relatively broad particle size is used, more time is needed for signal stabilization and a more pronounced particle size segregation effect is observed, resulting in poor signal stability. Hence, the use of a reasonably narrow particle size range is recommended. All powdered materials should be kept in a dry environment to avoid problems related to particle aggregation and adsorption on the internal walls of the CPI system due to the presence of moisture.

Matrix effect studies are essential for evaluation of any direct sampling technique. In CPI plasma emission spectrometry the matrix composition influences both sample delivery to the plasma and plasma processes, *i.e.* analyte vaporization, atomization and excitation. Taking into account the limited energy of a MIP discharge, it can be assumed that the large mass of the sample matrix consumes a large part of the plasma energy and only volatile elements can be efficiently determined by this system. However, this is not true if the analytes are localized on the surface of the solid particles. This is the case when the analytes are preconcentrated using solid-phase extraction. In general, matrix matching and particle size matching are imperative for accurate determinations. The presence of a matrix in the plasma environment causes spectral interferences that can be resolved only by wavelength selection, improved optical resolution or computer processing of the spectral data.

Matrix effects have to be recognized in any case and an adequate analytical procedure should be developed. For determination of aqueous fluorine with external calibration, four sets of powdered standards are prepared by dropping the standard fluoride solution onto four different support materials: calcium hydrogen phosphate, silica, carbon and alumina. The slope ratio of the calibration graphs is 3:2:2:1, respectively. It is obvious that the differences are due to matrix effects.[67] However, for the determination of volatile trace elements (Cu, Mg, Si, Zn) in the range 0–7 $\mu g\,g^{-1}$ in a number of metal oxides (Al_2O_3, MgO, La_2O_3, SnO_2) by CPI-Ar-MIP-OES, it is possible to calculate the single calibration graph for each analyte using a set of alumina-based reference materials. In comparison with alumina, insignificant differences are observed in the matrix effects of other metal oxides.[64]

In various spectroscopic techniques based on the direct introduction of powdered material to the excitation or ionization source, chemical modifiers are used to stimulate the vaporization of analytes by conversion into volatile chlorides, fluorides or oxides. When analyzing silicate materials by the CPI-MIP-OES technique, the addition of several per cent of silver chloride or ammonium fluoride has led to a two- to six-fold enhancement of the emission intensity for Mg, Mn, Fe and Cu, even though the residence time of the sample in the plasma was calculated to be about 10 ms.[71]

Sometimes, severe matrix effects can be reduced by modification of the matrix composition. The determination of zinc in titanium oxide by CPI-MIP-OES is difficult owing to spectral interferences from titanium emission lines and a high background level. The addition of several per cent of silica to the titania sample reduces both Ti and the background emission due to plasma cooling, improving the determination of zinc.[66]

One of the major considerations in solid sampling is the method of calibration. For the determination of trace elements in various materials by the CPI technique, several methods of calibration are useful, including external calibration using solid standards and single-standard addition. The most reliable results can be achieved using calibration with solid reference materials if only the matrix composition of the sample and the reference materials are almost identical. Matching similar particle size ranges is necessary for improved accuracy.

Despite calibration with solid reference materials, for the determination of major and trace contents of analytes, external calibration is adequate using calibration standards prepared by the dropping technique (the aliquots of analyte standard solution are dropped on a suitable powdered support with matrix matching and then dried). For multi-element determination of trace metals, the standard addition technique with aqueous standard solutions dropped on the powdered sample is strongly recommended. However, it should be noted that the use of dropping techniques raises problems related to the different chemical form in which the analyte is added and in that which it is present in the sample. Moreover, differences in vaporization efficiency are possible between surface impurities (analytes added) and bulk traces (analyte originally present in the sample).

The single calibration graph approach based on powdered reference materials is adequate for the determination of Cu, Mg, Si and Zn (detection limits around $100\,ng\,g^{-1}$) in Al_2O_3, SnO_2, La_2O_3 and MgO of spectrographic grade. Calibration functions were established using four aluminium oxide standards, as described above. The external calibration method with Al_2O_3 powdered reference materials was used for the determination of the Cu and Mg contents. For other elements the multi-element standard addition was applied using the dropping technique. The aliquots of standard solution were dropped on the CRM alumina weights and then thoroughly dried and homogenized.[64]

External calibration was applied for the determination of some metals (Co, Mo, V, Ni) at the per cent level in alumina- and silica-based catalysts by CPI-MIP-OES. Calibration standards were prepared by dropping the aliquots of single-element standard solution on the alumina or silica supports identical with those used for preparation of the catalysts. The slopes of the calibration graphs were similar to those obtained for calibration graphs based on powdered reference materials.[66]

Two approaches were considered for the external calibration in the direct determination of total fluorine in geological materials by CPI-MIP-OES. The first, rather expensive, possibility is to use a set of solid certified reference materials, which are available for a wide range of fluorine concentrations in coals and soils. The second approach is to prepare a set of calibration standards by dropping a standard fluoride solution on a suitable powdered support with matrix matching, *i.e.* carbon support for coal samples and silica for soils.[62,63,67]

The use of external calibration with the aid of synthetic standards obtained by multi-element sorption of analytes on suitable support materials is attractive for the determination of trace elements by CPI-MIP-OES coupled with preconcentration by solid-phase extraction (SPE). This provides an identical preparation procedure for samples and standards, including sorption efficiency. Additionally, matrix matching and particle size matching are easily attainable. The procedure is used for heavy metals determination in water samples, for precious metals in platinum ore and soil, as well as for the determination of aqueous fluoride in mineral water.[69]

References

1. *Sample Introduction in Atomic Spectroscopy*, ed. J. Sneddon, Elsevier, Amsterdam, 1999.
2. R. Sing, *Spectrochim. Acta, Part B*, 1999, **54**, 411–441.
3. S. A. Darke and J. F. Tyson, *Microchem. J.*, 1994, **50**, 310–336.
4. J. M. Carey and J. A. Caruso, *Crit. Rev. Anal. Chem.*, 1992, **23**, 397–439.
5. R. E. Russo, X. Mao, H. Liu, J. Gonzalez and S. S. Mao, *Talanta*, 2002, **57**, 425–451.
6. J. P. Matousek, B. J. Or and M. Selby, *Prog. Anal. At. Spectrosc.*, 1984, **7**, 275–314.
7. H. Matusiewicz, *Spectrochim. Acta Rev.*, 1990, **13**, 47–68.
8. U. Engel, A. Kehden, E. Voges and J. A. C. Broekaert, *Spectrochim. Acta, Part B*, 1999, **54**, 1279–1289.
9. D. J. C. Helmer and J. P. Walters, *Appl. Spectrosc.*, 1984, **38**, 392–398.
10. Y. N. Pak and S. R. Koirtyohann, *J. Anal. At. Spectrom.*, 1994, **9**, 1305–1310.
11. M. M. Mohamed, T. Uchida and S. Minami, *Appl. Spectrosc.*, 1989, **43**, 794–800.
12. L. R. Layman and G. M. Hieftje, *Anal. Chem.*, 1975, **47**, 194–202.
13. A. T. Zander and G. M. Hieftje, *Anal. Chem.*, 1978, **50**, 1257–1260.
14. F. Leis and K. Laqua, *Spectrochim. Acta, Part B*, 1978, **33**, 727–740.
15. F. Leis and K. Laqua, *Spectrochim. Acta, Part B*, 1979, **34**, 307.
16. T. Ishizuka and Y. Uwamino, *Anal. Chem.*, 1980, **52**, 125–129.
17. J. Uebbing, A. Ciocan and K. Niemax, *Spectrochim. Acta, Part B*, 1992, **47**, 601–610.
18. A. Ciocan, J. Uebbing and K. Niemax, *Spectrochim. Acta, Part B*, 1992, **47**, 611–617.
19. L. Hiddemann, J. Uebbing, A. Ciocan, O. Dessenne and K. Niemax, *Anal. Chim. Acta*, 1993, **283**, 152–159.
20. A. Ciocan, L. Hiddemann, J. Uebbing and K. Niemax, *J. Anal. At. Spectrom.*, 1993, **8**, 273–278.
21. F. Leis, H. E. Bauer, L. Prodan and K. Niemax, *Spectrochim. Acta, Part B*, 2001, **56**, 27–35.
22. K. X. Yang, J. X. Zhou, X. Hou and R. G. Michel, *J. Korean Phys. Soc.*, 1999, **35**, 133–142.
23. D. Cleveland, P. Stchur, X. Hou, K. X. Yang, J. Zhou and R. G. Michel, *Appl. Spectrosc.*, 2005, **59**, 1427–1444.
24. K. Kitagawa, A. Mizutani and M. Yanagisawa, *Anal. Sci.*, 1989, **5**, 539–544.
25. M. M. Abdillahi, *Appl. Spectrosc.*, 1993, **47**, 366–374.
26. G. Kaiser, D. Götz, P. Schoch and G. Tölg, *Talanta*, 1975, **22**, 889–899.
27. G. Volland, P. Tschöpel and G. Tölg, *Spectrochim. Acta, Part B*, 1981, **36**, 901–917.
28. D. G. Mitchell, K. M. Aldous and E. Canelli, *Anal. Chem.*, 1977, **49**, 1235–1238.

29. K. Chiba, M. Kurosawa, K. Tanabe and H. Haraguchi, *Chem. Lett.*, 1984, 75–78.
30. F. L Fricke, O. Rose, Jr. and J. A. Caruso, *Talanta*, 1976, **23**, 317–320.
31. U. Richts, J. A. C. Broekaert, P. Tschöpel and G. Tölg, *Talanta*, 1991, **38**, 863–869.
32. I. Atsuya, H. Kawaguchi, C. Veillon and B. L. Vallee, *Anal. Chem.*, 1977, **49**, 1489–1491.
33. H. Kawaguchi and D. S. Auld, *Clin. Chem.*, 1975, **21**, 591–594.
34. N. W. Barnett, *Anal. Chim. Acta*, 1987, **198**, 309–314.
35. R. G. Stahl, L. Brett and K. J. Timmins, *J. Anal. At. Spectrom.*, 1989, **4**, 1415–1425.
36. H. Matusiewicz and M. Kopras, *J. Anal. At. Spectrom.*, 2003, **18**, 273–278.
37. H. Matusiewicz, I. A. Brovko, R. E. Sturgeon and V. T. Luong, *Appl. Spectrosc.*, 1990, **44**, 736–739.
38. J. P. Matousek, B. J. Orr and M. Selby, *Spectrochim. Acta, Part B*, 1986, **41**, 415–429.
39. J. A. C. Broekaert and F. Leis, *Mikrochim. Acta*, 1985, **86**, 261–272.
40. G. Heltai, J. A. C. Broekaert, P. Burba, F. Leis, P. Tschöpel and G. Tölg, *Spectrochim. Acta, Part B*, 1990, **45**, 857–866.
41. Q. Jin, H. Zhang, W. Yang, Q. Jin and Y. Shi, *Talanta*, 1997, **44**, 1605–1614.
42. J. Yang, J. Zhang, C. Schickling and J. A. C. Broekaert, *Spectrochim. Acta, Part B*, 1996, **51**, 551–562.
43. J. H. Runnels and J. H. Gibson, *Anal. Chem.*, 1970, **42**, 1569.
44. S. Hanamura, B. W. Smith and J. D. Winefordner, *Anal. Chem.*, 1983, **55**, 2026–2032.
45. H. Lehnert, T. Twiehaus, D. Rieping, W. Buscher and K. Cammann, *Analyst*, 1998, **123**, 637–640.
46. E. H. Evans, J. A. Caruso and R. D. Satzger, *Appl. Spectrosc.*, 1991, **45**, 1478–1484.
47. T. Kawano, A. Nishide, K. Okutsu, H. Minami, Q. Zhang, S. Inoue and I. Atsuya, *Spectrochim. Acta, Part B*, 2005, **60**, 327–331.
48. H. Abdalla, K. C. Ng and J. D. Winefordner, *J. Anal. At. Spectrom.*, 1991, **6**, 211–213.
49. H. Abdalla, K. C. Ng and J. D. Winefordner, *Spectrochim. Acta, Part B*, 1991, **46**, 1207–1214.
50. H. Abdalla, K. C. Ng and J. D. Winefordner, *Anal. Chim. Acta*, 1992, **264**, 327–332.
51. W. R. L. Masamba, B. W. Smith and J. D. Winefordner, *Appl. Spectrosc.*, 1992, **46**, 1741–1744.
52. M. W. Wensing, B. W. Smith and J. D. Winefordner, *Anal. Chem.*, 1994, **66**, 531–535.
53. A. M. Pless, B. W. Smith, M. A. Bolshov and J. D. Winefordner, *Spectrochim. Acta, Part B*, 1996, **51**, 55–64.
54. A. M. Pless, A. Croslyn, M. J. Gordon, B. W. Smith and J. D. Winefordner, *Talanta*, 1997, **44**, 39–46.

55. A. D. Besteman, G. K. Bryan, N. Lau and J. D. Winefordner, *Microchem. J*, 1999, **61**, 240–246.
56. J. M. Gehlhausen and J. W. Carnahan, *Anal. Chem.*, 1991, **63**, 2430–2434.
57. H. C. Hoare and R. A. Mostyn, *Anal. Chem.*, 1967, **39**, 1153–1155.
58. R. M. Dagnall, D. J. Smith, T. S. West and S. Greenfield, *Anal. Chim. Acta*, 1971, **54**, 397–406.
59. R. Guevremont and K. N. De Silva, *Spectrochim. Acta, Part B*, 1986, **41**, 875–888.
60. W. Kessler and F. Gebhardt, *Glastechn. Ber.*, 1967, **40**, 194–200.
61. W. Kessler, *Glastechn. Ber.*, 1971, **44**, 479–483.
62. K. Jankowski, A. Jackowska and P. Łukasiak, *Anal. Chim. Acta*, 2005, **540**, 197–205.
63. K. Jankowski, Y. Jun, K. Kasiura, A. Jackowska and A. Sieradzka, *Spectrochim. Acta, Part B*, 2005, **60**, 369–375.
64. K. Jankowski, A. Jackowska, P. Łukasiak, M. Mrugalska and A. Trzaskowska, *J. Anal. At. Spectrom.*, 2005, **20**, 981–986.
65. K. Jankowski, A. Jackowska and M. Mrugalska, *J. Anal. At. Spectrom.*, 2007, **22**, 386–391.
66. K. Jankowski and A. Jackowska, *Trends Appl. Spectrosc.*, 2007, **6**, 17–25.
67. K. Jankowski, A. Jackowska and A. Tyburska, *Spectrosc. Lett.*, 2010, **43**, 91–100.
68. H. Takahara, M. Iwasaki and Y. Tanibata, *IEEE Trans. Instrum. Meas.*, 1995, **44**, 819–823.
69. T. Okamoto and Y. Okamoto, *JPl. Fusion Res. Ser.*, 2009, **8**, 1330–1333.
70. Y. Duan, Y. Su, Z. Jin and S. P. Abeln, *Anal. Chem.*, 2000, **72**, 1672–1679.
71. A. Jackowska, PhD thesis, Warsaw University of Technology, 2006 (in Polish).

Optimization of the MWP-OES System

9.1 What do we Optimize?

9.1.1 Sample Introduction System-related Parameters

In optical emission spectrometry (OES) the sample analysis is preceded by optimization of the measurement parameters. For MWP-OES it is quite complicated, especially that some elements of the apparatus (*e.g.* sample introduction system, discharge tube, optical system and others) are replaceable, and each element shows specific requirements and operating parameters. The most common parameters that are optimized, depending on the design of the instrument, are gas flow rates, MW power, integration time and photomultiplier tube (PMT) voltage. Other variable parameters may include pump speed and viewing position. The appropriate selection of the sample-introduction system is a critical step when particular analytical figures of merit are required. The analytical wavelength is certainly an important variable parameter of many instruments. Selection of the wavelength is considered in the next section of this chapter. The microwave-induced plasma (MIP) will have quite a few adjustments, depending on the control of the device desired. Unfortunately, they are usually interdependent to a large extent. The optimum nebulizer gas flow rate, for example, will depend on the diameter of the nebulizer orifice, the overall design of the discharge tube and also to some extent on the applied power.

Even though significant advances have been made in the MWP-OES technique, there is still a certain amount of interaction between the analyst and the instrument that is required for MWP-OES analyses. The skill required of the analyst can vary widely, depending on the types of samples to be analyzed and the sophistication of both the method and the instrumentation used.

RSC Analytical Spectroscopy Monographs No. 12
Microwave Induced Plasma Analytical Spectrometry
By Krzysztof J. Jankowski and Edward Reszke
© The Royal Society of Chemistry 2011
Published by the Royal Society of Chemistry, www.rsc.org

Some main variables are related to the sample introduction system. To give an example, the performance of the nebulizer has a considerable effect on the analytical performance of the whole spectrometer, such as measurement sensitivity, precision, *etc.* In the case of microwave plasmas characterized by limited energy, relatively large differences in the excitation efficiency of particular elements are found. Moreover, excitation can be accompanied by chemical reactions that effect the behaviour of an element in the high-temperature region. Therefore, for multi-element determination a problem appears in the selection of an element appropriate for optimization of the compromised experimental conditions. The optimization of experimental conditions for determination of zinc and phosphorous with the use of an ultrasonic nebulizer is an example.[1] The nebulizer gas flow rate can be a critical parameter because it largely determines the residence time of the analyte species in the plasma zone. The longer the residence time, the more time the analyte has to be atomized, excited and ionized. On the other hand, a trade-off between maximizing the analyte residence time in the discharge and the amount of sample loaded into the plasma takes place. For an element that has a high ionization potential, a long residence time would be desired. Thus a lower nebulizer gas flow rate might be used to obtain maximum sensitivity for phosphorus. On the other hand, for elements such as sodium and potassium that emit strong atomic lines and are easily ionized, a faster flow rate might be used.

9.1.2 Source-related Parameters

Applied power is the main source-related parameter. The power–signal relationships are complicated by interrelationships with other operating parameters, such as cavity design, dimensions of the discharge tube, viewing position relative to the spectrometer entrance slit, nature of the plasma gas and its flow rate, discharge pressure, sample size and the wavelength selected. In addition, the coupling efficiency of the MW power to the discharge must be controlled. The response of the GC-MIP-OES system operating at atmospheric pressure is correlated to the applied power level in either a positive, negative or a constant manner, depending, rather unpredictably, on the nature of the analyte and compound and the selection of the spectral line. Different power–signal relationships have been observed with different discharge pressures. In other studies, determination of the sensitivity of volatile compounds was found to be independent of the power.[2–5]

For solution nebulization MWP-OES the effect of applied power is greatest at lower nebulizer flow rates and tends to decrease as the flow rate increases. The effect is also more pronounced for helium than for argon.[6–8] The emission intensities for many spectral lines seem to be nearly independent of the plasma power.

The other source-related parameter is the plasma height. When the plasma is viewed side-on, the selection of the viewing position corresponds to the adequate plasma height. For the axial viewing mode, numerous studies suggest that for attaining the maximum emission signal for atom and ion lines the plasma should be at least 3 cm long to provide sufficient analyte residence time. Accordingly, cavities of 2 and 3 cm cavity depth have been constructed to obtain physically longer plasmas.[9–11] However, the optimum plasma length is limited to several centimetres owing to self-absorption, as observed for the surfatron.

9.1.3 Spectrometer-related Parameters

Although the MWP is an efficient excitation source and a large number of spectral lines are produced for each element, only a few lines can provide the sensitivity required for trace analysis. The sensitivities and detection limits towards a given element are often governed by spectral interferences of contaminants in the plasma gases, or by the elemental composition of the sample. The selectivity of the measurement to various elements is largely determined by the spectral region where the lines or bands are monitored. Typically, there is a compromise between selectivity and sensitivity. The spectral resolution of the optical system significantly influences these figures of merit.

Both the integration time and PMT voltage are crucial to achieve comparable sensitivity and reproducibility of measurements. Unstable or weakly emitting sources require much longer measurement times than stable or intense sources. Transient systems are generally less stable than continuous systems. For steady-state measurements, integration times typically vary from 0.5 s to 3 s and for transient signals they are typically about 0.1–1 s, but for special applications can be as short as 1 ms.

Before dealing with optimizing the parameters, we should first establish what the criteria are for assessing the optimum. For multi-element analysis we want to achieve low detection limits for trace elements, high reproducibility for major elements and a short analysis time. For optimizing the detection limits, the use of the signal-to-background ratio as a measured value is more suitable than the signal intensity.

The optimization of experimental conditions and analytical parameters is an important part of performing reliable and precise MWP-OES analysis. Even for standard applications, where it is possible to predict the optimal measurement conditions for different instruments, the operator should try to understand the influence of the different parameters on the analytical results. Very often, compromises have to be made between the analytical performance and the time of the analysis. For multi-element analysis the compromise conditions should be chosen in order to balance the detection limits among the elements. Sometimes, the use of such compromise conditions for single-element analysis is sufficient and it greatly facilitates a more tedious approach in which the power, the carrier-gas flow rate and the observation height are optimized for each element separately.

9.2 Sequence for Optimizing the Parameters

In many cases the choice of plasma gas is possible at the initial stage of the optimization. Argon, helium or nitrogen are the most common plasma gases for MWPs. Additionally, doping the plasma gas with small amounts of oxygen, hydrogen or nitrogen can be considered for specific applications in order to improve atomization efficiency for analyte-containing molecules, to reduce background emission or to reduce carbon deposition on the discharge tube walls when introducing organic samples.[12–17] Once the plasma gas is selected the plasma impedance matching can be performed under experimental conditions similar to those applied in the optimized analytical procedure, including the operation of the sample introduction system. MIP sources are tuned to the minimum reflected power.

One of the parameters optimized each time when working out the conditions of analysis is the viewing position, *i.e.* selection of the plasma area from which radiation is directed to the optical system of the spectrometer. This is connected with the fact that various elements show maximum emission in various plasma regions. For radial viewing the observation position is related to the plasma height from which the radiation is detected, while for axial viewing this is a selected position along the diameter of the discharge tube.[12,18–24] Sometimes, during multi-element determination, it is necessary to establish a compromise viewing position.

Spatial emission distribution has been studied for different plasma gases and different shapes of discharge tube. Differences were found in the emission distribution, depending on the location of the viewing position with respect to the discharge tube axis and also depending on the measurement conditions, including the matrix composition. In the case of an argon toroidal plasma, the spatial emission distribution of various elements is little dependent on the location of the viewing position, contrary to the plasma of a filament shape. From the three elements considered, only calcium reaches a maximum of emission on-axis of a filament Ar-MIP sustained in the TE_{101} cavity, whereas for aluminum and manganese the maxima occur in the plasma surrounding zone, although considerable on-axis emission intensity is also observed.[25] Studies of matrix effects caused by the presence of an element of low ionization potential, *e.g.* sodium, show that its presence in plasma causes, for many elements, an increase in emission intensity and a shift of the emission maximum towards the plasma centre as a result of various possible mechanisms.[25–27]

There are several features common to the MWP-OES methodology used in any analysis. Sample preparation, instrument calibration and wavelength selection are part of every MWP-OES analysis. For some analyses, the analyst must also consider interference corrections and optimization of the operating parameters. Before making analytical measurements using the instrument, the analyst should take the necessary steps to determine that the instrument is set up and functioning properly.

The sample preparation procedures for non-routine samples include determination of approximate detection limits for preselected lines, checking for spectral interferences from suspected concomitants and checking for blanks with the use of synthetic solutions, computing a dilution factor, when necessary, for minimizing the salt content in a sample introduced when the expected analyte concentration is possibly larger than 100 times the detection limit. The instrument calibration consists of measuring a few calibration standards in the concentration range that includes the expected analyte concentration in the sample (the acid concentration should be matched as well as sample matrix when necessary).

The preliminary steps for setting up instrument parameters include selection of a suitable pre-integration time as well as an integration time for calibration and sample analysis, and adjustment of detectors for expected signals.

The selection of the most suitable analytical line is a key part of MW-OES. The choice of appropriate wavelengths for trace analysis is sometimes difficult owing to the lack of truly comprehensive tables of MWP line intensities. The basic criterion is that the wavelengths be free from spectral interferences. When freedom from spectral interferences is not possible, lines whose emission intensities can be corrected for spectral interferences should be used. The use of several spectral lines for each element will probably also gain in interest. In some cases, more than one line can be chosen, to cover different composition ranges or use in different matrices. Careful line selection will lead to linear calibration curves with little scatter and low detection limits.

The optimization procedure for MWP-OES system equipped with a dual-flow torch usually concerns four operating conditions essential for the overall analytical performance of the method, namely the applied power level, the carrier gas flow rate, the plasma gas flow rate and the sample flow rate. The key parameters for assessing the suitability of the operating conditions are the signal-to background ratio (SBR) or the signal-to-noise ratio (SNR). The "one variable at a time" method is very popular although it suffers from some limitations, including dependence of the final result on the choice of starting conditions and the lack of information about interrelations between variables. Simplex optimization is generally considered the most efficient way to find the best conditions if there are several interrelated variables affecting the operation of the MWP and the analytical performance of the system.

The simplex or multi-simplex methodology has been used for optimization of a number MWP-OES systems and analytical procedures.[3–8,28–31] SNRs for chlorine and bromine as key parameters and factorial design and response surface methodology were used for optimization of the GC-MIP-OES system for the determination of halogenated organic compounds.[3] The results were compared to work done by others using the "one variable at a time" method. Although the final detection limits were similar for both optimization methods, the factorial analysis and response surface method proved to be faster and more reliable. For optimization of the multistage determination of mercury, including cold vapour generation, trapping and MIP-OES determination, both univariate and simplex methods as well as two-level full and fractional factorial design were applied. Separate optimization procedures were carried out for each stage.[5]

9.3 Relation between Analytical Signal and Aerosol (Sample) Parameters

Theoretical considerations on the relation between the signal intensity and the analyte concentration in the sample have been widely discussed, including by

Winefordener's group.[32,33] The relationship between the efficiency of the measurement (ε_m) and the efficiency of detection (ε_d) for a given spectrometric technique can be estimated. The efficiency of the measurement is defined as the probability that a given atom in the sample is detected above the background noise in the observation region. This relationship is given by:

$$\varepsilon_m = \varepsilon_{SR}\,\varepsilon_s\,\varepsilon_t\,\varepsilon_d$$

where ε_{SR} is the sample-related overall efficiency, including the efficiencies of sample nebulization, aerosol transport, vaporization, atomization or ionization and excitation processes, ε_s is the spatial probing efficiency and ε_t is the temporal probing efficiency.

The ε_m and ε_d values for atomic spectroscopic methods are the means of comparing the potential detection power of individual methods and of determining whether any given atomic method has achieved its maximum capability. It should be noted that the efficiency of measurement is also identical to the sensitivity of measurement.

For the case of MWP-OES, the ε_m estimated by Winefordner *et al.*[32] for hypothetical atomic resonance lines at 200, 500 and 800 nm ranged from 5×10^{-6} to 10^{-9} and is comparable for glow discharge (GD)-, MWP- and ICP-OES. This means that approximately 10^5–10^9 atoms at 5000 K will produce 1 count, assuming residence times of atoms within the plasma is about 1 ms. However, for low-flow MWPs the residence time may be even 5–10 times better, leading to improvement of the ε_m value.

The dependence of SBR on the aerosol generation and transport efficiency was shown for three nebulizers coupled to MIP-OES by Jankowski *et al.*[1,34] One can see from Figure 9.1. that the measurement sensitivity changes are closely related to changes of the aerosol concentration with an increase in carrier gas flow rate (influencing the residence time of the sample in the plasma) and a certain optimal range of aerosol concentration occurs at which the maximum SBR is achieved.

9.4 Optimizing Plasma Parameters for Trace Analysis

During a standard optimization procedure the operating conditions are determined on the basis of spectroscopic measurements. Sometimes, as a criterion the emission intensity of the introduced element is taken into account. A better choice is the SBR, which is directly connected to the measurement sensitivity. This is of particular importance when low analyte concentrations are determined. It is good practice to minimize the background signal by optimization of the instrumental parameters after choice of the excitation source, the plasma gas and the appropriate sample introduction system, where possible.

It has been shown that SBR levels may be maximized by optimization of certain physical parameters, such as discharge tube dimensions and the geometry of the viewed image.[19] For a helium MIP the SBR is constant for any

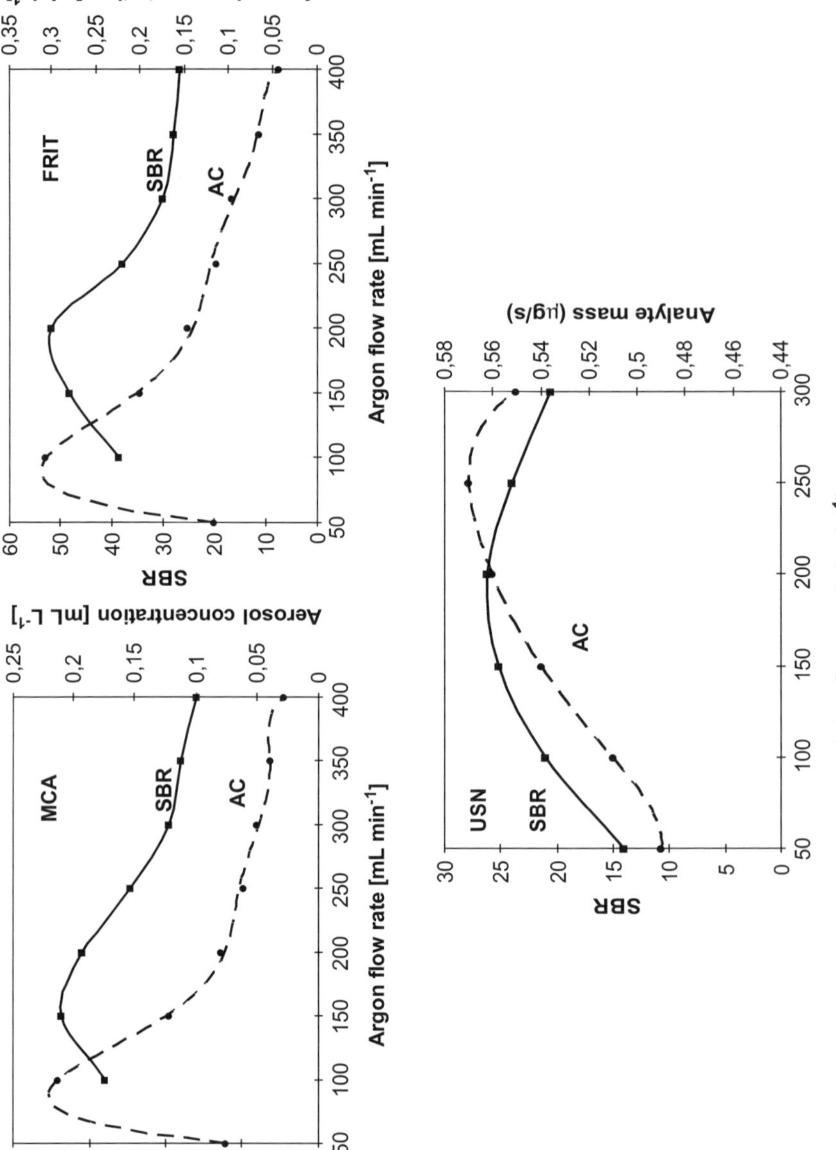

Figure 9.1 Correlation between SBR by MIP-OES and aerosol concentration for three nebulizers: MCA - multicapillary array nebulizer, FRIT - Frit nebulizer, USN - ultrasonic nebulizer.

viewing position along the diameter of the discharge tube. Pulse operation of the MIP has been reported to improve the SBR.[35] The plasma gas flow rate is the most important parameter determining the SBR.[11]

By using an oxygen-sheathed Ar-MPT-OES, the background emission and noise are significantly reduced, while the emission intensities of tested elemental lines increase or decrease slightly in comparison with a pure argon MPT-OES.[16] The improvements of SBR and SNR were both very significant. As a result, the detection limit and determination precision were improved.

The SNR is a quantitative measure of the influence of noise on the analysis and is readily related to the precision and the limit of detection. An identification and characterization of the factors that influence the SNR, their effects and magnitudes, allows improved results to be obtained by rational optimization of an analytical method. Goode and Kimbrough[36,37] discussed factors influencing the SNR in a microwave plasma detector for gas chromatography. Factors examined in this study included the various gas flow rates, optical parameters of the monochromator and the electronic measurement parameters, including those used in a wavelength modulation system. Also, data smoothing has been shown as a mean for SNR enhancement.[38]

The analytical capabilities of MWP sources depend not only on the characteristics of the plasma gas and the operating conditions, but also on the coupling device used for plasma generation. Important progress has been made in this last factor since the first resonant cavities were developed, as discussed in Chapters 2 and 3.

9.5 Instrument Tests

Improvements in the accuracy of measurements by MWP-OES can be achieved by careful control of certain instrumental parameters. Besides preparing samples carefully and selecting the proper components of the instrumental setup and operating parameters, the operator should make sure that the instrument is maintained properly. The major components of a MWP-OES system comprise the excitation source, the sample introduction device, the optical system and the detection system. Each part of the whole MWP-OES system can degrade the analytical performance.

The number of emission counts for a given line of a plasma gas component (Ar, He, H, OH) is useful for trend analysis. It should be noted that the user should generally wait for an MWP-OES system to warm up properly before performing these tests, typically 20–30 min. Then, the measurement can be used for drift diagnostics.

The improper performance of both a sample introduction system and microwave coupling are the main causes of poor precision and accuracy for an MWP-OES analysis. In general, the analyst should keep in mind that anything that prevents a stable operation can be a source of error for analyses. Keeping the discharge tube and sample introduction system clean is important in assuring a stable flow of sample to the plasma. If samples with a complex matrix have been run, it is necessary to take the system apart for a more thorough cleaning by running a blank solution for several minutes. The impedance matching should be checked from day to day.

Some general guidelines regarding instrument maintenance and performance verification may be helpful in assuring that an MWP-OES instrument will provide acceptable analytical results. The analytical performance characteristics that are of concern are the drift, the precision, the accuracy, the long-term stability, the selectivity expressed by practical resolution, the robustness, the background equivalent concentration and the linear dynamic range. This information can be extracted from measurements of line profiles, line intensities, the relative standard deviation of the signal or background, the SBRs and the ionic to atomic line intensity ratios.

A set of general tests which have been proposed to verify the performance of an ICP-OES instrument can be applied to MWP-OES as well.[39–43] Some simple tests can be performed on a daily basis. Other testing procedures are more tedious and are used occasionally as diagnostic tests. These experiments allow verification of practical resolution, efficiency of ionization and excitation, light absorption, *etc.* For checking the practical resolution of the spectrometer coupled with a MWP system, the Cd(II) 228 nm and Ba(II) 455 nm lines can be used for the UV and Vis spectral ranges, respectively.[42]

Plasma robustness may be verified by the Mg(II) 279/Mg(I) 285 nm line ratio or the broadening of the H_β line. The Mg atom-to-ion line ratio provides an indirect estimate of the efficiency of atomization and ionization.[43] The value of this ratio depends on the sample composition and sample introduction system used. For Ar- and He-MIPs sustained in a TE_{101} cavity and for 3 sample introduction systems examined (solution nebulization with and without desolvation, and continuous powder introduction), the ratio varies as 2.5-5-, 1.6-3 and 3-19, respectively.[44] In general, for MIP-OES the value of the ratio is lower than that for ICP-OES; however, the value obtained for the CPI of a carbon matrix sample is close to that obtained for laser ablation (LA)-ICP-OES.[45] Interestingly, for He-MIP-OES sustained in the three-phase rotating field system and solution nebulization without desolvation, the ratio exceeds 4.5.[46]

References

1. K. Jankowski, D. Karmasz, A. Ramsza and E. Reszke, *Spectrochim. Acta, Part B*, 1997, **52**, 1813–1823.
2. A. J. McCormack, S. C. Tong and W. D. Cooke, *Anal. Chem.*, 1965, **37**, 1470–1476.
3. M. Caetano, R. E. Golding and E. A. Key, *J. Anal. At. Spectrom.*, 1992, **7**, 1007–1011.
4. J. Sanz, A. De Diego, J. C. Raposo and J. M. Madariaga, *Anal. Chim. Acta*, 2003, **486**, 255–267.

5. M. Murillo, N. Carrion, J. Chirinos, A. Gamiero and E. Fassano, *Talanta*, 2001, **54**, 389–395.

6. J. J. Urh and J. W. Carnahan, *Anal. Chem.*, **57**, 1253–1255.

7. H. Matusiewicz, M. Ślachciński, M. Hidalgo and A. Canals, *J. Anal. At. Spectrom.*, 2007, **22**, 1174–1178.

8. H. Matusiewicz, B. Golik and A. Suszka, *Chem. Anal. (Warsaw)*, 1999, **44**, 559–566.

9. A. Bollo-Kamara and E. G. Codding, *Spectrochim. Acta, Part B*, 1981, **36**, 973–982.

10. M. Wu and J. W. Carnahan, *Appl. Spectrosc.*, 1992, **46**, 163–168.

11. K. Jankowski, A. Jackowska, A. P. Ramsza and E. Reszke, *J. Anal. At. Spectrom.*, 2008, **23**, 1234–1238.

12. G. Heltai, J. A. C. Broekaert, F. Leis and G. Tölg, *Spectrochim. Acta, Part B*, 1990, **45**, 301–311.

13. H. Matusiewicz, *Spectrochim. Acta, Part B*, 1992, **47**, 1221–1227.

14. M. Ohata, H. Ota, M. Fushimi and N. Furuta, *Spectrochim. Acta, Part B*, 2000, **55**, 1551–1564.

15. K. G. Michlewicz, J. J. Urh and J. W. Carnahan, *Spectrochim. Acta, Part B*, 1985, **40**, 493–499.

16. Q. Jin, W. Yang, F. Liang, H. Zhang, A. Yu, Y. Cao, J. Zhou and B. Xu, *J. Anal. At. Spectrom.*, 1998, **13**, 377–384.

17. T. H. Risby and Y. Talmi, *CRC Crit. Rev. Anal. Chem.*, 1983, **14**, 231–265.

18. S. R. Goode and J. N. Emily, *Spectrochim. Acta, Part B*, 1994, **49**, 31–45.

19. M. Selby, R. Rezaaiyaan and G. M. Hieftje, *Appl. Spectrosc.*, 1987, **41**, 749–761.

20. M. Ohata and N. Furuta, *J. Anal. At. Spectrom.*, 1997, **12**, 341–347.

21. H. Tanabe, H. Haraguchi and K. Fuwa, *Spectrochim. Acta, Part B*, 1983, **38**, 49–60.

22. D. Kollotzek, P. Tschöpel and G. Tölg, *Spectrochim. Acta, Part B*, 1982, **37**, 91–96.

23. Y. N. Pak and S. R. Koirtyohann, *Appl. Spectrosc.*, 1991, **45**, 1132–1142.

24. L. J. Galante, M. Selby, D. R. Luffer, G. M. Hieftje and M. Novotny, *Anal. Chem.*, 1988, **60**, 1370–1376.

25. K. Jankowski, *Spectrochim. Acta, Part B*, 2002, **57**, 853–863.

26. J. P. Matousek, B. J. Orr and M. Selby, *Spectrochim. Acta, Part B*, 1986, **41**, 415–419.

27. M. Selby, R. Rezaaiyaan and G. M. Hieftje, *Appl. Spectrosc.*, 1987, **41**, 761–771.

28. G. M. Greenway and N. W. Barnett, *J. Anal. At. Spectrom.*, 1989, **4**, 783–787.

29. A. Delgado, A. Usobiaga, A. Prieto, O. Zuloaga, A. de Diego and J. M. Madariaga, *J. Sep. Sci.*, 2008, **31**, 768–774.

30. A. M. Gonzalez and P. C. Uden, *J. Chromatogr.*, 2000, **898**, 201–210.

31. H. Matusiewicz and M. Ślachciński, *Microchem. J.*, 2006, **82**, 78–85.

32. J. D. Winefordner, E. P. Wagner II and B. W. Smith, *J. Anal. At. Spectrom.*, 1996, **11**, 689–702.

33. J. D. Winefordner, G. A. Petrucci, C. L. Stevenson and B. W. Smith, *J. Anal. At. Spectrom.*, 1994, **9**, 131–143.
34. K. Jankowski, D. Karmasz, L. Starski, A. Ramsza and A. Waszkiewicz, *Spectrochim. Acta, Part B*, 1997, **52**, 1801–1812.
35. M. M. Mohamed and Z. F. Ghatass, *Fresenius' J. Anal. Chem.*, 2000, **368**, 449–455.
36. S. R. Goode and L. K. Kimbrough, *Spectrochim. Acta, Part B*, 1987, **42**, 309–322.
37. S. R. Goode and L. K. Kimbrough, *J. Anal. At. Spectrom.*, 1988, **3**, 915–918.
38. K. J. Mulligan, M. Zerezhgi and J. A. Caruso, *Spectrochim. Acta, Part B*, 1983, **38**, 369–375.
39. J. M. Mermet and E. Poussel, *Appl. Spectrosc.*, 1995, **49**, 12A–18A.
40. E. Poussel, J. M. Mermet and O. Samuel, *Spectrochim. Acta, Part B*, 1993, **48**, 743–755.
41. M. Carre, E. Poussel and J. M. Mermet, *J. Anal. At. Spectrom.*, 1992, **7**, 791–797.
42. J. M. Mermet, *J. Anal. At. Spectrom.*, 1987, **2**, 681–686.
43. J. M. Mermet, *Anal. Chim. Acta*, 1991, **250**, 85–94.
44. A. Jackowska, PhD Thesis, Warsaw University of Technology, 2006 (in Polish).
45. A. C. Ciocan, X. L. Mao, O. V. Borisov and R. E. Russo, *Spectrochim. Acta, Part B*, 1998, **53**, 463–470.
46. K. Jankowski, A. P. Ramsza, E. Reszke and M. Strzelec, *J. Anal. At. Spectrom.*, 2010, **25**, 44–47.

Analytical Performance of MWP-OES

10.1 Introduction

The plasma techniques of optical emission spectrometry (OES), including the microwave plasma (MWP) technique, have a number of advantages due to which they belong to the most popular analytical methods. When utilizing a MWP as an excitation source coupled with solution nebulization, over 70 elements can be determined at the trace level [detection limits (DLs) from 0.1 to 4000 ppb, exceptionally].[1–6] A unique feature of a helium plasma is the ability to detect non-metals at the picogram level of absolute DL.

MWPs coupled with analyte vapour-generating techniques lead to matrix-free analyte excitation, but as a result of adding trapping techniques very highly selective and sensitive determination methods can be realized. The low DLs for metals in the form of their volatile species and for non-metals in the case of helium discharges, together with the low cost of operation, makes MWPs especially attractive. Besides being able to determine a large number of elements over a wide range of concentrations, a major advantage of the plasma OES technique is that many elements can be determined easily in the same analytical run. Moreover, this technique is characterized by high precision and accuracy of measurement, as well as a wide linear dynamic range, which facilitates fitting of the analyte concentration in the sample to the measurement range. When applying MWP-OES, relatively small spectral interferences are also observed.

A linearity of response from 2.5 to 5 orders of magnitude can be expected, depending upon the element to be determined and the particular plasma configuration. The wide linear dynamic range (LDR) is an important feature of the plasma OES method. In the case of ICPs, it can reach even 5 orders of magnitude. In MWPs, the LDR is somewhat smaller, typically 3–4 orders of

---

RSC Analytical Spectroscopy Monographs No. 12
Microwave Induced Plasma Analytical Spectrometry
By Krzysztof J. Jankowski and Edward Reszke
© The Royal Society of Chemistry 2011
Published by the Royal Society of Chemistry, www.rsc.org

magnitude.[1–6] Ng i Shen[7] obtained a linear dependence of the $\log I = f \log C$ relationship for several elements in the 0.1–$100\,\text{mg}\,\text{L}^{-1}$ range. Similar results were obtained when determining analytical curves for several elements in groundwater.[8] However, for some elements a relatively narrow linear range (about two orders of magnitude) is obtained.[7–9] In the case of alkali metals, the upper linearity range is limited by the occurrence of self-absorption. For the GC-MIP method, and other methods in which the sample is introduced to plasma in a gaseous form, linearity for the calibration curve can be achieved for even five orders of magnitude.

The precision and accuracy of the MWP-OES system are considered sufficient for most trace elemental analyses. Even in the presence of interferences, modern signal compensation techniques allow the analyst to perform analyses with remarkable accuracy. Typical values of precision thus obtained for MWP-OES with solution nebulization, at concentrations well away from the detection limit, are between 1 and 4%.[1–6] The precision of MWP-OES is usually determined by taking multiple measurements of the analyte concentration during the continuous nebulization of a dilute, synthetic solution. Better precision can be obtained through the use of longer integration times and signal processing techniques.

10.2   Interferences in MWP-OES

In OES, the interferences caused by the matrix can be divided into two categories: spectral interferences and non-spectral interferences. Spectral interferences can be a problem with any radiation source, in particular when spectrometers with low resolving power are used. Fortunately, argon and helium MWPs produce a relatively simple and less intense background emission compared to other plasmas. The most important are the spectral interferences from continuum, molecular and atomic emissions, which are generally more significant in the Vis than in the UV region.

Spectral interferences arise mainly from impurities in the plasma gases and from atmospheric species (CO_2, H_2O, N_2). Interferences from molecular bands and line structures of species such as OH radicals and N_2 molecules are well known. To differentiate these molecular species from the atomic species of interest, the use a high-resolution spectrometer is required. This is discussed in Chapter 5.

Different categories of background interference encountered in MWP-OES include the direct wavelength coincidence from an interfering emission line or band, the wing overlap or line-broadening type of partial overlapping of the line of interest with an interfering line situated nearby, and the presence of an elevated or depressed background continuum. A molecular band structure may arise from components in the analytical sample or may be part of the background structure under the analyte emissions. Each of these interferences has its own correction techniques similar to those applied in other OES methods. Elimination of the wing overlap interference is usually only possible by improvement in the resolution. Mathematical models can be used to correct for this type of interference. The only certain solution, however, is to select an interference-free wavelength for the selected element. The line-broadening type of partial overlapping interference can be corrected for by measurement of the background on either side of the wavelength of interest.

10.2.1 Non-spectral Interferences in Microwave Plasmas

Although a MWP offers many advantages as a source for OES, there are several obvious limitations. For example, a low-power MWP suffers from substantial matrix interferences and is strongly affected by sample loading. The non-spectral interferences include chemical interferences and physical interferences. They can occur in the plasma, or during the process of sample nebulization, desolvation and transportation. Basically, the matrix effects in an MWP-OES are similar to those observed in other plasma OES techniques. Physical interferences occurring in the liquid sampling systems include nebulizer interferences and desolvation interferences. Nebulizer interferences are related to changes in the aerosol generation efficiency owing to differences in the volatility or surface tension of the solution. Pneumatic nebulization can be affected by the high salt content of an aqueous sample, which affects the nebulization efficiency and hence the sensitivity of the determination. Desolvation interferences occur due to the use of desolvation devices and are related to solvent vaporization.

The quenching effect is one of the physical interferences that is commonly encountered with MWP sources. This effect depends somewhat on the nature of the injected species, and is especially pronounced for water and molecular species due to the large quantity of microwave energy dissipated in decomposing these species. Obviously, such a limitation constrains the sample size used with a MWP. The presence of water causes a decrease in the analyte emission significantly, irrespective of the type of MWP in use. If the sample matrix is composed of high volatile compounds, then this affects the vaporization rate and analyte excitation efficiency.[10,11] This is especially true with halogen ionic emission, which can be readily observed when applying a desolvation system.[12]

Chemical interferences are connected with the fact that the analyte vaporization, atomization and excitation processes are accompanied by chemical reactions involving other sample components.[1,2,13] Stable-compound formation interferences are defined as interferences due to the formation of stable species occurring in the plasma that degrade the atomization efficiency of the element being determined. The decrease in calcium emission in the presence of phosphates[14,15] or aluminum[16] are classical examples in OES. The emission from calcium in a MWP has been shown to be affected only slightly or not at all by the presence of phosphate, whereas other anionic species have no effect. In the case of MWPs, this kind of interference can be eliminated by the addition of strontium as a spectroscopic buffer.[17] The formation of stable oxides of the analyte also appears not to be significant for MWPs.[18–20]

Ionization interferences are usually related to matrix effects caused by easily ionizable elements (EIEs).They have been extensively studied for various MWPs and sample introduction techniques.[1–3,13,17–20] In general, the presence of matrix EIEs causes a shift of the ionization equilibrium, whose magnitude depends on the ionization potential of both the analyte and the EIE as well as the wavelength selected (atomic or ionic). For a solution nebulization (SN)-MWP-OES system operating with or without desolvation, large discrepancies between ionization interferences have been observed by various investigators, indicating the complexity of this phenomenon. The processes occurring in MIPs were studied by measurement of the emission of selected 35 elements, drawing attention especially to the character and magnitude of the matrix effects. The results obtained were interpreted on the basis of the spectro-chemical parameters of the elements and probable mechanisms of their exci-tation.[18] In order to quantify the effect the signal enhancement factor was introduced, being a ratio of the maximum signal obtained for a given element over the whole EIE concentration range studied for the signal measured for the solution without matrix addition. Most of the atomic lines are enhanced in the presence of sodium, whereas all of the ionic lines tested are reduced. This can be explained by the shift of the ionization equilibrium. Four groups of elements were selected regarding differences in the effect observed. Only for the B, V, Mo, Ti and Zr signals was a decrease observed, with an increase in the dis-sociation energy of the metal–oxygen bond for particular elements. The pre-sence of refractory elements in a sample matrix sometimes leads to both vaporization and excitation interference problems. This study concluded that atomic lines observed in the MIP can be divided into *soft* and *hard* lines, similar to ICP-OES,[21] depending on the value of the enhancement factor measured.

The presence of the EIE causes a strong enhancement of the analyte emission in the Beenakker cavity system, but has a much smaller influence on the analyte emission in the surfatron system even at higher concentrations of EIEs. For the Beenakker cavity source, in the absence of EIEs the analyte does not appear to penetrate well into the hot region of the plasma. In the case of the surfatron, the central plasma channel appears to allow greater penetration of sample material into the plasma even in the absence of EIEs.[22] A reduced matrix effect for microwave plasma torches (MPTs) has been found in comparison with the surfatron. However, owing to the low power operation, it was still severe.

Another common interference in MWPs is the pronounced enhancement of analyte emission when alkali halide salts are present.[23,24] If alkali metals are introduced to the plasma in the form of chlorides, then an additional effect of increase in emission is observed, connected with the formation of volatile chlorides of the elements being determined.[18,25] A very high increase in emis-sion for some elements (1000-fold for Mn)[24] can be observed, especially when thermal vaporization is used as the method of sample introduction to the MIP.

To explain the nature of matrix effects caused by elements of low ionization potential, several reaction mechanisms in plasma have been suggested: a shift of the ionization equilibrium, enhanced collisional excitation, increased analyte penetration,[26] a solvent vaporization effect, an effect of a concomitant anion, and others.[13,27–30] Attention has been drawn also to the effect of the matrix on the nebulization yield and aerosol transport to the plasma.[31,32] Generally, none of the mechanisms explain sufficiently the phenomena observed, which indicates a considerable effect of several factors on the excitation conditions.

The presence of EIEs in plasma causes not only the occurrence of chemical interferences but also changes in the spatial emission distribution over the discharge tube cross-section, connected with increased analyte penetration to the plasma or ambipolar diffusion.[13,22,33–38] By addition of EIEs, a decrease in argon emission intensity on moving away from the plasma centre is much faster than that for the analyte, which allows minimization of spectral interference by selection of an appropriate location of the viewing position. For the interference originating from the OH band, no similar relationship was found.[38]

Because of improved sample tolerance, the atomic emission from a MPT is not affected significantly by the acidity of the sample.[39] This provides much flexibility and convenience for sample pretreatment, as well as much ruggedness of the determination.

Atsuya *et al.*[25] examined the effect of hydrochloric acid on the emission of manganese and copper, and found that HCl apparently enhances the role of potassium chloride. They argued that potassium chloride permits the volatilization of the sample from the filament as chloride.

A relatively low tolerance of MWP to organic sample loading was discussed in previous chapters. However, when a sufficiently small amount of the sample is introduced to the GC-MIP-OES, it has the advantage of almost uniform elemental response for each halogen, irrespective of analyte molecular structure.[40] Despite quenching effects, carbon deposits on the discharge tube walls might be a problem. When analyzing organic-based solutions, it may be necessary to remove carbon deposits periodically from the discharge tube. Dopant gases (O_2, N_2, H_2) are sometimes added to MWPs, usually acting as scavengers of hydrocarbon species and carbon deposit eliminators.[41,42]

10.3   Calibration Strategies

MWP-OES is a comparative analytical method and therefore needs calibration. External calibration by measuring the emitted intensities for a set of standard solutions with known composition is the most popular technique. In fact, calibration is the most important part of the preparation for analysis as the final analytical result can never be more reliable than the calibration. The relationship is then used to calculate the analyte concentration in an unknown sample on the basis of measured intensities. Some care must be taken when implementing, in order to avoid mistakes.

Usually, the sample is supplied as a liquid solution because of its homogeneity, ease of handling and the possibility of preparing calibration standards. The type of calibration used plays a key role in the minimization of matrix effects. Calibration with synthetic standards without matrix matching often leads to inaccurate results, especially when the concentration of the matrix becomes high. A simple dilution of the samples can be sufficient to reduce the magnitude of the matrix effects, but at the expense of the power of detection. An important concept related to sample preparation and interference correction is that of matrix matching. Matrix matching involves preparing calibration solutions whose major components match those of the sample solution. While matrix matching certainly involves matching the solvents, it also involves matching the concentrations of acids and other major solutes. Unfortunately, the sample composition should be approximately known. When preparing the blank solution to be used in the calibration process, it is recommended that the blank be matrix matched identically with the standard solutions to be used.

The technique of standard additions can be used when the above techniques cannot solve the practical problems. On-line standard addition is more convenient and consumes less amount of sample compared with off-line standard addition, and may be able to eliminate matrix and spectral interferences.[43,44] Internal standardization is rarely applied in MWP-OES and may also be a source of additional errors, even though the background level produced by a MWP is usually low. Hence, the use of standard addition is certainly the most reliable way to correct for non-spectral interferences because of accurate matrix matching. Standard addition, however, is time consuming as it requires the preparation of numerous samples and, in most cases, additional dilutions.

Although well-investigated MWPs can be used for OES, the broad use of the approach for routine work has not been realized up to now. It is possible, in MWP-OES, to obtain linear calibrations over two, three or more orders of magnitude. The standards included in the calibration process should cover the entire composition range that will be used for the analysis. Although extrapolation of calibration curves is possible in MWP-OES, because of its mainly linear calibration curves, and is sometimes necessary owing to a lack of suitable reference materials, extrapolation should be avoided whenever possible because of the increased uncertainties of the results. Calibration functions based on net line signals should be straight lines through the origin. The scatter of the individual measurements about the line of best fit is usually negligible. The precision with which a net line signal is measured for the standard gives a good estimate of the precision with which concentrations are determined.

The existence of a background signal at the analytical wavelength may be detected by analysis of a blank. When this approach is not possible, alternatives include the analysis of reference materials and synthetic samples, care being taken to ensure that the reagents employed are not contaminated.

It is difficult to give a fixed value for the number of calibration standards that is necessary and sufficient for a good calibration. Uncertainties in calibration tend to decrease as the number of samples is increased, but the time and cost of calibration both increase. In fact the number depends on the expected quality and the covered composition ranges. We suggest a minimum of five calibration standards per calibration curve in a standard case, for a linear calibration curve with only two regression parameters (slope and intercept). Because calibration curves are generally linear over three to four orders of magnitude in MWP-OES, it is usually necessary to measure only two standards and a blank, to recalibrate the MWP instrument.

When calibrating for analysis of solid samples with high contents of different elements and when significant interferences are to be expected, the number of samples should be increased. When large composition ranges are to be covered, again the number of calibration samples should be increased. However, in particular cases the matrix effects can be surprisingly similar over a large composition range and can be covered with a single calibration function.[45]

10.4 General Analytical Characteristics of MWP-OES

As indicated above, the MWP-OES technique is applicable to the determination of a large number of elements. The DLs for these elements are generally in the $\mu g\,L^{-1}$ range, with absolute detection abilities as low as the picogram level.

To date, the application of MWPs in atomic spectrometry to practical analyses is still very limited due to the lack of commercial instrumentation available, excluding gas samples, where it has served particularly well as an element-selective detector for gas chromatography.[46–49] The G2350A detector offers as a standard the possibility of determining seven non-metals, with DLs ranging from 1 to $150\,pg\,s^{-1}$ and linear dynamic ranges covering 3–4 orders of magnitude.

In recent years the MPT-OES instrument was developed and then commercialized. The analytical performance of this instrument has been reported by Yang *et al.*[4] The DLs for 56 elements determined with the use of various sampling techniques varied from 0.1 to $83\,\mu g\,L^{-1}$. The LDRs cover 2–4 orders of magnitude and the precision achieved varies typically from 1 to 3% relative standard deviation (RSD). Comparison of the ICP-OES technique with solution nebulization with both Ar-MPT-OES and oxygen-sheathed Ar-MPT-OES revealed that the MPT gave lower detection limits for 8 from 15 and for 5 from 10 elements determined.[50]

A high-power nitrogen MIP based on the Okamoto cavity has been commercialized as an ionization source for mass spectrometry. However, it was examined also as an excitation source. For As, Sb, Se, Bi, Te, Sn and Pb it offers DLs in the range $0.45–3\,mg\,L^{-1}$ and $0.86–102\,\mu g\,L^{-1}$ when coupled with solution nebulization and the hydride generation technique, respectively, and LDRs from 3 to 4 orders of magnitude.[51] However, these DLs are relatively high in comparison with ICP-OES, ICP-MS and AAS.

Recently, an instrument for continuous emission monitoring based on MWPs has been offered for analyzing airborne aerosols, both dry and wet. The EPD1 monitor is able to determine at least 12 elements at the sub-ppb level when introduced as liquid aerosols and, in the case of small solid particles, delivery DLs from 0.088 to $4.3\,\mu g\,m^{-3}$ are attained.[52–54] Two particle size analyzers based on MWPs have been commercialized in Japan; however, the analytical performance was not specified.[55–58]

10.5 Comparison of Different MWP-based Techniques

Many factors contribute to the DL attainable with a MWP-OES system. The nature of the matrix used, the operating pressure, the plasma gas type and flow rate, the applied microwave power and the volume of plasma viewed all have their effect. However, of major importance is the choice of the MWP source and the sample introduction technique. Differences in the measurement sensitivity for MWP-OES techniques are connected with different measurement conditions. The DLs achieved for selected elements by various solution nebulization and MWP-OES systems are listed in Table 10.1.

In early work concerning plasma methods, opinions can be found that a MIP is characterized by the smallest precision of measurement.[62] This finds justification in the instability of the discharge itself, owing to the non-symmetrical microwave coupling. However, the most recent designs of resonators allow a considerable improvement of this state of matter. Studies on the nebulization of solutions show that the quality of the aerosol also has a considerable effect on the background noise. Nebulizers compatible with low-power MIPs (see Chapter 7) enable generation of fine aerosols, due to which the short-term precision is 2–4%.[8] For MPT-OES with solution nebulization, RSD values between 1% and 5% are reported.[4,50] However, CMP-OES usually suffers problems of higher emission backgrounds than other MWP-OES systems and has limited precision (5–10%).[2,59–61]

MWP-OES is a useful and sensitive method of microsample analysis when interfaced with electrothermal vaporization (ETV). In Table 10.2 are presented

Table 10.1 Comparison of detection limits ($ng\,mL^{-1}$) by SN-MWP-OES for various MWP sources.[2,4,8,59–61]

Element	LP-MIP[a]	MP-MIP[b]	MPT side-on	MPT end-on	N_2-MIP	CMP[c]	TE_{101}	TEM
Al	1400	970	5.3	–	12	3	110	18
Ba	180	–	16	5.2	–	3100	110	20
Ca	40	–	0.15	–	3.1	2	3	0.5
Cd	0.7	1.5	6.7	2.5	24	6	5	7
Cr	90	6.0	10	2.4	24	9	50	14
Cu	15	1.8	16	3.3	2.3	90	20	7
Fe	650	6.2	4.7	–	13	12	60	27
Mg	63	3.9	3.1	0.49	2.2	2	7	3
Mn	18	2.1	2.4	–	6.9	–	25	10
Mo	420	–	37	4.1	180	–	500	–
Na	2	–	1.4	–	46	0.05	0.9	1.5
Ni	–	10	39	11	23	0.7	90	28
Pb	139	33	103	16.9	80	10	60	15
Ti	–	–	45	–	4.3	4	70	–
V	91	–	20	3.1	8.8	–	170	–
Zn	420	3.4	7.8	3.0	5.8	5000	15	8

[a]Low-power MIP.
[b]Medium-power MIP.
[c]Capacitive microwave plasma.

Table 10.2 Comparison of detection limits (ng mL^{-1}) by ETV-MWP-OES for various analytical systems.[2,4,63–66]

	ETV-MIP[a]	*TS-MIP*[b]	*TB-MIP*[c]	*GF-TE$_{101}$*[d]	*ETV-MPT*	*DSI-CMP*[e]
Ag	–	–	–	4.0	3.0	12.0
As	120	600	–	120	–	–
Cd	0.2	20	500	8.0	(0.15)	4.0
Cu	30	50	650	24	(0.03)	7.0
Fe	500	–	500	55.0	–	–
Mn	–	–	–	11.0	135	16.0
Pb	–	300	100	56	12.0	12.0
Zn	400	20	400	–	(0.23)	5.0

[a]Electrothermal vaporization MIP.
[b]Tantalum strip MIP.
[c]Tungsten boat MIP.
[d]Graphite furnace TE$_{101}$.
[e]Direct sample insertion CMP.

Table 10.3 Comparison of detection limits (ng mL^{-1}) by HG-MWP-OES for various MWP sources.[2,4,67–69]

	TM$_{010}$	*Surfatron*	*MPT*	*N$_2$-MIP*	*TEM*	*TE$_{101}$*
As	0.32	1.3	5.4	3.0	4.0	1.0
Ge	0.04	–	–	–	7.0	3.0
Hg	0.5	0.9	1.4	–	6.0	2.6
Sb	6.1	0.39	2.5	1.9	2.0	5.0
Se	40	1.2	8.8	0.9	4.0	6.4
Sn	1.4	–	5.9	52.3	–	–

the DLs of some elements achieved for various techniques of electrothermal vaporization.

Coupling of the hydride generation system with a MWP provides a very sensitive spectroscopic method. Several factors have an influence on this: the gaseous sample is introduced to the plasma, which does not overload the plasma, and relatively small plasma gas-flow rates can be used for MWPs, which leads to a small dilution of the sample and prolongs its residence time in the plasma region. Various MWP devices have been examined, but also different HG arrangements that should be taken into account when comparing the DLs showed in Table 10.3.

10.6 Microwave Plasmas *versus* Other Plasma Sources

An evaluation of the analytical performance of the MIP-OES technique in relation to the most popular instrumental methods can be found in the reviews of Broekaert and Tölg.[70,71] A general comparison of various spectroscopic techniques has been made by Winefordner *et al.*[72] Compared with other types of plasma sources, the MWP offers some attractive characteristics, such as its unique features of high excitation efficiency for metal and non-metal elements, the capability of working with various gases, simplicity, and low cost for instrumentation and maintenance. The MWP is a powerful alternative source for elemental determination, and has been extensively applied in analytical atomic spectrometry. A MWP shows analytical possibilities similar to an ICP and, owing to low exploitation costs, can be a technique competitive to ICPs. A somewhat different character of sample excitation mechanisms in MWPs provides favourable conditions for determining non-metals such as nitrogen, chlorine or fluorine. In addition, the MWP has the capability of exciting the halogens well enough so that they can be quantified at the $\mu g\,mL^{-1}$ level at their UV and Vis lines.

In Table 10.4 are presented exemplary DLs for selected elements, obtained when analyzing aqueous solutions by applying nebulization as the technique of sample introduction and different kinds of plasma as the excitation source and, additionally, flame atomic absorption spectrometry (FAAS). It is generally assumed that in plasma emission techniques the best detection of elements is obtained for ICP methods. The remaining two types of plasma, MIP and direct current plasma (DCP), allow detection with somewhat smaller sensitivity, whereas FAAS is in many cases decisively the least sensitive.

More detailed studies have shown that MWP-OES successfully competes with ICP-OES. For a number of elements the MWP has shown superior performance

Table 10.4 Comparison of detection limits (ng mL^{-1}) by various spectroscopic techniques.[2,62,73–75]

Element	MIP-OES	ICP-OES	DCP-OES	MINDAP-OES[a]	FAAS
Ag	10	7	–	–	2
Al	60	23	15	13	20
As	30	53	–	–	150
B	10	4.8	10	–	6000
Ba	100	1.3	–	15	10
Ca	10	0.2	0,5	1.2	1
Cd	0.12	2.5	–	1.7	2
Cr	25	6	1	–	3
Cu	9	5.4	1	–	1
Fe	8	4.6	5	280	10
K	2	–	10	5.4	1
Mg	0.6	0.15	0.1	13	0.1
Mn	6	1.4	2	–	2
Na	1.8	29	–	0.29	0.2
Ni	35	10	1	5.3	2
P	90	76	100	–	–
Pb	10	42	13	84	10
Ti	6	3,8	–	–	90
V	80	5	–	47	20
Zn	10	1.8	4	120	1

[a]Microwave-induced Nitrogen Discharge at Atmospheric Pressure OES.

Table 10.5 Comparison of absolute detection limits (pg) by various spectrometric techniques coupled with ETV.[74,76,77]

Element	ETV-MIP-OES	ETV-ICP-OES	GF-AAS
Ag	1.6	1.0	6.0
Al	13	7.5	60
As	120	200	60
Br	100	n.d.	n.d.
Ca	1.0	0.002	45
Cd	2.8	1.0	1.8
Cu	0.13	0.35	45
Fe	10	20	300
Hg	0.5	10	–
K	0.4	550	15
Mg	0.3	0.01	1.0
Ni	100	4.5	150
P	660	100	15
Pb	0.6	4.0	0.7
Se	16	6.0	20
Ti	18	6.0	60 000
Zn	0.4	0.25	0.1

to the argon ICP.[8,50] Comparing nitrogen high-power MIP-OES with argon ICP-OES, Ohata and Furuta[61] found that the ICP provides equivalent or superior results. Similar detection power has been shown for Al, Cu, Pb and Sn. However, an N_2-MIP atomic mass spectrometer working with an Okamoto cavity offers comparable analytical performance to an argon ICP-AMS.

A comparison of the analytical performance of the ETV technique coupled with common spectrometric techniques is presented in Table 10.5. In general, the MIP method provides equivalent DLs or better for some elements, including non-metals (here represented by Br).

Matusiewicz and Sturgeon[78] determined hydride-forming elements by trapping and vaporization in a graphite furnace (GF) with subsequent detection by plasma emission spectrometry or atomic absorption spectrometry. The DLs of elements for the GF-MIP-OES method are slightly worse than those achieved when using the ICP plasma or the GF-AAS method.

Although an ICP source can provide good sensitivity for elemental analysis, it is very difficult to integrate an ICP into a small, low-cost instrument because of its complex structure and relatively large heavy-power system, as well as its high flow rates of supporting gases. MWPs can be sustained at fairly low power and low gas flow rates, making them a desirable source for on-site, real-time waste stream monitoring.

References

1. A. E. Croslyn, B. W. Smith and J. D. Winefordner, *CRC Crit. Rev. Anal. Chem.*, 1997, **27**, 199–255.

2. Q. Jin, Y. Duan and J. A. Olivares, *Spectrochim. Acta, Part B*, 1997, **52**, 131–161.
3. J. A. C. Broekaert and U. Engel, in *Encyclopedia of Analytical Chemistry*, ed. R.A. Meyers, Wiley, Chichester, 2000, pp. 9613–9667.
4. W. Yang, H. Zhang, A. Yu and Q. Jin, *Microchem. J*, 2000, **66**, 147–170.
5. K. Jankowski, *Chem. Anal. (Warsaw)*, 2001, **46**, 305–327.
6. K. Wagatsuma, *Appl. Spectrosc. Rev.*, 2005, **40**, 229–243.
7. K. C. Ng and W. L. Shen, *Anal. Chem.*, 1986, **58**, 2084–2087.
8. K. Jankowski, *J. Anal. At. Spectrom.*, 1999, **14**, 1419–1423.
9. P. G. Brown, D. L. Haas, J. M. Workman, J. A. Caruso and F. L. Fricke, *Anal. Chem.*, 1987, **59**, 1433–1436.
10. L. R. Layman and G. M. Hieftje, *Anal. Chem.*, 1975, **47**, 194–202.
11. R. K. Skogerboe and G. N. Coleman, *Anal. Chem.*, 1976, **48**, 611A–622A.
12. Y. Okamoto, *Jpn. J. Appl. Phys.*, 1999, **38**, L338–L341.
13. J. P. Matousek, B. J. Orr and M. Selby, *Spectrochim. Acta, Part B*, 1986, **41**, 415–429.
14. G. L. Long and L. D. Perkins, *Appl. Spectrosc.*, 1987, **41**, 980–985.
15. C. I. M. Beenakker, B. Bosman and P. W. J. M. Boumans, *Spectrochim. Acta, Part B*, 1978, **33**, 373–381.
16. G. F. Larson and V. A. Vassel, *Anal. Chem.*, 1976, **48**, 1161–1166.
17. J. O. Berman and K. Bostrum, *Anal. Chem.*, 1979, **51**, 516–520.
18. K. Jankowski and M. Dreger, *J. Anal. At. Spectrom.*, 2000, **15**, 269–276.
19. J. J. Urh and J. W. Carnahan, *Anal. Chem.*, 1985, **57**, 1253–1255.
20. Z. Zhang and K. Wagatsuma, *Spectrochim. Acta, Part B*, 2002, **57**, 1247–1257.
21. M. W. Blades and G. Horlick, *Spectrochim. Acta, Part B*, 1981, **36**, 861–883.
22. M. Selby, R. Rezaaiyaan and G. M. Hieftje, *Appl. Spectrosc.*, 1987, **41**, 761–771.
23. J. F. Alder and M. T. C. da Cuhna, *Can. J. Spectrosc.*, 1980, **25**, 32.
24. H. Kawaguchi and B. L. Vallee, *Anal. Chem.*, 1975, **47**, 1029–1034.
25. I. Atsuya, H. Kawaguchi, C. Veillon and B. L. Vallee, *Anal. Chem.*, 1977, **49**, 1489–1491.
26. D. D. Nygaard and T. R. Gilbert, *Appl. Spectrosc.*, 1981, **35**, 52–56.
27. M. W. Blades and G. Horlick, *Spectrochim. Acta, Part B*, 1981, **36**, 881–900.
28. M. R. Tripković and I. D. Holclajtner-Antunović, *J. Anal. At. Spectrom.*, 1993, **8**, 349–359.
29. G. D. Rayson and G. M. Hieftje, *Spectrochim. Acta, Part B*, 1986, **41**, 683–697.
30. M. H. Miller, D. Eastwood and M. S. Hendrick, *Spectrochim. Acta, Part B*, 1984, **39**, 13–56.
31. R. K. Skogerboe and K. W. Olson, *Appl. Spectrosc.*, 1978, **32**, 181–187.
32. K. O'Hanlon, L. Ebdon and M. Foulkes, *J. Anal. At. Spectrom.*, 1997, **12**, 329–331.
33. K. Tanabe, H. Haraguchi and K. Fuwa, *Spectrochim. Acta, Part B*, 1983, **38**, 49–60.
34. M. Ohata and N. Furuta, *J. Anal. At. Spectrom.*, 1997, **12**, 341–347.

35. Y. N. Pak and S. R. Koirtyohann, *Appl. Spectrosc.*, 1991, **45**, 1132–1142.
36. M. Selby, R. Rezaaiyaan and G. M. Hieftje, *Appl. Spectrosc.*, 1987, **41**, 749–761.
37. J. P. Matousek, B. J. Orr and M. Selby, *Spectrochim. Acta, Part B*, 1984, **38**, 231–239.
38. K. Jankowski, *Spectrochim. Acta, Part B*, 2002, **57**, 853–863.
39. Q. Jin, H. Zhang, Y. Wang, X. Yuan and W. Yang, *J. Anal. At. Spectrom.*, 1994, **9**, 851–856.
40. N. A. Stevens and M. F. Borgerding, *Anal. Chem.*, 1998, **70**, 4223–4227.
41. A. Besner and J. Hubert, *J. Anal. At. Spectrom.*, 1988, **3**, 381–385.
42. T. Maeda, K. Wagatsuma and Y. Okamoto, *Anal. Bioanal. Chem.*, 2005, **382**, 1152–1158.
43. Y. Israel and R. M. Barnes, *Anal. Chem.*, 1984, **56**, 1188–1191.
44. D. Ye, H. Zhang, J. Yu and Q. Jin, *Chem. Res. Chin. Univ.*, 1995, **16**, 1871.
45. K. Jankowski, A. Jackowska, P. Łukasiak, M. Mrugalska and A. Trzaskowska, *J. Anal. At. Spectrom.*, 2005, **20**, 981–986.
46. P. C. Uden, *J. Chromatogr. A*, 1995, **703**, 393–416.
47. *Selective Detectors*, ed. R. E. Sievers, Wiley, New York, 1995.
48. L. L. P. van Stee and U. A. T. Brinkman, *Trends Anal. Chem.*, 2002, **21**, 618–626.
49. L. L. P. van Stee and U. A. T. Brinkman, *J. Chromatogr. A*, 2008, **1186**, 109–122.
50. Q. Jin, W. Yang, F. Liang, H. Zhang, A. Yu, Y. Cao, J. Zhou and B. Xu, *J. Anal. At. Spectrom.*, 1998, **13**, 377–384.
51. A. Matsumoto and T. Nakahara, *Can. J. Anal. Sci. Spectrosc.*, 2004, **49**, 334–345.
52. *Real-time Air Particulate Monitor EPD1*, data sheet, Elemetric Instruments, 2004.
53. Y. Duan, Y. Su, Z. Jin and S. P. Abeln, *Rev. Sci. Instrum.*, 2000, **71**, 1557–1563.
54. Y. Duan, Y. Su, Z. Jin and S. P. Abeln, *Anal. Chem.*, 2000, **72**, 1672–1679.
55. H. Takahara, M. Iwasaki and Y. Tanibata, *IEEE Trans. Instrum. Meas.*, 1995, **44**, 819–823.
56. S. Tamura, T. Kikuchi, H. Takahara, M. Mishima and Y. Fujii, *Polar Meteorol. Glaciol.*, 2001, **15**, 124–132.
57. K. Kobayashi, A. Sato, T. Homma and T. Nagatomo, *Jpn. J. Appl. Phys.*, 2005, **44**, 1027–1030.
58. H. Saitoh, K. Kawahara, S. Ohshio, A. Nakamura and N. Nambu, *Sci. Technol. Adv. Mater.*, 2005, **6**, 205–209.
59. D. L. Haas and J. A. Caruso, *Anal. Chem.*, 1984, **56**, 2014–2019.
60. L. Zhao, D. Song, H. Zhang, Y. Fu, Z. Li, C. Chen and Q. Jin, *J. Anal. At. Spectrom.*, 2000, **15**, 973–978.
61. M. Ohata and N. Furuta, *J. Anal. At. Spectrom.*, 1998, **13**, 447–453.
62. A. T. Zander, *Anal. Chem.*, 1986, **58**, 1139A–1149A.
63. K. Chiba, M. Kurosawa, K. Tanabe and H. Haraguchi, *Chem. Lett.*, 1984, 75–78.

64. F. L. Fricke, O. Rose, Jr. and J. A. Caruso, *Talanta*, 1976, **23**, 317–320.
65. Q. Jin, H. Zhang, W. Yang, Q. Jin and Y. Shi, *Talanta*, 1997, **44**, 1605–1614.
66. J. Yang, J. Zhang, C. Schickling and J. A. C. Broekaert, *Spectrochim. Acta, Part B*, 1996, **51**, 551–562.
67. H. Matusiewicz and M. Kopras, *J. Anal. At. Spectrom.*, 2003, **18**, 1415–1425.
68. H. Tao and A. Miyazaki, *Anal. Sci.*, 1991, **7**, 55–59.
69. T. Nakahara, *Anal. Sci.*, 2005, **21**, 477–484.
70. J. A. C. Broekaert and G. Tölg, *Fresenius' Z. Anal. Chem.*, 1987, **326**, 495–509.
71. J. A. C. Broekaert, *Anal. Chim. Acta*, 1987, **196**, 1–21.
72. J. D. Winefordner, E. P. Wagner II and B. W. Smith, *J. Anal. At. Spectrom.*, 1996, **11**, 689–702.
73. R. K. Winge, V. J. Peterson and V. A. Fassel, *Appl. Spectrosc.*, 1979, **33**, 206–209.
74. B. Welz, *Atomic Absorption Spectrometry*, Verlag Chemie, Weinheim, 1985.
75. R. D. Deutsch, J. P. Keilsohn and G. M. Hieftje, *Appl. Spectrosc.*, 1985, **39**, 531–534.
76. J. M. Carey and J. A. Caruso, *CRC Crit. Rev. Anal. Chem.*, 1992, **23**, 397–439.
77. H. Matusiewicz, *Spectrochim. Acta Rev.*, 1990, **13**, 47–68.
78. H. Matusiewicz and R. E. Sturgeon, *Spectrochim. Acta, Part B*, 1996, **51**, 377–397.

Analytical Applications of MWP-OES

11.1 Microwave Plasma Spectroscopic Techniques: Overview of Practical Uses

MWPs find wider and wider application in analytical spectrometry techniques. In comparison with other excitation sources, this kind of plasma bears several important features that constitute their attractiveness. These allow achieving a high efficiency of excitation, especially in the case of a helium plasma. Low detection limits (DLs) for non-metals, and especially for halogens reached in helium MWP-OES, are the benchmark of this excitation source. However, the determination of almost all elements at trace levels is possible. The versatility of MWP-OES/MS makes it a good analytical technique for a wide variety of applications. This versatility is not only due to the large number of elements that can be determined rapidly at trace levels, but also to the wide variety of spectroscopic techniques that can be applied using MWPs. In the case of analytical spectrometry, a large number of different MWPs have been examined and new plasma sources are constantly being designed and improved. Because the tolerance of early MWPs to the introduction of liquid samples was relatively low, and since all MWPs suffer more or less from matrix effects, sample introduction techniques have been carefully studied when coupled with MWPs. Consequently, a wide variety of sample introduction techniques have been developed for use with MWPs. This resulted in introducing MWPs in many spectroscopic techniques covering absorption, emission and fluorescence spectrometry, mass spectrometry and so-called coupled techniques, among which GC-MIP-OES is recognized as a matured and powerful technique in the classification of analytical techniques utilizing MWPs.

In Table 11.1 are collected examples of various spectroscopic techniques involving MWPs. The first group covers techniques in which MWPs are utilized

RSC Analytical Spectroscopy Monographs No. 12
Microwave Induced Plasma Analytical Spectrometry
By Krzysztof J. Jankowski and Edward Reszke
© The Royal Society of Chemistry 2011
Published by the Royal Society of Chemistry, www.rsc.org

Table 11.1 MWP analytical spectrometric techniques.

Sample introduction technique	Spectrometric technique	Acronym	Ref
Gas sampling	Optical emission	GS-MWP-OES	1,2
Electrothermal vaporization	Optical emission	ETV-MWP-OES	3,4
Solution nebulization	Optical emission	SN-MWP-OES	5–10
Hydride generation	Optical emission	HG-MWP-OES	11–14
Cold vapour generation	Optical emission	CV-MWP-OES	15,16
Chemical vapour generation	Optical emission	CVG-MWP-OES	17,18
Laser ablation	Optical emission	LA-MWP-OES	19–21
Spark ablation	Optical emission	SA-MWP-OES	22,23
Direct sample insertion	Optical emission	DSI-MWP-OES	24
Continuous powder introduction	Optical emission	CPI-MWP-OES	25,26
Flow injection	Optical emission	FIA-MWP-OES	27–29
Gas chromatography	Optical emission	GC-MWP-OES	30–35
Supercritical fluid chromatography	Optical emission	SFC-MWP-OES	36,37
Liquid chromatography	Optical emission	LC-MWP-OES	38,39
High-performance liquid chromatography	Optical emission	HPLC-MWP-OES	40
Capillary zone electrophoresis	Optical emission	CZE-MWP-OES	41
Electrothermal vaporization	Atomic absorption	ETV-MWP-AAS	42
Solution nebulization	Atomic absorption	SN-MWP-AAS	43,44
Gas chromatography	Atomic absorption	GC-MWP-DDL-AAS	45,46
Solution nebulization	Atomic fluorescence	SN-MIP-AFS	47–49
Electrothermal vaporization	Atomic fluorescence	ETV-MIP-LIF	50
Gas sampling	Mass spectrometry	GS-MIP-MS	51,52
Hydride generation	Mass spectrometry	HG-MIP-MS	53
Chemical vapour generation	Mass spectrometry	CVG-MIP-MS	54
Electrothermal vaporization	Mass spectrometry	ETV-MWP-MS	55
Solution nebulization	Mass spectrometry	SN-MWP-MS	56–60
Solution nebulization	Mass spectrometry	SN-MWP-TOFMS	61
Gas chromatography	Mass spectrometry	GC-MWP-MS	62,63
Supercritical fluid chromatography	Mass spectrometry	SFC-MWP-MS	64
High-performance liquid chromatography	Mass spectrometry	HPLC-MWP-MS	65,66
Solution nebulization	Cavity ringdown spectroscopy	SN-MWP-CRDS	67

as an excitation and radiation source for multi-element analysis by means of OES, where various methods of sample introduction are used depending on the form of the sample studied. From among these applications of MWPs an important place is played by coupled techniques, consisting of combining different chromatographic techniques, in these cases with MWP-OES. Especially, GC-MIP-OES has an established position in the group of modern methods of trace analysis and speciation studies.[30–35] This technique is utilized for the analysis of trace amounts of organic compounds containing heteroatoms such as N, S, Cl, Br, F, P and Si, as well as organometallic compounds.[30–32] Advantage is taken of both the high resolution of the chromatographic method and the high sensitivity (for many elements detection is achieved at the level of pg s^{-1}) and selectivity of detection. OES with a MWP has also been applied as a detection method in several types of liquid chromatography, capillary electrophoresis and flow injection analysis.

In the second group of spectroscopic methods taking advantage of microwave plasma, MWPs fulfil exclusively the role of atomizers. This type of plasma has been used in atomic absorption spectrometry,[42–46] atomic fluorescence[47–50] and cavity ringdown spectroscopy.[67] Several MWP systems are efficient ionization sources and have found application in mass spectrometry.[51–66] The use of low-energy microwave plasmas for the fragmentation of organic molecules to study their structure by mass spectrometry is an interesting example.[65,68,69] All the above-mentioned non-emission spectrometric techniques will be discussed in more detail in the next chapter.

In this chapter, MWP-OES applications have been grouped into four generalized categories: environmental and water, clinical, industrial and geological. While an exhaustive review of the application areas of MWP-OES is beyond the scope of this book, some examples of MWP-OES applications are discussed to give the reader an idea of the types of analyses where this technique has been used successfully. Examples of using MWPs in the analysis of real environmental materials are described, as well as possibilities of its application in the analysis of industrial samples and in the control of technological processes. Detailed information about specific applications can be found in a number of reviews.[3,4,29–34,70–79] Selected chapters in books,[80–84] while not exhaustive references, provide detailed information on some particular applications.

11.1.1 Types of Analyses

Obtaining qualitative information, *i.e.* what elements are present in the sample, involves identifying the presence of emission at the wavelengths characteristic of the elements of interest. In general, at least three spectral lines of the element are examined to be sure that the observed emission can be indeed classified as that belonging to the element of interest in order to minimize uncertainty of identification caused by occasional spectral line interferences from other elements. Fortunately, the relatively small number of intense emission lines available for most elements by MWP-OES allows one to reduce problems with spectral interferences.

At present, spectrometric methods are the most common tools used by the analytical chemist to perform elemental analysis. In the case of atomic emission and mass spectrometries, new plasma sources are being designed and improved. The multi-element determination capability is the key feature of these methods.

As shown in Chapter 10, MWP-OES offers moderate DL values for most elements including, advantageously, non-metals.

Numerous studies have revealed the great potential of GC-MWP-OES for the determination of the empirical formula of organic compounds. The need for the quantification of non-metals limits the choice of detectors to those operating with an He plasma, and in particular to MWP-OES. This would require a detector capable of responding simultaneously to a number of elements with a high degree of selectivity and sensitivity. The most important requirement of the detector is that its response should be proportional to the quantity of samples entering it. If the eluting solute molecules were decomposed completely forming atomic species, then the empirical formula of the solutes would be obtained. The validity of empirical formulas obtained by this technique and also methods for formula calculations have been given by Valente and Uden.[85]

Initially, very optimistic prospects concerning quantitative determination of interelement ratios were drawn.[86–88] However, later workers[85,89–94] have shown that the elemental response from chemically different compounds depends on a number of operating conditions and may vary considerably. Hence, the identification of a compound of completely unknown origin would be difficult. Nevertheless, the concept of compound-independent calibration has been raised and is still being developed.[95–97] Yie-ru *et al.*[90] published the results of the determination of C to H ratios for several groups of homologous hydrocarbons, which showed the accuracy being unaffected by the different reference compounds adopted. Kovacic and Ramus[98] shown the compound independence of GC-MPD elemental response factors for a wide range of aliphatic, aromatic and heterocyclic compounds, and relative standard deviations (RSDs) between 3 and 6% for C, Cl, F, N and O for determination of number of atoms in a given compound. Empirical formulas for polychlorinated biphenyls (PCBs), thiols and diols were reported within 5–10% at low analyte levels,[86] while the accuracy of Se to C ratios was determined to within $\pm10\%$ for a number of organoselenium compounds.[99] Stuff *et al.*[100] have found the GC-MPD technique an excellent tool for screening samples for chemical warfare related materials.

Speciation of metals and organometals has become a challenging research area in environmental studies and food analyses. The combination of gas chromatography with detection by the microwave plasma detector has probably been the most successful commercialized hyphenated technique used for speciation studies. Some recent applications of GC-MWP coupling for speciation studies will be discussed at the end of this chapter.

GC-MIP-OES has also been used as a complementary method in the determination of molecular formulae. By combining information from both mass spectrometric and gas chromatography–atomic emission detection, elemental composition and molecular structure of individual compounds can be obtained. If sample organic molecules separated by GC and subsequently introduced into the MIP can be reproducibly fragmented into polyatomic ions and determined as such by mass spectrometry, and their constituent atoms reproducibly excited in the parallel run with the use of plasma emission spectrometry, then the molecular formula can be established even when different isomers of the same compound are present in the sample. Several groups have shown that low-pressure MIP is an effective ionization source for soft fragmentation of organic compounds.[69,101] Hooker and DeZwaan[102] combined MIP-OES and mass spectral data in molecular formula determinations of large molecules. Clarkson and Cooke[103] used these complementary techniques for identification of an unusual volatile component in processed tobacco.

A recent highly interesting application of MWP-OES is the analysis of dispersion, mixing, coating and composition of powdered materials, including nanoparticles. Moreover, the MWP-OES system should be applicable to the identification and characterization of individual microparticles, thus becoming a very useful tool for fundamental nanotechnology research. This will be discussed in Section 11.4.

11.2 Selected Applications of MWP-OES in Environmental Analysis

The main field of application of emission spectrometry involving microwave plasmas are analytical problems connected with control and environment protection. The potential of GC-MIP-OES for determinations of pesticide and herbicide residues has proved to be very great for over 40 years.[104,105] The plasma emission detector is less sensitive than the electron capture detector (ECD); however, it is much more selective and can be readily used for simultaneous multi-element qualitative and quantitative measurement. Simultaneous multi-wavelength monitoring enables the determination of several groups of pesticides containing different heteroatoms during one chromatographic run. Samples with a complex matrix, *e.g.* environmental and biological, contain many constituent compounds that greatly complicate the interpretation of their corresponding chromatograms obtained with the use of ECD. The MPD is able to reduce, and in many cases eliminate, these interferences. Because of the selectivity of the detector, no sample matrix effects are shown in the chromatogram.[31] A database of over 400 pesticides has been established for screening environmental and biological samples for sulfur-, nitrogen-, phosphorus- and chlorine-containing pesticide residues by GC-MIP-OES by Cook *et al.*,[106] while Stan and Linkerhägner[107] have collected a similar number of data for a German multi-method approved for pesticide residue analysis in foodstuffs. The determination of pesticides in water by GC-MIP-OES was reported by Viñas *et al.*[108] The low-power helium MIP has been proposed as an ionization source for mass spectrometric detection of pesticides separated by GC.[63]

The second flagship of MWP-OES applications is the determination of volatile halogen- and sulfur-containing compounds in environmental water samples, including halomethanes, dioxins, PCBs and others.[109–116] For a quasi-continuous monitoring, analyzers for the element-specific detection of adsorbable haloorganics in wastewater have been extensively investigated; these use element-selective detection of the halogens by microwave-induced helium plasma spectrometry.[117–121]

Other examples of environmental MWP-OES applications include various water quality analyses as required by ISO regulations and the U.S. Environmental Protection Agency (EPA),[122] and the determination of heavy metals in ocean water.[6,123] A critical evaluation of the usefulness of the MIP-OES method for multi-element analysis in ground and drinking water has been presented. Relatively low DLs for a majority of the 31 elements studied, especially for alkali metals and alkali earth metals, were obtained. Ultrasonic nebulizers are useful for improving sensitivity for such analyses. Although the DLs achieved for MIP-OES are comparable to those offered by ICP-OES in ISO regulation 11885,[124] only for about 50% of the studied elements do both techniques fulfil the EPA programme requirements to a similar degree. It has been shown that, by the MIP-OES method, common elements like Fe, Mn, Zn, Cu, Pb, Cr, Na, K, Mg and Ca can be directly determined, which in many cases is sufficient for the needs of typical monitoring of impurities in water. For eight elements, in the MIP-OES method a greater sensitivity is achieved than in the ICP-OES method. Especially alkali metals can be determined at a much lower content level. Also, such important elements as cadmium, lead or silver are better detected in a microwave plasma. For the next eight out of 31 elements the DLs in MIPs and ICPs are comparable, whereas for the remaining 15 elements, better possibilities of analysis are provided by an ICP. This concerns especially the determination of metals forming stable oxides, such as W, Mo, Zr and Ti. For analyses requiring very low detection levels, it may be necessary to perform some kind of preconcentration step prior to the analysis.[125]

The ETV-MIP-OES technique has been applied to the analysis of a large variety of agricultural and food materials. Types of samples include soils, plant materials, foods, animal tissues and body fluids. Most agricultural and food materials are generally not in the form of dilute aqueous solutions, nor they are readily soluble in distilled water. Therefore, analyses of powdered microsamples of these materials by MIP-OES were carried out. In order to determine the accuracy of the ETV-MIP-OES technique, an analysis of certified reference materials was performed, especially those of vegetable and animal origin.[3,4,126,127]

11.3 Selected Applications of MWP-OES in Clinical Analysis

Determinations of essential, toxic and therapeutic trace elements are important in medical research laboratories, as well as in clinical and pharmaceutical laboratory environments. Unfortunately, procedures developed with the use of spectrometric techniques are rarely applied to routine practice. However, the determination of trace elements in urine, blood or hair is a challenging task as such and can be used to demonstrate the robustness and sensitivity of the technique developed.

Many clinical samples are either too small or contain elemental concentrations too low for MWP-OES analysis using conventional solution nebulization. In these cases, it is often necessary to turn to alternative sample introduction techniques, such as ultrasonic nebulization, electrothermal vaporization or hydride generation, or preconcentration techniques such as analyte trapping or solid-phase extraction. Moreover, severe matrix effects are usually observed in the analysis of clinical samples, which must be solved when elaborating an analytical procedure.

Examples of MWP-OES (and MIP-MS) analyses of clinical samples include determinations of As, F, Ge, Hg, Ni and Se in urine;[128–132] Ag, As, Ca, Cu, Fe, Ge, Hg, Mg, Mn, Na, Ni, Pb, Sr and Zn in hair;[8,127,131–136] As, Ca, Cd, Cr, Cu, Fe, Ge, Hg, K, Li, Mg, Mn, Na, Ni, Pb, Se and Zn in blood;[53,130–132,134–141] Ca, Cd, Cr, Cu, Fe and Ni in different tissues of rat;[141] rare earth elements in the organs of mice;[142] as well as speciation of Hg in blood[143] and As, Se and Sn in urine.[144–147]

The second type of possible clinical applications of MWP analytical spectrometries is the screening of pharmaceuticals and their metabolites, as well as quantitative analysis of both target metabolites and unexpected compounds. Jordan *et al.*[148] proposed an analytical procedure for the determination of catechol derivatives in human urine extracts *via* derivatization with a borate ester and GC-MIP-OES. A similar procedure has been proposed by Luffer and Novotny[149] for the determination of catechols, monosaccharides and cortisone in urine with the use of SFC-MIP-OES. The DL for boron in catechol borate was $25\,pg\,s^{-1}$ and the molar selectivity was over 5000. Quimby *et al.*[150] reported the ability of GC-MPD to detect the isotope ^{13}C selectively over ^{12}C as a tool for the determination of ^{13}C-labelled compounds and their metabolites in urine. They utilized the molecular emission from intense C bands in the vacuum ultraviolet region. A shift of about 0.4 nm was observed between the ^{13}C and ^{12}C band heads. A DL of $7\,pg\,s^{-1}$ for the ^{13}C compound and a selectivity of 2500 was reported.

MWP-based spectrometric techniques were also used for diagnostic purposes in individual medical studies. Tateyama *et al.*[151] investigated correlations of calcium accumulation in arteries, veins and other tissues in single humans with the use of MIP-OES. The utility of HPLC-MIP-MS for the analysis of underivatized amino acids has been demonstrated for a synthetic mixture of 20 amino acids and a cytochrome *c* digest by Kwon and Moini.[65] Mercury and selenium were determined in blood and urine of workers previously exposed to mercury vapour at a chloroalkali plant in order to evaluate the clinical significance of this exposure.[130] The assessment of toxic metal exposure following a water pollution incident has been made by Powell *et al.*[134] They determined heavy metal content in human blood and hair.

11.4 Selected Applications of MWP-OES in Industrial Analysis

The area of industrial analysis is quite a broad topic, covering many different types of MWP-OES applications. Possible significant applications of MIPs include the determination of semiconductor grade gases and materials,[1,152] polymers,[153,154] plasticizers,[155] petroleum products and many other industrial compounds of interest.[156–158] Other important industrial MWP-OES applications include analyses of industrial waste products[159–161] and analyses of dust and other airborne particulates, as well as continuous emission monitoring of gas streams. A particular advantage associated with metal analysis by MWP-OES is the relatively low potential of spectral interferences, since many metals exhibit relatively simple emission spectra. This spectrometric technique provides spectral simplification and improved signal-to-background ratios up to six-fold over the emission observed from the spark discharge. The emission from a MWP is linear with concentration for many lines of various elements determined. Some representative applications of MWP-OES for the analysis of metals and related materials include the analysis of steels for Al, B, Cr, Cu, Fe, Mg, Mn, Mo, Ni, Si and V,[22,162,163] the determination of Fe, Ni, Pb and Sn in brass[163] and the determination of Al, Cr, Cu, Fe, Mg, Mn, Si and Zn in aluminum.[22,163,164]

A special application of microwave-induced plasmas deals with the sizing and analysis of powder particles by atomizing them in a microwave plasma and observing the atomic spectra. The single-particle introduction system for a helium MIP has been developed and commercialized.[26,165] The analyzer system can simultaneously measure the three-dimensional structure of individual particles, their composition and size. Analyzing the obtained spectra with a time resolution below $100\,\mu s$, the chemical composition is determined from the set of wavelength emissions, the number of particulates from the number of peaks, the particle size from the peak height, and the structure from the synchronizing emission. The technique can be applied for analyses of fine powdered composites and nanomaterials,[166–169] but also for analysis of particles in exhaust gases, dusts, airborne particles and aerosols.[169–171]

The other specific application of MWPs is the determination of the chemical structure of polymers and their empirical formulae. Polymers are decomposed by controlled pyrolysis and the resulting products are analyzed by GC-AED and MS. The method was applied to a number of polymers, including polyethylene, polysiloxanes, epoxy resins and others.[172–174]

The analysis of various petroleum products for trace metal content is one of the more popular applications for organic analysis by MWP-OES. Some typical applications include the determination of organolead and organomanganese compounds and oxygen-containing additives in gasoline,[175–178] the determination of Fe, Ni and V porphyrins and other components in crude oils,[179–182] as well as the analysis of other petroleum related products.[183–186] Also considered under the category of organic analyses would be the introduction of organic effluents and vapours.[187]

A number of applications require real-time analysis of gases for volatile organic and inorganic compounds or fine particles, including air analysis, engine exhaust, hazardous-waste incinerators and process streams. Most general applications for volatiles involve collection of the samples, followed by off-line analysis by GC or HPLC. However, these methods typically require

30–60 min for the separation, which precludes real-time measurements. An alternative approach to these protocols is the use of continuous-emission monitors (CEMs).[188,189] Owing to the fact that MWPs can be operated with many gases, including molecular gases such as oxygen, nitrogen and air, they are ideally suitable for monitoring metal and non-metal pollutants directly in ambient air.[2,190] This makes them ideally suited for stack-gas or process-gas real-time monitoring.[52,191,192] More demanding applications include air or gaseous effluents analysis near industrial plants or near waste storage places, which requires portable instrumentation that can operate with batteries. Since the ICP-OES spectrometers are weighty and need a lot of energy and gas for plasma maintenance, they are very hard to use for on-site applications, but a very compact MWP instrument can be applied to on-site, real-time waste stream monitoring. A portable MWP-OES system operated with a dual-flow torch, where the internal gas is the sample gas, has been designed for continuous air analysis as well. The systems proved to be very robust and enabled monitoring of particulate matter in storage places of fly ash and in workplaces where ultrafine powders are handled. Duan *et al.*[171] described a MPT-based analyzer operated with argon or helium. In the case of helium, up to 40% of the central gas flow could be air, by which detection limits for metals in air of 0.08 (Cu) to 2.3 (Mn) $\mu g\,m^{-3}$ can be obtained. Timmermans *et al.*[193,194] monitored metals at the stack of a waste incinerator with a MPT-OES instrument and could use the emission signals for controlling the burning conditions in the incinerator on an on-line basis.

Siemens *et al.*[195] studied the MIP-OES system for the monitoring of trace amounts of mercury in flue gases using a non-mixed argon/nitrogen discharge. The DL of $8\,\mu g\,m^{-3}$ of Hg in nitrogen was lower than the recommended threshold limit value. Vermaak *et al.*[196] utilized an air MIP-OES for the detection of gaseous lead in exhaust gases. Baumann and Heumann[197] determined organobromine compounds, hydrogen bromide and tetraalkyllead compounds in exhaust gases. The hydrogen bromide was firstly converted into 2-bromocyclohexanol; then the organobromine compounds were adsorbed on a Tenax GC column and the lead compounds on a Porapak N column for preconcentration and evolved by means of thermal desorption to the GC-MIP-OES system. The contents of the compounds studied in the exhaust gases were from a few to several hundred $\mu g\,m^{-3}$.

Collection of airborne particulates sometimes requires use of air filtering techniques. Reamer *et al.*[198] studied the content of the total gaseous lead (TGL) and total particle lead (TPL) in exhaust gases, tunnel air and laboratory air. The sample was transferred through a system in which particulate lead was collected on a filter and gaseous lead on an appropriate adsorbent. The compounds determined were released by means of a freeze-drying system to the GC-MIP-AES instrument. The TGL and TPL contents measured were 20–1000 and 1000–55 000 $ng\,Pb\,m^{-3}$, respectively. Serravallo and Risby[199] studied the determination of vinyl chloride in air by means of He-MIP-OES under reduced pressure on the basis of the chlorine emission measurement for the 479.45 nm line.

11.5 Selected Applications of MWP-OES in Geological Analysis

Geological applications of MWP analytical spectrometry involve determinations of toxic, trace and major constituents of various ores, coals, rocks, soils and related materials. The major use of MWP-OES/MS in this field is for determining the origins of the formation and distribution of toxic inorganic and organometallic compounds in soils, coals and sediments.[200] GC is usually interfaced with analytical pyrolysis for characterization of soils and sediments to provide information in organic geochemistry. MIP-OES was applied to study sulfur-containing pyrolysis products of sediments,[201,202] while both MIP-OES and MIP-MS have been used for the detection of nitrogen-containing products.[203,204] In recent years the methods for the determination of organic contaminants in soils and sediments have been extensively developed.[205,206] The technique is also used for prospecting purposes. Other applications of MWP-OES for analysis of geological samples include the determination of elements in coals, soils and sediments.[207–209] The direct determination of total fluorine by CPI-MIP-OES has been developed. A DL between 3 and 6 μg g^{-1} was achieved for fluorine determination in coal and soil[25,208] and in soil, fluorite or phosphogypsum,[209] depending on the matrix composition. The CPI-MIP-OES system has been proposed for the determination of noble metals in ores and soils. For the separation and preconcentration of analytes, as well as for preparing calibration standards, solid-phase extraction on activated carbon was utilized.[210]

11.6 Selected Applications of MIP-OES in Speciation Studies

There are many applications of MWPs reported in the literature which span the field of speciation. They are collected and discussed in a numerous reviews.[30–32,35,71,72,74,76,77,211–214] In this chapter we only mention a few examples to show the potential of MWP applications. Without doubt the GC-MIP-OES system is one of techniques of choice for speciation studies.[71,215] Owing to the excellent selectivity and relatively high sensitivity offered by atomic emission detection and the use of helium MIPs, very clear chromatograms can be obtained even in the analysis of complex environmental and biological matrices. Liu *et al.*[216] have presented an analytical procedure for the separation and speciation analysis of 15 different alkyltin compounds that can occur in soils and marine sediments. In another study they determined simultaneously a number of organic tin, lead and mercury compounds present in environmental samples using CGC-MIP-OES. The linearities of the respective calibration graphs were around 3–4 orders of magnitude.[217] Rodriguez Pereiro *et al.*[218,219] also speciated alkyllead, alkylmercury and alkyltin compounds in a number of environmental samples by multi-capillary GC-MIP-OES. They confirmed the accuracy of the method with the use of certified reference materials: sediment and fish tissue. Simultaneous multi-element speciation of Hg, Sn and Pb organic compounds by *in situ* derivatization, cold trapping and GC-AED has been developed by Reuther *et al.*,[220] providing DLs below $1\,ng\,L^{-1}$.

Costa-Fernandez *et al.*[221] studied the speciation of mercury and arsenic in waters and urine using HPLC-CV-MIP-OES and HPLC-HG-MIP-OES, respectively, with a flow injection mode. The DLs for the organoarsenic compounds varied from 1 to $6\,ng\,mL^{-1}$, while the DL for inorganic Hg was $0.15\,ng\,mL^{-1}$ and that for methylmercury was $0.35\,ng\,mL^{-1}$. In later studies, Dietz *et al.*[222] used cold trapping for preconcentration of mercury species and the DLs achieved were from 0.001 to $0.006\,ng\,mL^{-1}$. In the speciation studies of lead in snow from Greenland, Łobiński *et al.*[223] used a separation procedure to attain 1250-fold preconcentration of the analytes. The content of lead compounds in snow was from as low as 0.02 to $0.48\,pg\,g^{-1}$. The success of numerous applications of GC-MIP-OES has resulted in the development of automatic speciation analyzers based on this technique.[224–226]

References

1. S. Kirschner, A. Golloch and U. Telgheder, *J. Anal. At. Spectrom.*, 1994, **9**, 971–974.
2. B. W. Pack, G. M. Hieftje and Q. Jin, *Anal. Chim. Acta*, 1999, **383**, 231–241.
3. J. A. C. Broekaert and F. Leis, *Mikrochim. Acta*, 1985, **86**, 261–272.
4. H. Matusiewicz, *Spectrochim. Acta Rev.*, 1990, **13**, 47–68.
5. G. L. Long and L. D. Perkins, *Appl. Spectrosc.*, 1987, **41**, 980–985.
6. P. G. Brown, D. L. Haas, J. M. Workman, J. A. Caruso and F. L. Fricke, *Anal. Chem.*, 1987, **59**, 1433–1436.
7. K. Jankowski, D. Karmasz, L. Starski, A. Ramsza and A. Waszkiewicz, *Spectrochim. Acta, Part B*, 1997, **52**, 1801–1812.
8. H. Matusiewicz, M. Ślachciński, M. Hidalgo and A. Canals, *J. Anal. At. Spectrom.*, 2007, **22**, 1174–1178.
9. Y. Okamoto, *Anal. Sci.*, 1991, **7**, 283–288.
10. L. Zhao, D. Song, H. Zhang, Y. Fu, Z. Li, C. Chen and Q. Jin, *J. Anal. At. Spectrom.*, 2000, **15**, 973–978.
11. E. Bulska, J. A. C. Broekaert, P. Tschöpel and G. Tölg, *Anal. Chim. Acta*, 1993, **276**, 377–384.
12. N. W. Barnett, *Spectrochim. Acta, Part B*, 1987, **42**, 859–864.
13. H. Tao and A. Miyazaki, *Anal. Sci.*, 1991, **7**, 55–59.
14. R. Pereiro, M. Wu, J. A. C. Broekaert and G. M. Hieftje, *Spectrochim. Acta, Part B*, 1994, **49**, 59–73.
15. G. Kaiser, D. Götz, P. Schoch and G. Tölg, *Talanta*, 1975, **22**, 889–899.
16. Y. Nojiri, A. Otsuki and K. Fuwa, *Anal. Chem.*, 1986, **58**, 544–547.
17. W. Drews, G. Weber and G. Tölg, *Fresenius' Z. Anal. Chem.*, 1989, **332**, 862–865.
18. N. W. Barnett, *J. Anal. At. Spectrom.*, 1988, **3**, 969–972.

19. J. Uebbing, A. Ciocan and K. Niemax, *Spectrochim. Acta, Part B*, 1992, **47**, 601–611.
20. T. Ishizuka and Y. Uwamino, *Anal. Chem.*, 1980, **52**, 125–129.
21. D. Cleveland, P. Stchur, X. Hou, K. X. Yang, J. Zhou and R. G. Michel, *Appl. Spectrosc.*, 2005, **59**, 1427–1444.
22. Y. N. Pak and S. R. Koirtyohann, *J. Anal. At. Spectrom.*, 1994, **9**, 1305–1310.
23. L. R. Layman and G. M. Hieftje, *Anal. Chem.*, 1975, **47**, 194–202.
24. A. M. Pless, B. W. Smith, M. A. Bolshov and J. D. Winefordner, *Spectrochim. Acta, Part B*, 1996, **51**, 55–64.
25. K. Jankowski and A. Jackowska, *Trends Appl. Spectrosc.*, 2007, **6**, 17–25.
26. H. Takahara, M. Iwasaki and Y. Tanibata, *IEEE Trans. Instrum. Meas.*, 1995, **44**, 819–823.
27. Y. Madrid, M. Wu, Q. Jin and G. M. Hieftje, *Anal. Chim. Acta*, 1993, **277**, 1–8.
28. H. Zhang, D. Ye, J. Zhao, J. Yu, R. Men, Q. Jin and D. Dong, *Microchem. J.*, 1996, **53**, 69–78.
29. Q. Jin, W. Yang, F. Liang, H. Zhang, A. Yu, Y. Cao, J. Zhou and B. Xu, *J. Anal. At. Spectrom.*, 1998, **13**, 377–384.
30. E. Bulska, *J. Anal. At. Spectrom.*, 1992, **7**, 201–210.
31. J. J. Sullivan, *Trends Anal. Chem.*, 1991, **10**, 23–26.
32. R. Łobiński and F. C. Adams, *Trends Anal. Chem.*, 1993, **12**, 41–49.
33. B. F. Scott and P. L. Wylie, *Chem. Plant Prot.*, 1995, **12**, 33–57.
34. P. C. Uden, *Trends Anal. Chem.*, 1987, **6**, 238–246.
35. Y. K. Chau and P. T. S. Wong, *Fresenius' J. Anal. Chem.*, 1991, **339**, 640–645.
36. D. R. Luffer, L. J. Galante, P. A. David, M. Novotny and G. M. Hieftje, *Anal. Chem.*, 1988, **60**, 1365–1369.
37. G. K. Webster and J. W. Carnahan, *Anal. Chem.*, 1992, **64**, 50–55.
38. L. Zhang, J. W. Carnahan, R. E. Winans and P. H. Neill, *Anal. Chem.*, 1989, **61**, 895–897.
39. L. J. Galante, D. A. Wilson and G. M. Hieftje, *Anal. Chim. Acta*, 1988, **215**, 99–109.
40. D. Kollotzek, D. Oechsle, G. Kaiser, P. Tschöpel and G. Tölg, *Fresenius' Z. Anal. Chem.*, 1984, **318**, 485–489.
41. Y. Liu and V. Lopez-Avila, *J. High Resolut. Chromatogr.*, 1993, **16**, 717–720.
42. Y. Duan, X. Li and Q. Jin, *J. Anal. At. Spectrom.*, 1993, **8**, 1091–1096.
43. K. C. Ng, R. S. Jensen, M. J. Brechmann and W. C. Santos, *Anal. Chem.*, 1988, **60**, 2818–2821.
44. Y. Duan, M. Hou, Z. Du and Q. Jin, *Appl. Spectrosc.*, 1993, **47**, 1871–1879.
45. A. Zybin, C. Schnürer-Patschan and K. Niemax, *J. Anal. At. Spectrom.*, 1995, **10**, 563–567.
46. J. Koch, A. Zybin and K. Niemax, *Appl. Phys. B*, 1998, **67**, 475–479.
47. L. D. Perkins and G. L. Long, *Appl. Spectrosc.*, 1988, **42**, 1285–1289.
48. Y. Duan, X. Du, Y. Li and Q. Jin, *Appl. Spectrosc.*, 1995, **49**, 1079–1085.
49. V. Rigin, *Anal. Chim. Acta*, 1993, **283**, 895–901.

50. Y. Oki, H. Uda, C. Honda, M. Maeda, J. Izumi, T. Morimoto and M. Tanoura, *Anal. Chem.*, 1990, **62**, 680–683.
51. P. G. Brown, T. M. Davidson and J. A. Caruso, *J. Anal. At. Spectrom.*, 1988, **3**, 763–769.
52. M. E. Cisper, A. W. Garrett, Y. X. Duan, J. A. Olivares and P. H. Hemberger, *Int. J. Mass Spectrom.*, 1998, **178**, 121–128.
53. M. Ohata, T. Ichinose, N. Furuta, A. Shinohara and M. Chiba, *Anal. Chem.*, 1998, **70**, 2726–2730.
54. Y. Duan, M. Wu, Q. Jin and G. M. Hieftje, *Spectrochim. Acta, Part B*, 1995, **50**, 1095–1108.
55. T. Kawano, A. Nishida, K. Okutsu, H. Minami, Q. Zhang, S. Inoue and I. Atsuya, *Spectrochim. Acta, Part B*, 2005, **60**, 327–331.
56. J. T. Creed, T. M. Davidson, W. L. Shen, P. G. Brown and J. A. Caruso, *Spectrochim. Acta, Part B*, 1989, **44**, 909–924.
57. L. K. Olson and J. A. Caruso, *Spectrochim. Acta, Part B*, 1994, **49**, 7–30.
58. K. Oishi, T. Okumoto, T. Iino, M. Koga, T. Shirasaki and N. Furuta, *Spectrochim. Acta, Part B*, 1994, **49**, 901–914.
59. Y. Okamoto, *J. Anal. At. Spectrom.*, 1994, **9**, 745–749.
60. M. Huang, A. Hirabayashi, T. Shirasaki and H. Koizumi, *Anal. Chem.*, 2000, **72**, 2463–2467.
61. Y. Su, Y. Duan and Z. Jin, *Anal. Chem.*, 2000, **72**, 2455–2462.
62. W. C. Story and J. A. Caruso, *J. Anal. At. Spectrom.*, 1993, **8**, 571–575.
63. A. M. Zapata and A. Robbat, Jr., *Anal. Chem.*, 2000, **72**, 3102–3108.
64. L. K. Olson and J. A. Caruso, *J. Anal. At. Spectrom.*, 1992, **7**, 993–998.
65. J. Y. Kwon and M. Moini, *J. Am. Soc. Mass Spectrom.*, 2001, **12**, 117–122.
66. A. Chatterjee, Y. Shibata, H. Tao, A. Tanaka and M. Morita, *J. Chromatogr. A*, 2004, **1042**, 99–106.
67. C. Wang, S. P. Koirala, S. T. Scherrer, Y. Duan and C. B. Winstead, *Rev. Sci. Instrum.*, 2004, **75**, 1305–1313.
68. E. Poussel, J. M. Mermet, D. Deruaz and C. Beaugrand, *Anal. Chem.*, 1988, **60**, 923–927.
69. L. K. Olson, W. C. Story, J. T. Creed, W. L. Shen and J. A. Caruso, *J. Anal. At. Spectrom.*, 1990, **5**, 471–475.
70. J. P. Matousek, B. J. Orr and M. Selby, *Prog. Anal. At. Spectrosc.*, 1984, **7**, 275–314.
71. R. Łobiński and F. C. Adams, *Spectrochim. Acta, Part B*, 1997, **52**, 1865–1903.
72. R. Łobiński, *Appl. Spectrosc.*, 1997, **51**, 260A–278A.
73. W. Yang, H. Zhang, A. Yu and Q. Jin, *Microchem. J.*, 2000, **66**, 147–170.
74. B. Rosenkranz and J. Bettmer, *Trends Anal. Chem.*, 2000, **19**, 138–156.
75. K. Jankowski, *Chem. Anal. (Warsaw)*, 2001, **46**, 305–327.
76. L. L. P. van Stee, U. A. T. Brinkman and H. Bagheri, *Trends Anal. Chem.*, 2002, **21**, 618–626.
77. J. C. A. Wuilloud, R. G. Wuilloud, A. P. Vonderheide and J. A. Caruso, *Spectrochim. Acta, Part B*, 2004, **59**, 755–792.

78. J. A. C. Broekaert and V. Siemens, *Spectrochim. Acta, Part B*, 2004, **59**, 1823–1839.
79. T. Nakahara, *Anal. Sci.*, 2005, **21**, 477–484.
80. *Selective Detectors*, ed. R. E. Sievers, Wiley, New York, 1995.
81. R. P. W. Scott, *Tandem Techniques*, Wiley, New York, 1997.
82. J. A. C. Broekaert and U. Engel, in *Encyclopedia of Analytical Chemistry*, ed. R. A. Meyers, Wiley, Chichester, 2000, pp. 9613–9667.
83. R. L. Grob and E. F. Barry, *Modern Practice of Gas Chromatography*, Wiley-Interscience, Hoboken, NJ, 2004.
84. J. A. C. Broekaert, *Analytical Atomic Spectrometry with Flames and Plasmas*, VCH, Weinheim, 2005.
85. A. L. P. Valente and P. C. Uden, *Analyst*, 1990, **115**, 525–529.
86. C. I. M. Beenakker, *Spectrochim. Acta, Part B*, 1977, **32**, 173–187.
87. R. M. Dagnall, T. S. West and P. Whitehead, *Analyst*, 1973, **98**, 647–654.
88. W. R. McLean, D. L. Stanton and G. E. Penketh, *Analyst*, 1973, **98**, 432–442.
89. K. S. Brenner, *J. Chromatogr.*, 1978, **167**, 365–380.
90. Y. Huang, Q. Ou and W. Yu, *J. Anal. At. Spectrom.*, 1990, **5**, 115–120.
91. J. T. Anderson and B. Schmid, *Fresenius' J. Anal. Chem.*, 1993, **346**, 403–409.
92. G. Becker, A. Colmsjö and C. Östman, *Anal. Chim. Acta*, 1997, **340**, 181–189.
93. D. F. Gurka, S. Pyle and R. Titus, *Anal. Chem.*, 1997, **69**, 2411–2417.
94. E. S. Chernetsova, A. I. Revelsky, D. Durst, T. G. Sobolevsky and I. A. Revelsky, *Anal. Bioanal. Chem.*, 2005, **382**, 448–451.
95. J. T. Anderson, *Anal. Bioanal. Chem.*, 1993, **373**, 344–355.
96. F. Augusto, E. Carasek da Rocha, L. Montero and A. L. P. Valente, *J. Braz. Chem. Soc.*, 1998, **9**, 17–21.
97. Y. Juillet, E. Gibert, A. Begos and B. Bellier, *Anal. Bioanal. Chem.*, 2005, **383**, 848–856.
98. N. Kovacic and T. L. Ramus, *J. Anal. At. Spectrom.*, 1992, **7**, 999–1005.
99. R. Bos and N. W. Barnett, *J. Anal. At. Spectrom.*, 1997, **12**, 733–741.
100. J. R. Stuff, W. R. Creasy, A. A. Rodriguez and H. D. Durst, *J. Microcolumn Sep.*, 1999, **11**, 644–651.
101. E. Poussel, J. M. Mermet, D. Deruaz and C. Beaugrand, *Anal. Chem.*, 1988, **60**, 923–927.
102. D. B. Hooker and J. DeZwaan, *Anal. Chem.*, 1989, **61**, 2207–2211.
103. P. Clarkson and M. Cooke, *Anal. Chim. Acta*, 1996, **335**, 253–259.
104. C. A. Bache and D. J. Lisk, *J. Assoc. Off. Anal. Chem.*, 1967, **6**, 1246–1250.
105. J. L. Bernal, M. J. del Nozal, M. T. Martin and J. J. Jimenez, *J. Chromatogr. A*, 1996, **754**, 245–256.
106. J. Cook, M. Engel, P. Wylie and B. Quimby, *J. AOAC Int.*, 1999, **82**, 313–326.
107. H. J. Stan and M. Linkerhägner, *J. Chromatogr. A*, 1996, **750**, 369–390.

108. P. Viñas, N. Campillo, I. Lopez-Garcia, N. Aguinaga and M. Hernandez-Cordoba, *J. Chromatogr. A*, 2002, **978**, 249–256.
109. M. L. Bruce and J. A. Caruso, *Appl. Spectrosc.*, 1985, **39**, 942–949.
110. A. H. Mohamad, M. Zerezghi and J. A. Caruso, *Anal. Chem.*, 1986, **58**, 469–471.
111. I. Rodriguez, M. I. Turnes, M. C. Mejuto and R. Cela, *J. Chromatogr. A*, 1996, **721**, 297–304.
112. J. Hajšlova, P. Cuhra, M. Kempny, J. Poustka, K. Holadova and V. Kocourek, *J. Chromatogr. A*, 1995, **699**, 231–239.
113. S. Pedersen-Bjergaard, S. I. Semb, E. M. Brevik and T. Greibrokk, *J. Chromatogr. A*, 1996, **723**, 337–347.
114. B. Rosenkranz, C. B. Breer, W. Buscher, J. Bettmer and K. Cammann, *J . Anal. At. Spectrom.*, 1997, **12**, 993–996.
115. B. D. Quimby, M. F. Delaney, P. C. Uden and R. M. Barnes, *Anal. Chem.*, 1979, **51**, 875–880.
116. C. Gerbersmann, R. Łobiński and F. C. Adams, *Anal. Chim. Acta*, 1995, **316**, 93–95.
117. H. Frischenschlager, M. Peck, C. Mittermayr, E. Rosenberg and M. Grasserbauer, *Fresenius' J. Anal. Chem.*, 1997, **357**, 1133–1141.
118. K. Cammann, L. Lendero, H. Feuerbacher and K. Balschmiter, *Fresenius' J. Anal. Chem.*, 1983, **316**, 194–200.
119. B. Koschuh, M. Montes, J. F. Camuña, R. Pereiro and A. Sanz-Medel, *Microchim. Acta*, 1998, **129**, 217–223.
120. T. Twiehaus, S. Evers, W. Buscher and K. Cammann, *Fresenius' J. Anal. Chem.*, 2001, **371**, 614–620.
121. B. Rosenkranz and J. Bettmer, *Trends Anal. Chem.*, 2000, **19**, 138–156.
122. K. Jankowski, *J. Anal. At. Spectrom.*, 1999, **14**, 1419–1423.
123. K. C. Ng and W. Shen, *Anal. Chem.*, 1986, **58**, 2084–2087.
124. *Water Quality – The Determination of 33 Elements by Inductively Coupled Plasma Atomic Emission Spectroscopy*, ISO 11885, International Organization for Standardization, 1996.
125. K. Jankowski, J. Yao, K. Kasiura, A. Jackowska and A. Sieradzka, *Spectrochim. Acta, Part B*, 2005, **60**, 369–375.
126. J. M. Carey and J. A. Caruso, *CRC Crit. Rev. Anal. Chem.*, 1992, **23**, 397–439.
127. J. Yang, J. Zhang, C. Schicling and J. A. C. Broekaert, *Spectrochim. Acta, Part B*, 1996, **51**, 551–562.
128. K. Chiba, K. Yoshida, K. Tanabe, M. Ozaki, H. Haraguchi, J. D. Winefordner and K. Fuwa, *Anal. Chem.*, 1982, **54**, 761–764.
129. H. W. Kuo, W. G. Chang, Y. S. Huang and J. S. Lai, *Bull. Environ. Contam. Toxicol.*, 1999, **62**, 677–684.
130. D. G. Ellingsen, R. I. Holland, Y. Thomassen, M. Landro-Olstad, W. Frech and H. Kjuus, *Br. J. Ind. Med.*, 1993, **50**, 745–752.
131. A. Shinohara, M. Chiba and Y. Inaba, *J. Anal. Toxicol.*, 1999, **23**, 625–631.
132. W. Drews, G. Weber and G. Tölg, *Anal. Chim. Acta*, 1990, **231**, 265.
133. H. Matusiewicz, *Spectrochim. Acta, Part B*, 2002, **57**, 485–494.

134. J. J. Powell, S. M. Greenfield, R. P. H. Thompson, J. A. Cargnello, M. D. Kendall, J. P. Landsberg, F. Watt, H. T. Delves and I. House, *Analyst*, 1995, **120**, 793–798.
135. F. E. Lichte and R. K. Skogerboe, *Anal. Chem.*, 1972, **44**, 1321–1323.
136. F. E. Lichte and R. K. Skogerboe, *Anal. Chem.*, 1972, **44**, 1480–1482.
137. M. S. Black and R. E. Sievers, *Anal. Chem.*, 1976, **48**, 1872–1874.
138. A. Shinohara, M. Chiba and Y. Inaba, *Anal. Sci.*, 1998, **14**, 713–717.
139. A. D. Besteman, G. K. Bryan, N. Lau and J. D. Winefordner, *Microchem. J*, 1999, **61**, 240–246.
140. A. D. Besteman, N. Lau, D. Y. Liu, B. W. Smith and J. D. Winefordner, *J. Anal. At. Spectrom.*, 1996, **11**, 479–481.
141. M. M. Mohamed and Z. F. Ghatass, *Fresenius' J. Anal. Chem.*, 2000, **368**, 449–455.
142. A. Shinohara, M. Chiba and Y. Inaba, *Anal. Sci.*, 2001, **17**(suppl), i1539–i1541.
143. E. Bulska, H. Emteborg, D. C. Baxter, W. Frech, D. Ellingsen and Y. Thomassen, *Analyst*, 1992, **117**, 657–663.
144. A. Chatterjee, Y. Shibata, H. Tao, A. Tanaka and M. Morita, *Anal. Chim. Acta*, 2001, **436**, 253–263.
145. A. Chatterjee, Y. Shibata, H. Tao, A. Tanaka and M. Morita, *J. Anal. At. Spectrom.*, 1999, **14**, 1853–1859.
146. A. Chatterjee, Y. Shibata, H. Tao, A. Tanaka and M. Morita, *Anal. Chem.*, 2000, **72**, 4402–4412.
147. G. A. Zachariadis and E. Rosenberg, *Talanta*, 2009, **78**, 570–576.
148. S. W. Jordan, I. S. Krull and S. B. Smith Jr, *Anal. Lett.*, 1982, **15**, 1131–1148.
149. D. R. Luffer and M. V. Novotny, *J. Microcolumn Sep.*, 2005, **3**, 39–46.
150. B. D. Quimby, P. C. Dryden and J. J. Sullivan, *Anal. Chem.*, 1990, **62**, 2509–2512.
151. Y. Tateyama, Y. Takano, Y. Tohno, Y. Moriwake, S. Tohno, M. Hashimoto and T. Araki, *Biol. Trends Elem. Res.*, 2000, **74**, 211–221.
152. K. B. Starowieyski, A. Chwojnowski, K. Jankowski, J. Lewinski and J. Zachara, *Appl. Organomet. Chem.*, 2000, **14**, 616–622.
153. J. Koch, M. Miclea and K. Niemax, *Spectrochim. Acta, Part B*, 1999, **54**, 1723–1735.
154. K. Jankowski, *Microchem. J*, 2001, **70**, 41–49.
155. K. Jankowski, A. Jerzak, A. Sernicka-Poluchowicz and L. Synoradzki, *Anal. Chim. Acta*, 2001, **440**, 215–221.
156. H. Matusiewicz, *Microchim. Acta*, 1993, **111**, 71–82.
157. K. Jankowski, *Talanta*, 2001, **54**, 855–862.
158. C. S. Cerbus and S. J. Gluck, *Spectrochim. Acta, Part B*, 1983, **38**, 387–397.
159. S. Dugenest, H. Casabianca and M. F. Grenier-Loustalot, *Analusis*, 1999, **27**, 75–81.
160. K. Jankowski, *Anal. Chim. Acta*, 1995, **317**, 365–369.
161. Y. Talmi and A. W. Andren, *Anal. Chem.*, 1974, **46**, 2122–2126.

162. D. J. C. Helmer and J. P. Walters, *Appl. Spectrosc.*, 1984, **38**, 392–398.
163. U. Engel, A. Kehden, E. Voges and J. A. C. Broekaert, *Spectrochim. Acta, Part B*, 1999, **54**, 1279–1289.
164. L. Hiddemann, J. Uebbing, O. Dessenne and K. Niemax, *Anal. Chim. Acta*, 1993, **283**, 152–159.
165. E. M. S. Frame, Y. Takamatsu and T. Suzuki, *Spectroscopy*, 1996, **11**, 17–22.
166. H. Saitoh, K. Kawachara, S. Ohshio, A. Nakamura and N. Nambu, *Sci. Technol. Adv. Mater.*, 2005, **6**, 205–209.
167. K. Kobayashi, A. Sato, T. Homma and T. Nagatomo, *Jpn. J. Appl. Phys.*, 2005, **44**, 1027–1030.
168. T. Okamoto and Y. Okamoto, *J. Plasma Fusion Res. Ser.*, 2009, **8**, 1330–1334.
169. M. Mishima, T. Suzuki and H. Takahara, *Yokogawa Tech. Rep.*, 2001, **31**, 5–10.
170. S. Tamura, T. Kikuchi, H. Takahara, M. Mishima and Y. Fujii, *Polar Meteorol. Glaciol.*, 2001, **15**, 124–132.
171. Y. Duan, Y. Su, Z. Jin and S. P. Abeln, *Anal. Chem.*, 2000, **72**, 1672–1679.
172. H. J. Perpall, P. C. Uden and R. L. Deming, *Spectrochim. Acta, Part B*, 1987, **42**, 243–251.
173. U. Fuchshieger, H. J. Grether and M. Grasserbauer, *Fresenius' J. Anal. Chem.*, 1994, **349**, 283–288.
174. R. Oguchi, A. Shimizu, S. Yamashita, K. Yamaguchi and P. Wylie, *J. High Resolut. Chromatogr.*, 1991, **14**, 412–416.
175. I. Rodriguez Pereiro and R. Łobiński, *J. Anal. At. Spectrom.*, 1997, **12**, 1381–1385.
176. Y. K. Chau, F. Yang and M. Brown, *Appl. Organomet. Chem.*, 1997, **11**, 31–37.
177. Y. Zeng, J. A. Seeley, T. M. Dowling, P. C. Uden and M. Y. Khuhawar, *J. High Resolut. Chromatogr.*, 1992, **15**, 669–676.
178. S. R. Goode and C. L. Thomas, *J. Anal. At. Spectrom.*, 1994, **9**, 73–78.
179. B. D. Quimby, P. C. Dryden and J. J. Sullivan, *J. High Resolut. Chromatogr.*, 1991, **14**, 110–116.
180. Y. Zeng and P. C. Uden, *J. High Resolut. Chromatogr.*, 1994, **17**, 217–222.
181. Y. Zeng and P. C. Uden, *J. High Resolut. Chromatogr.*, 1994, **17**, 223–229.
182. A. H. Hegazi, J. T. Andersson and M. S. El-Gayar, *Fuel Process. Technol.*, 2003, **85**, 1–19.
183. G. A. Depauw and F. Froment, *J. Chromatogr. A*, 1997, **761**, 231–247.
184. N. Kolbe, O. van Reihnberg and J. T. Andersson, *Energy Fuels*, 2009, **23**, 3024–3031.
185. C. Bradley and J. W. Carnahan, *Anal. Chem.*, 1988, **60**, 858–863.
186. J. Zhang, L. Li, J. Zhang, Q. Zhang, Y. Yang and Q. Jin, *Petrol. Sci. Technol.*, 2007, **25**, 443–451.
187. K. B. Olsen, J. C. Evans, D. S. Sklarew, J. S. Fruchter, D. C. Girvin and C. L. Nelson, *Environ. Sci. Technol.*, 1990, **24**, 258–263.
188. P. L. Lemieux, J. V. Ryan, N. B. French, W. J. Haas, Jr., S. J. Priebe and D. B. Burns, *Waste Manage.*, 1998, **18**, 385–391.

189. D. J. Butcher, *Microchem. J.*, 2000, **76**, 55–72.

190. M. Seeling and J. A. C. Broekaert, *Spectrochim. Acta, Part B*, 2001, **56**, 1747–1760.

191. P. P. Woskov, D. Y. Rhee, P. Thomas, D. R. Cohn, J. E. Surma and C. H. Titus, *Rev. Sci. Instrum.*, 1996, **67**, 3700–3707.

192. P. P. Woskov, K. Hadidi, P. Thomas, K. Green and G. Flores, *Waste Manage.*, 2000, **20**, 395–402.

193. E. A. H. Timmermans and J. J. A. M. van der Mullen, *Spectrosc. Eur.*, 2003, **15/5**, 14–21.

194. E. A. H. Timmermans, F. P. J. de Groote, J. Jonkers, A. Gamero, A. Sola and J. J. A. M. van der Mullen, *Spectrochim. Acta, Part B*, 2003, **58**, 823–836.

195. V. Siemens, T. Harju, T. Laitinen, K. Larjava and J. A. C. Broekaert, *Fresenius' J. Anal. Chem.*, 1995, **351**, 11–18.

196. H. Vermaak, O. Kujirai, S. Hanamura and J. D. Winefordner, *Can. J. Spectrosc.*, 1986, **31**, 95–99.

197. H. Baumann and K. G. Heumann, *Fresenius' Z. Anal. Chem.*, 1987, **327**, 186–192.

198. D. C. Reamer, W. H. Zoller and T. C. O'Haver, *Anal. Chem.*, 1978, **50**, 1449.

199. F. A. Serravallo and T. H. Risby, *Anal. Chem.*, 1976, **48**, 673–676.

200. T. Hamasaki, H. Nagase, Y. Yoshioka and T. Sato, *Crit. Rev. Environ. Sci. Technol.*, 1995, **25**, 45–91.

201. S. J. Rowland, R. Evens, L. Ebdon and A. W. G. Reeves, *Anal. Proc.*, 1992, **29**, 10–11.

202. J. A. Seeley, Y. Zeng, P. C. Uden, T. I. Eglinton and I. Ericson, *J. Anal. At. Spectrom.*, 1992, **7**, 979–985.

203. J. S. Sinninghe Damste, T. I. Eglinton and J. W. de Leeuw, *Geochim. Cosmochim. Acta*, 1992, **56**, 1743–1751.

204. P. Read, H. Beere, L. Ebdon, M. Leizers, M. Hetheridge and S. Rowland, *Org. Geochem.*, 1997, **26**, 11–17.

205. N. Campillo, R. Peñalver and M. Hernandez-Cordoba, *J. Chromatogr. A*, 2007, **1173**, 139–145.

206. P. Canosa, R. Montes, J. P. Lamas, M. Garcia-Lopez, I. Orriols and I. Rodriguez, *Talanta*, 2009, **79**, 598–602.

207. J. M. Gehlhausen and J. W. Carnahan, *Anal. Chem.*, 1991, **63**, 2430–2434.

208. K. Jankowski, A. Jackowska and M. Mrugalska, *J. Anal. At. Spectrom.*, 2007, **22**, 386–391.

209. K. Jankowski and M. I. Szynkowska, unpublished data.

210. K. Jankowski, A. Jackowska and P. Łukasiak, *Anal. Chim. Acta*, 2005, **540**, 197–205.

211. I. Rodriguez Pereiro, V. O. Schmitt and R. Łobiński, *Anal. Chem.*, 1997, **69**, 4799–4807.

212. J. Szpunar and R. Łobiński, *Fresenius' J. Anal. Chem.*, 1999, **363**, 550–557.

213. J. Szpunar, *Analyst*, 2000, **125**, 963–988.

214. J. Szpunar, *Analyst*, 2005, **130**, 442–465.
215. S. Aguerre, G. Lespes, V. Desauziers and M. Potin-Gautier, *J. Anal. At. Spectrom.*, 2001, **16**, 263–269.
216. Y. Liu, V. Lopez-Avila, M. Alcaraz and W. F. Beckert, *J. High Resolut. Chromatogr.*, 1994, **17**, 527–536.
217. Y. Liu, V. Lopez-Avila, M. Alcaraz and W. F. Beckert, *Anal. Chem.*, 1994, **66**, 3788–3796.
218. I. Rodriguez Pereiro, A. Wasik and R. Łobiński, *Chem. Anal. (Warsaw)*, 1997, **42**, 799–808.
219. I. Rodriguez Pereiro and A. Carro Diaz, *Anal. Bioanal. Chem.*, 2002, **372**, 74–90.
220. R. Reuther, L. Jaeger and B. Allard, *Anal. Chim. Acta*, 1999, **394**, 259–269.
221. J. M. Costa-Fernandez, F. Lunzer, R. Pereiro-Garcia, A. Sanz-Medel and N. Bordel-Garcia, *J. Anal. At. Spectrom.*, 1995, **10**, 1019–1025.
222. C. Dietz, Y. Madrid and C. Camara, *J. Anal. At. Spectrom.*, 2001, **16**, 1397–1402.
223. R. Łobiński, C. F. Boutron, J. P. Candelone, S. Hong, J. Szpunar-Łobiński and F. C. Adams, *Anal. Chem.*, 1993, **65**, 2510–2515.
224. J. Sanz Landaluze, A. De Diego, J. C. Raposo and J. M. Madariaga, *Anal. Chim. Acta*, 2003, **486**, 255–267.
225. S. Slaets and F. C. Adams, *Anal. Chim. Acta*, 2000, **414**, 141–149.
226. R. Feldhaus, W. Buscher, E. Kleine-Benne and P. Quevauviller, *Trends Anal. Chem.*, 2002, **21**, 356–365.

Non-emission Microwave Plasma Spectroscopic Techniques and Tandem Sources

12.1 Microwave Plasma Atomic Absorption Spectrometry

12.1.1 Instrumental Setup

As shown in the previous chapter, microwave plasmas (MWPs) can also serve as atom or ion reservoirs in a number of non-emission spectroscopic techniques. Among them, microwave plasma mass spectrometry (MWP-MS) has gained widespread acceptance. We will present shortly the analytical potential of the MWP atomization and ionization sources with respect to their characteristics and analytical applications. In atomic absorption spectrometry (AAS), the incident light of a wavelength characteristic of the element of interest passes through the atomic vapour produced in an atomization cell. A portion of the light absorbed by the atoms of that element is then measured and used to determine the concentration of that element in the sample. The instrumental setup, including a MWP as the atom source, is shown in Figure 12.1.

Considering the physical characteristics of MWPs, previously discussed in Chapters 1 and 5, one can conclude that in a MWP the ground state atoms are overpopulated. This leads to an increase of the light absorption efficiency. However, in most MWPs the absorption pathlength is not very long. Despite this, characteristic concentrations attainable by AAS with the use of typical microwave induced plasma (MIP) or capacitive microwave plasma (CMP) sources[1–3] are comparable with those for inductively coupled plasma AAS (ICP-AAS). In further studies the surfatron has been predominantly used to maintain a plasma, because in this case the plasma can be made long enough

RSC Analytical Spectroscopy Monographs No. 12
Microwave Induced Plasma Analytical Spectrometry
By Krzysztof J. Jankowski and Edward Reszke
© The Royal Society of Chemistry 2011
Published by the Royal Society of Chemistry, www.rsc.org

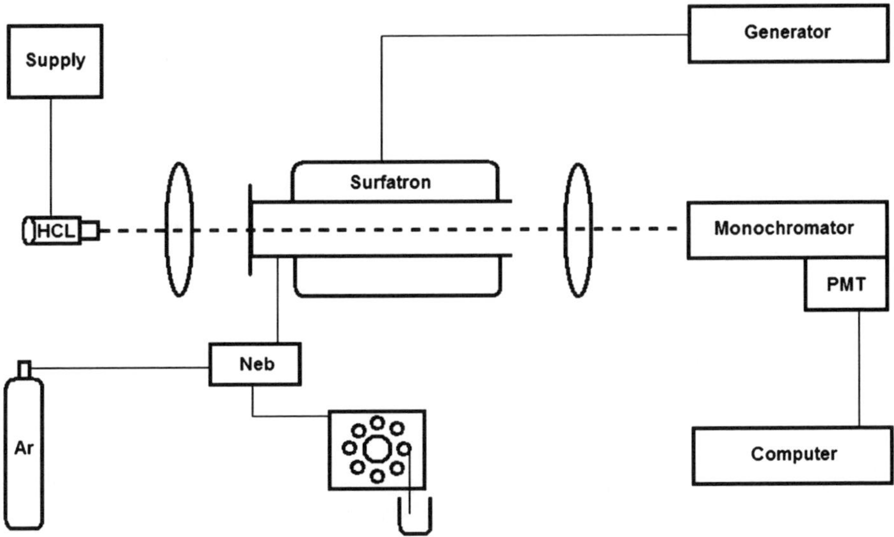

Figure 12.1   A schematic diagram of the MWP-AAS instrumental arrangement.

for AAS measurements.[4–6] This results in a substantial improvement of the analytical performance of the method. Since a MWP is generally operated at relatively low power and a much lower plasma gas flow rate than other plasma sources, and produces lower overall background, the use of a MWP as an atomizer for AAS should offer advantages of high sensitivity. The low aerosol flow rate makes a relatively long residence time for analyte atoms in the absorption volume. Moreover, traditional AAS primarily makes use of atomic resonance lines which are usually the prominent lines in MIP optical emission spectrometry (MIP-OES) spectra. A comparison of the characteristic concentrations for a number of AAS methods is given in Table 12.1.[4,5]

These data demonstrate that a MWP has potential as an atomization cell for AAS. The characteristic concentrations for MWP-AAS are several times better than those for flame AAS and even 10 to 10 000 better than those for ICP-AAS. The most significant problems faced in ICP-AAS are the reduced sensitivity and lower detection power due to the relatively high plasma gas flow rate, the low absorption volume and the high temperature that favours the production of ionic species. On the other hand, the performance of MWP-AAS is hindered by the susceptibility of low-power MWPs to solvent loading, especially for the surfatron as discussed in the Chapter 7. To improve further the analytical performance of MIP-AAS with solution nebulization, a high-efficiency desolvation system or an electrothermal vaporization technique has been employed. Detailed studies have shown an improvement in the detection capability for MIP-AAS by using a desolvation system.[6] The MWP-AAS linear dynamic ranges are usually around three orders of magnitude. The other studies have shown that the utilization of an atom trapping technique leads to a further

Table 12.1 Comparison of characteristic concentrations ($ng\,mL^{-1}$) for some elements obtained by AAS with different atomizers.

Element	Wavelength (nm)	MIP (surfatron)	ICP	Flame
Ag	328.1	1	4000	30
Ca	422.7	7	200	20
Cd	228.8	1	2000	90
Co	240.7	20	2000	40
Cu	324.7	6	2000	25
Fe	248.3	15	2000	50
Mg	285.3	5	70	3
Mn	279.5	7	800	20
Ni	232.0	30	300	50
Zn	213.9	1	10000	10

improvement in the sensitivity of MIP-AAS measurements, which could be even better than that of graphite furnace AAS (GF-AAS).[7]

To show analytical applications of the MIP-AAS arrangement, an electro-thermal vaporization (ETV) device has been coupled with the spectrometric system.[4] The characteristic concentrations reported for ETV-MIP-AAS are in the nanogram per millilitre range or higher. To examine the feasibility of the recommended method, some water samples were analyzed without any pre-treatment by using the method of standard addition. The results were in good agreement with those obtained by conventional flame AAS. However, wider applications of MWP-AAS suffer from somewhat more severe matrix effects than flame AAS.

The second application of MWPs in AAS consists of several instrumental arrangements in which a reduced-pressure MIP is used as an atomizer for diode laser AAS. This technique has been widely investigated by Niemax's group.[8–14] The use of low-pressure MWPs leads to a homogeneous plasma distribution in the whole atomization cell and a lowered background. The technique is based on wavelength modulation diode laser AAS in modulated low-pressure MIPs. The double modulation laser AAS eliminates not only flicker noise from the laser as well as from the plasma but also etalon effects that limit the detection in wavelength modulation diode laser atomic absorption spectrometry (WM-LAAS). For GC-MWP-OES the best detection limits (DLs) for chlorine are typically about $7\,pg\,s^{-1}$. This value is much larger than the DL for chlorine obtained by double modulation laser atomic absorption spectrometry (DM-LAAS) in MIPs and direct current plasmas (DCPs) (0.12 and $0.25\,pg\,s^{-1}$, respectively).[9] The method has been applied to the determination of chlorinated hydrocarbons in oil and chlorophenols in plant extracts.[15,16] The capability of the diode laser atomic absorption spectrometry (DLAAS) technique coupled with laser ablation has been demonstrated as a method for the analysis of polymer materials. The stoichiometry of a number of poly (vinyl chloride) samples has been reproduced and a detection limit for chlorine of $85\,\mu g\,g^{-1}$ has been obtained.[17,18]

12.2   Microwave Plasma Atomic Fluorescence Spectrometry

In atomic fluorescence spectrometry (AFS), a light source, such as that used for AAS, is used to excite atoms through radiative absorption transitions. When these selectivity excited atoms decay through radiative transitions to lower levels, their emission is measured to determine the concentration, much the same as in OES. Initially, Beenakker-type cavities were used for sample atomization.[19,20] However, this caused some limitations for AFS measurement since the plasma was located inside the discharge tube. The flame-like plasma of the microwave plasma torch (MPT) discharge proved to be beneficial for atomic fluorescence excitation and measurement where the fluorescence is viewed side-on.[21–24] Both ultrasonic and pneumatic nebulization with desolvation are usually used for coupling with MWP-AFS, as shown in Figure 12.2.

In general, an atomization cell for AFS should efficiently produce ground state atoms of the element being determined as well as having a high quantum efficiency and long residence time for the sample and also present a low background signal. The temperature of the tail flame of the MPT is relatively high and its size is relatively large in comparison with the outlet of the discharge tube placed in a conventional low-power MIP cavity. Also, the number density of the ground state is significantly greater. This makes the tail flame useful for observing the fluorescence, although it produces relatively higher background radiation than MIPs. In AFS, the optimal viewing position is usually higher than that used for OES. For either argon or helium low-power MIPs,[19,20] the fluorescence is viewed side-on at a position of 4–5 mm and 8 mm above the top of the cavity, respectively. For argon MPT-AFS the optimum observation

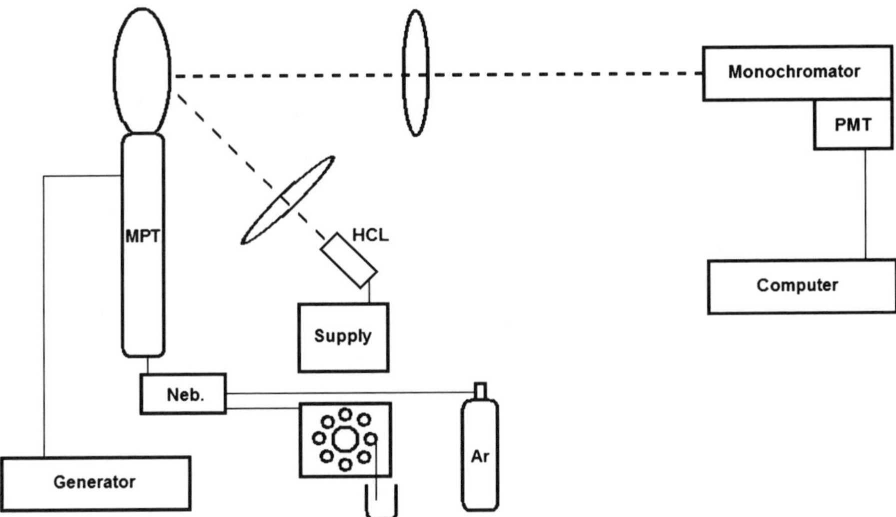

Figure 12.2   A schematic diagram of the MWP-AFS instrumental arrangement.

height is from approximately 24–26 mm above the top of the torch, depending on the element being determined.[22] This is much more beneficial for multi-element determination. In addition, low maintenance costs and a convenient operation are important for analytical applications. Apparently, the MPT provides a more convenient operation than traditional MIPs. To perform multi-element determinations, a compromise observation height is generally applied. Since in the case of the MPT the optimum observation height for different elements is almost identical, this does not lead to substantial sensitivity losses for individual elements. However, it is difficult to detect a large number of elements simultaneously using AFS, because the possibility of the spatial arrangement of several light sources and detectors is limited, especially in the case of MIPs.

Perkins and Long[19,20] used a low-power Ar or He MIP as the atomization source and hollow cathode lamps (HCLs) or a xenon arc lamp as the incident light source for AFS measurements. The DLs for the elements studied were in the low $\mu g\,mL^{-1}$ to high $ng\,mL^{-1}$ range. Their study showed that the DLs with He MIP-AFS are 4–50 times better than those obtained with Ar MIP-AFS. However, the DLs with Ar MPT-AFS are much better than those obtained with Ar MIP-AFS, as shown by the data in Table 12.2. Nevertheless, they are only comparable or worse than those obtained by Ar ICP-AFS. The linear dynamic ranges span more than five orders of magnitude for HCL-MIP-AFS or three for pulsed HCL-MPT-AFS.

Reductive carbonylation by means of simultaneous treatment of the sample with carbon monoxide and sodium borohydride is used for improvement of the determination of iron-group metals by MWP-AFS. The carbonyls released are collected by cold trapping and separated by GC before wide bandwidth AFS monitoring.[25] Absolute DLs for Fe, Co and Ni are 5, 20 and 3 pg, respectively. The method can be applied to the analysis of various materials, including semiconductors, ceramics, chemical reagents and environmental samples.

Table 12.2 MWP-AFS detection limits for some elements (3σ, $ng\,mL^{-1}$).

Element	Wavelength (nm)	HCL Ar-MIP[19]	HCL He-MIP[20]	Xe arc Ar-MIP[19]	HCL Ar-MPT[22]
Ag	328.1	60	15	105	42
Al	396.2	1050	120	1500	–
Ba	553.5	30	12	60	–
Ca	422.7	30	2.6	30	–
Co	240.7	1500	29	1500	18
Cr	357.9	3000	60	3000	–
Fe	248.3	900	45	1500	60
K	766.5	30	1.5	75	–
Li	670.8	30	1.8	75	–
Mg	285.2	30	2.0	105	6.3
Mn	279.5	750	–	600	15
Na	589.0	15	0.15	75	–
Sr	460.7	30	7.5	120	–
Zn	213.9	60	1.8	60	1.2

A MIP cavity in combination with a tungsten coil vaporization system was used for atomization of sodium for its laser-induced fluorescence detection at the $pg\,mL^{-1}$ level in pure water samples.[26]

12.3 Microwave Plasma Mass Spectrometry

Plasma mass spectrometry from its early beginnings used MWPs as ionization sources, which are attractive alternatives to the argon ICP. The first investigation of the MWP as an ion source for coupling with MS was published in 1981 by Douglas and French,[27] but the first commercial nitrogen MWP mass spectrometer appeared 10 years later.[28] MWP-MS was extensively studied during the last 20 years of the 20th century and the results are discussed in detail in some books and reviews.[29–34] Here, just a short summary will be given, along with an update covering the past 10 years.

MWP-MS has been demonstrated to be a powerful trace elemental analysis technique.[29–40] In general, replacing the ICP with the MWP with mass spectrometric detection brings the ability to utilize a variety of plasma gases and helps to minimize air entrainment into the plasma (especially for low-pressure MWPs arrangements), as well as to maintain a higher ion flux through the expansion stage. Other advantages of the MWP-MS include a decreased rate of sampling-orifice deterioration and reduced requirements for shielding electronic noise in comparison with ICPs. Additionally, low gas consumption MWPs allow arrangement of compact and technically simpler instruments and should provide lower DLs because of lowered analyte dilution. However, many of the difficulties associated with MWP-OES also occur with MWP-MS. In the MWP-MS technique, the analyte ions formed in the plasma are sent through a mass spectrometer where they are separated according to their mass-to-charge ratios (m/e). The number of ions of the same m/e value are then measured for both qualitative and quantitative purposes. A representative instrumental arrangement for elemental analysis by MWP-MS is shown in Figure 12.3.

However, the MWP-MS setup can be modified. Various sample introduction techniques are used, including solution nebulization,[36,39,41–43] electrothermal vaporization,[44] hydride generation,[45] chemical vapour generation,[46] gas chromatography,[47–52] supercritical fluid chromatography[53] and HPLC.[54–57] In order to optimize the ionization efficiency of the source and to improve plasma stability, various MWP constructions,[27,37–39,58–60] discharge tube arrangements[31,37,39,44,49,61,62] and ionization modes[31,32,39,42,63–66] are used. Moreover, various m/e separation modes can be used for the detection of ions produced by MWPs.[31,34,39,42,47,67–69]

One of the shortcomings of conventional MIP cavities is the filament shape of the plasma, which is easily influenced by a change in the experimental conditions. Moreover, the sample does not penetrate the plasma very well, but rather travels along the plasma column. Thus, the annular plasmas generated in MPTs as well as nitrogen or helium plasmas sustained in the Okamoto cavity are found to be very promising ion sources for coupling with MS. In the MWP-

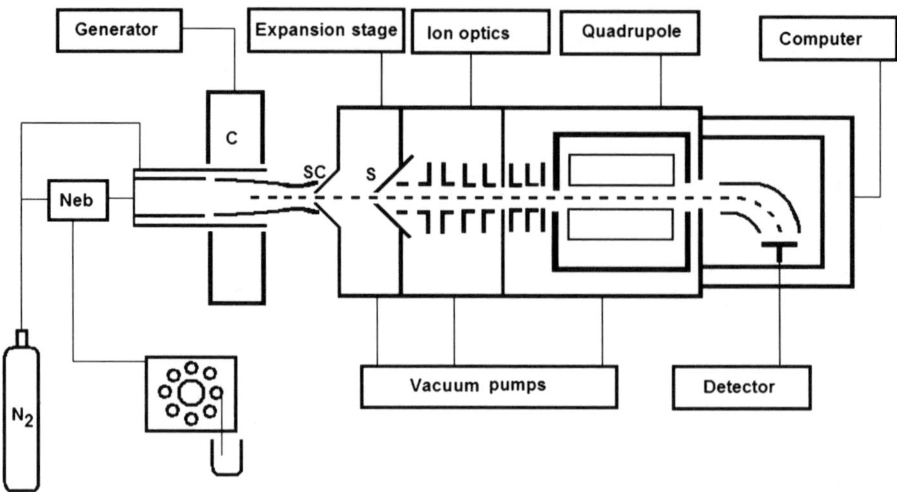

Figure 12.3   The MWP-MS instrumental arrangement: C, microwave cavity; SC, sampling cone; S, skimmer; neb, nebulizer.

MS system the interfacing orifice is less prone to clogging than is the case with the ICP-MS system;[27] however, the MWP-MS system is more susceptible to matrix effects.[35] In the reduced pressure (RP) MWP-MS system , the sampling cone can be replaced by the front plate of the cavity. Thus, the coupling of an RP-MWP to MS and the extraction of ions from the plasma to the ion optical system are easier and perhaps more efficient than for an atmospheric pressure MWP, and the air entrainment is eliminated.

The ionization efficiency of argon plasma for different elements has been shown to depend on their ionization potential and it cannot provide sufficient energy to efficiently ionize some of the non-metals with higher ionization potentials, especially halogens.[70] The second limitation is the presence of spectral isobaric and molecular interferences caused by either the Ar isotopes or Ar-containing polyatomic species. The helium MWP-MS has shown some favourable advantages, including a simpler background spectrum and the efficient production of monopositive ions from the hard-to-ionize halogens.[31] With the helium MIP-MS,[36,63,71] halogens in aqueous samples can be determined at the low to sub ppb level, while by using the He MPT-MS,[72] C, F, P, S, Cl, Br and I can be determined, with DLs ranging from $12\,\mathrm{ng\,mL^{-1}}$ to $1000\,\mathrm{ng\,mL^{-1}}$, as shown in Table 12.3.

In summary, N_2 MWP-MS based on the Okamoto cavity is the most matured and robust equivalent to ICP-MS for elemental analysis.[38,40] An annular-shaped plasma is formed which improves the analytical performance. No doubt, the utilization of MWPs for atomic MS should be regarded as a milestone in the historical development of MWPs. In general, the introduction of organic solvents into an MWP-MS system causes much less carbon deposition on the sampler and skimmer.[38,57] The use of N_2 MWP-MS is beneficial for environmental and clinical analysis for some elements, including As, Se, K, Ca

Table 12.3 Detection limits (3σ, $ng\,mL^{-1}$) measured in various MWP-MS systems.

Isotope	IP (eV)	Ar TM_{010}[27]	Ar surfatron[59]	He TM_{010}[36]	He MPT[72]	N_2-TM_{010}[73]	MP^a N_2-MIP[37]	HP^b N_2-MIP[39,40]	Ar ICP[29,40]
^{107}Ag	7.57	0.1	–	–	–	–	–	0.006	0.0017
^{75}As	9.81	–	4	0.10	1.2	–	0.32	0.02	0.004
^{138}Ba	5.21	–	7	–	0.11	11	0.004	0.0005	0.001
^{79}Br	11.84	–	–	0.18	15	–	–	–	1
^{40}Ca	6.11	–	–	2	–	22	0.24	0.003	n.m.^{c}
^{114}Cd	8.99	2	4	0.03	0.18	18	–	0.0005	0.002
^{35}Cl	13.01	–	–	39	41	–	–	–	5
^{59}Co	7.86	–	–	7	0.11	10	0.17	0.001	0.0002
^{52}Cr	6.76	0.1	–	–	–	–	0.44	0.001	0.014
^{63}Cu	7.72	0.1	–	0.07	0.14	–	–	0.005	0.009
^{56}Fe	7.87	–	–	33	0.20	–	6.8	0.001	0.077
^{127}I	10.45	–	–	0.04	12	–	–	–	0.01
^{39}K	4.34	–	–	17	–	–	0.48	0.002	n.m.^{c}
^{24}Mg	7.65	–	0.8	–	–	–	0.43	0.002	0.002
^{55}Mn	7.43	–	–	–	0.12	–	0.6	0.001	0.001
^{98}Mo	7.10	–	–	–	0.45	–	–	0.01	0.0008
^{60}Ni	7.63	8	–	–	0.21	–	–	0.007	0.002
^{208}Pb	7.42	1	2	2	0.12	–	0.008	0.0026	0.0008
^{80}Se	9.75	–	40	0.5	1.0	–	1.4	0.04	0.032
^{88}Sr	5.69	–	0.5	–	0.06	3	0.39	0.001	0.0003
^{51}V	6.74	0.02	–	0.02	–	–	0.31	0.001	0.0005

^{a}Medium pressure.
^{b}High pressure.
^{c}Not measurable.

and Fe, for which severe spectral interferences are observed in argon plasma MS.[74–79] Isotope dilution analysis has been widely applied to the determination of selenium,[80–85] and also some other elements by MWP-MS.[85,86]

A low-power low gas flow helium MWP-MS proved to be an excellent element-specific detector for GC, since helium is useful for the ionization of non-metals and is often used in GC as a carrier gas. Low DLs for non-metals and wide linear dynamic ranges up to four orders of magnitude in the positive ion mode have been reported.[29–31,47,49,51] A chemical vapour generation technique has been employed for S, Cl, Br and I determination with the He MPT-MS,[42,46] leading to improvement of DLs in the range from high $ng\,mL^{-1}$ to low $pg\,mL^{-1}$. In gas chromatography, the signals are very short and simultaneous measurements are preferred for multi-element detection without risking skew, as required for the determination of empirical formulae. Using a helium MIP ion source coupled with an orthogonal acceleration time-of-flight mass spectrometer, empirical formulae based on halogen-to-carbon ratios for both aliphatic and aromatic hydrocarbons can be determined. Absolute DLs for the halogenated hydrocarbons ranging from 160 to 330 fg (10^{-15} g) were obtained.[47] Various chromatographic techniques coupled with MWP-MS have been used for multi-element analysis and speciation studies of natural samples.[33,34,50,56,87–91] Using MWP-MS coupled to HPLC with the aid of a dual oscillating capillary nebulizer, underivatized amino acids can be directly determined.[55,92]

Most MWPs used as sources for analytical spectrometry have been mainly restricted to elemental analysis by OES and MS. However, low-pressure (LP) MIPs operating at MW power levels below 100 W can be used as molecular ionization sources for MS. The degree of fragmentation and atomization of the sample molecules can be controlled by modification of the plasma operating conditions, including both the pressure and power level, the choice of plasma gas and doping of the plasma gas with scavangers or by change of the plasma position with respect to the sample injector probe. Poussel *et al.*[93] used a LP surfatron MIP interfaced to a MS and obtained soft ionization and fragmentation. Heppner combined GC with LP MIP-MS and obtained fragmentation of organic compounds to a greater degree for higher power levels.[94] Markey and Abramson proposed soft fragmentation for the determination of radioactive carbon-containing compounds.[95] Olson *et al.*[96] used a LP MIP for fragmentation of organic compounds where the sample was introduced into the expansion stage of the mass spectrometer and thus to the tail flame of the plasma. In more recent studies, MWP sources operating at atmospheric pressure have also been used for soft fragmentation of various organic compounds.[43,97,98]

12.4 Microwave Plasma Cavity Ringdown Spectroscopy

Cavity ringdown spectroscopy (CRDS) has been explored as a sensitive spectroscopic technique for elemental analysis since 1997.[99] In principle, it consists of light absorption by ground state atoms. Utilizing various plasma sources (including ICPs and MWPs) as atomizers and different lasers, several elements have been determined and the DLs have ranged from $ng\,mL^{-1}$ to $pg\,mL^{-1}$. In this technique, the laser beam is introduced to the stable optical cavity between two mirrors where ground state atoms are generated by the atomization source. The intensity of the light trapped between these mirrors decays exponentially with time due to the finite mirror reflectivity, optical absorption and scattering. The absorbance corresponds to the ringdown lifetimes (exponential decays) of the same cavity when the blank and an analyte are introduced, respectively. Tunable lasers or continuous wave diode lasers are used for atomic absorption measurements. Owing to the much longer pathlength, CRDS provides orders of magnitude improvement in sensitivity over conventional AAS. As discussed in Section 12.1, an ICP source is not beneficial for atomic absorption measurements, in particular because they need ground level populations while the high temperatures in the ICP favour the production of atomic excited states as well as ions. By contrast, MWPs offer some attractive characteristics, including the low gas flow rate, low power level and relatively high atomization efficiency. Moreover, MWP-CRDS exhibits the unique feature that the decay baseline stability is not affected by the plasma.[100] To improve the portability and reduce the cost of the system, both compact MWPs and a continuous wave diode laser are used.[101] The possibility of the determination of strontium has been demonstrated and a DL of about $1\,pg\,mL^{-1}$ has been estimated. For mercury[102,103] and lead,[104] DLs of 0.4 and $0.8\,ng\,mL^{-1}$ were obtained, respectively. A further improvement in the detection limit would result from increasing the effective pathlength of the laser beam in the atomization cell. Possible solutions include the utilization of a tube-shaped MWP and the use of a reduced pressure MWP which fills the whole optical cavity.[105] When the latter was used for sustaining a helium plasma, the DL for fluorine was estimated to be $100\,ng\,g^{-1}$.

12.5   Tandem Sources and Miscellaneous

The low-power MIP is known to be susceptible to the introduction of large amounts of water aerosols as well as high concentrations of organics. However, the MWP has unique excitation and/or ionization conditions. Hence, it is common practice to couple it with energized sampling techniques according to the "combined" or "tandem source" principle.[106,107] The use of separate plasma sources for sample volatilization and excitation/ionization is known as the "tandem source" concept. Initially, the use of two plasma sources in series was implemented without differentiation of the atomization and excitation sections. Freeman and Hieftje[108] used two MIPs in series in order to increase the plasma length for the determination of elemental ratios of some halogenated organic compounds. An increase of analyte intensities was observed in comparison with a single MIP operation. In a series of papers, Leis *et al.*[109,110] described an improvement of the glow discharge (GD) source with microwave boosting. The use of a MIP for boosting the emission from the GD caused a significant enhancement of signal-to-background ratios for many analytical lines of some elements determined in steel, aluminum, copper and lead samples compared with a conventional GD. The MIP-MIP atomization/excitation tandem source for OES was explored by Ng and Chen[111] for determination of aqueous Ca, Cr, Mn and Sr; however, no essential improvement was observed in comparison with MIP-OES. With the use of an ICP-MIP atomization/excitation tandem source, efficient solution analysis can be performed.[112] By utilizing power modulation of the MIP source, the stability of the background signals in atomic emission could be improved and the DLs for a number of elements ranged from 0.5 to 40 ng mL^{-1}.[113] Glow discharge proved to be an effective primary source of tandem apparatus for analysis of gaseous, liquid and solid samples. The GD-MIP atomization/excitation tandem source was developed by Duan *et al.*[114] and an enhancement of signal measured was observed with increasing microwave power. More recently, a compact GD-MIP ion source was coupled with TOF-MS.[115] Finally, the microwave plasma torch–atmospheric sampling glow discharge modulated tandem source was investigated for sequential acquisition of both molecular fragmentation and atomic mass spectra.[116] Microwave radiation was also used for constructing "pre-vaporization sources", namely microwave powered thermospray nebulizers.[117,118] Although tandem sources show some promise for improving the analytical performance of spectrometric techniques, they still need further optimization to prove their attractiveness for analytical applications.

References

1. K. C. Ng, R. S. Jensen, M. J. Brechmann and W. C. Santos, *Anal. Chem.*, 1988, **60**, 2818–2821.
2. K. C. Ng and T. J. Garner, *Appl. Spectrosc.*, 1993, **47**, 241–243.
3. D. C. Liang and M. W. Blades, *Anal. Chem.*, 1988, **60**, 27–31.
4. Y. Duan, X. Li and Q. Jin, *J. Anal. At. Spectrom.*, 1993, **8**, 1091–1096.
5. Y. Duan, M. Hou, Z. Du and Q. Jin, *Appl. Spectrosc.*, 1993, **47**, 1871–1879.
6. Y. Duan, H. Zhang, M. Huo and Q. Jin, *Spectrochim. Acta, Part B*, 1994, **49**, 583–592.
7. H. Lu, Y. Ren, H. Zhang and Q. Jin, *Microchem. J.*, 1991, **44**, 86–92.
8. C. Schnürer-Patschan and K. Niemax, *Spectrochim. Acta, Part B*, 1995, **50**, 963–969.
9. A. Zybin, C. Schnürer-Patschan and K. Niemax, *J. Anal. At. Spectrom.*, 1995, **10**, 563–567.
10. V. Liger, A. Zybin, Y. Kuritsyn and K. Niemax, *Spectrochim. Acta, Part B*, 1997, **52**, 1125–1138.
11. J. Koch and K. Niemax, *Spectrochim. Acta, Part B*, 1998, **53**, 71–79.
12. K. Kunze, A. Zybin, J. Koch, J. Franzke, M. Miclea and K. Niemax, *Spectrochim. Acta, Part A*, 2004, **60**, 3393–3401.
13. A. Zybin, J. Koch, D. J. Butcher and K. Niemax, *J. Chromatogr. A*, 2004, **1050**, 35–44.

14. A. Zybin, J. Koch, H. D. Wizemann, J. Franzke and K. Niemax, *Spectrochim. Acta, Part B*, 2005, **60**, 1–11.
15. A. Zybin and K. Niemax, *Anal. Chem.*, 1997, **69**, 755–757.
16. J. Koch, A. Zybin and K. Niemax, *Appl. Phys. B*, 1998, **67**, 475–479.
17. J. Koch, M. Miclea and K. Niemax, *Spectrochim. Acta, Part B*, 1999, **54**, 1723–1735.
18. J. Koch, A. Zybin and K. Niemax, *Spectrosc. Spectr. Anal.*, 2000, **20**, 149–155.
19. L. D. Perkins and G. L. Long, *Appl. Spectrosc.*, 1988, **42**, 1285–1289.
20. L. D. Perkins and G. L. Long, *Appl. Spectrosc.*, 1989, **43**, 499–504.
21. Y. Duan, X. Kong, H. Zhang, J. Liu and Q. Jin, *J. Anal. At. Spectrom.*, 1992, **7**, 7–10.
22. Y. Duan, X. Du, Y. Li and Q. Jin, *Appl. Spectrosc.*, 1995, **49**, 1079–1085.
23. W. Yang, H. Zhang, A. Yu and Q. Jin, *Microchem. J.*, 2000, **66**, 147–170.
24. Z. B. Gong, F. Liang, P. Y. Yang, Q. H. Jin and B. L. Huang, *Spectrosc. Spectr. Anal.*, 2002, **22**, 63–66.
25. V. Rigin, *Anal. Chim. Acta*, 1993, **283**, 895–901.
26. Y. Oki, H. Uda, C. Honda, M. Maeda, J. Izumi, T. Morimoto and M. Tanoura, *Anal. Chem.*, 1990, **62**, 680–683.
27. D. J. Douglas and J. B. French, *Anal. Chem.*, 1981, **53**, 37–41.
28. K. Oishi, Y. Okamoto, M. Koga and H. Yamamoto, *Hitachi Hyoron*, 1991, **73**, 885.
29. H. Zhang, A. Montaser, B. S. Zimmer, N. P. Vela and J. A. Caruso, in *Inductively Coupled Plasma Mass Spectrometry*, ed. A. Montaser, Wiley-VCH, Weinheim, 1998, pp. 891–939.
30. E. H. Evans, J. J. Giglio, T. M. Castillano and J. A. Caruso, *Inductively Coupled and Microwave Induced Plasma Sources for Mass Spectrometry*, Royal Society of Chemistry, Cambridge, 1995.
31. L. K. Olson and J. A. Caruso, *Spectrochim. Acta, Part B*, 1994, **49**, 7–30.
32. B. S. Sheppard and J. A. Caruso, *J. Anal. At. Spectrom.*, 1994, **9**, 145–149.
33. N. P. Vela, L. K. Olson and J. A. Caruso, *Anal. Chem.*, 1993, **65**, 585A–597A.
34. S. J. Ray, F. Andrade, G. Gamez, D. McClenathan, D. Rogers, G. Schilling, W. Wetzel and G. M. Hieftje, *J. Chromatogr. A*, 2004, **1050**, 3–34.
35. D. J. Douglas, E. S. Quan and R. G. Smith, *Spectrochim. Acta, Part B*, 1983, **38**, 39–48.
36. J. T. Creed, T. M. Davidson, W. L. Shen, P. G. Brown and J. A. Caruso, *Spectrochim. Acta, Part B*, 1989, **44**, 909–924.
37. W. L. Shen, T. M. Davidson, J. T. Creed and J. A. Caruso, *Appl. Spectrosc.*, 1990, **44**, 1003–1010.
38. T. Shirasaki and K. Yasuda, *Anal. Sci.*, 1992, **8**, 375–376.

39. K. Oishi, T. Okumoto, T. Iino, M. Koga, T. Shirasaki and N. Furuta, *Spectrochim. Acta, Part B*, 1994, **49**, 901–914.
40. M. Ohata and N. Furuta, *J. Anal. At. Spectrom.*, 1998, **13**, 447–453.
41. M. Huang, A. Hirabayashi, T. Shirasaki and H. Koizumi, *Anal. Chem.*, 2000, **72**, 2463–2467.
42. Y. Duan, E. P. Chamberlin and J. A. Olivares, *Int. J. Mass Spectrom. Ion Process.*, 1996, **161**, 27–39.
43. W. L. Shen and R. D. Satzger, *Anal. Chem.*, 1991, **63**, 1960–1964.
44. E. H. Evans, J. A. Caruso and R. D. Satzger, *Appl. Spectrosc.*, 1991, **45**, 1478–1484.
45. W. C. Wetzel, J. A. C. Broekaert and G. M. Hieftje, *Spectrochim. Acta, Part B*, 2002, **57**, 1009–1023.
46. Y. Duan, M. Wu, Q. Jin and G. M. Hieftje, *Spectrochim. Acta, Part B*, 1995, **50**, 1095–1108.
47. B. W. Pack, J. A. C. Broekaert, J. P. Gruzowski, J. Poehlman and G. M. Hieftje, *Anal. Chem.*, 1998, **70**, 3957–3963.
48. G. O'Connor, S. J. Rowland and E. H. Evans, *J. Sep. Sci.*, 2002, **25**, 839–846.
49. W. C. Story and J. A. Caruso, *J. Anal. At. Spectrom.*, 1993, **8**, 571–575.
50. H. Suyani, J. Creed, J. Caruso and R. D. Satzger, *J. Anal. At. Spectrom.*, 1989, **4**, 777–782.
51. L. K. Olson, D. Heitkemper and J. A. Caruso, in *Element-Specific Chromatographic Detection by Atomic Emission Spectroscopy*, ed. , ACS Symposium Series 479, ACS, Washington, DC, 1992, pp. 288–308.
52. D. Heitkemper and J. A. Caruso, *J. Chromatogr. Libr.*, 1991, **47**, 49–73.
53. L. K. Olson and J. A. Caruso, *J. Anal. At. Spectrom.*, 1992, **7**, 993–998.
54. A. Chatterjee, Y. Shibata, H. Tao, A. Tanaka and M. Morita, *J. Chromatogr. A*, 2004, **1042**, 99–106.
55. J. Y. Kwon and M. Moini, *J. Am. Soc. Mass Spectrom.*, 2001, **12**, 117–122.
56. K. Kohda, T. Shirasaki, T. Yokokura and M. Takagi, *Chromatography*, 1998, **19**, 274–275.
57. D. Heitkemper, J. Creed and J. A. Caruso, *J. Chromatogr. Sci.*, 1990, **28**, 175–181.
58. M. Wu, Y. Duan, Q. Jin and G. M. Hieftje, *Spectrochim. Acta, Part B*, 1994, **49**, 901–914.
59. D. Boudreau and J. Hubert, *Appl. Spectrosc.*, 1993, **47**, 609–614.
60. P. Siebert, G. Petzold, A. Hellenbart and J. Müller, *Appl. Phys. A*, 1998, **67**, 155–160.
61. R. D. Satzger and T. W. Brueggemeyer, *Microchim. Acta*, 1989, **99**, 239–246.
62. K. Eberhardt, G. Buchert, G. Herrmann and N. Trautmann, *Spectrochim. Acta, Part B*, 1992, **47**, 89–94.

63. P. G. Brown, T. M. Davidson and J. A. Caruso, *J. Anal. At. Spectrom.*, 1988, **3**, 763–769.
64. M. E. Cisper, A. W. Garrett, Y. X. Duan, J. A. Olivares and P. H. Hemberger, *Int. J. Mass Spectrom.*, 1998, **178**, 121–128.
65. P. Španel, K. Dryahina and D. Smith, *Int. J. Mass Spectrom.*, 2007, **267**, 117–124.
66. R. S. Lysakowski Jr, R. E. Dessy and G. L. Long, *Appl. Spectrosc.*, 1989, **43**, 1139–1145.
67. Y. Su, Y. Duan and Z. Jin, *Anal. Chem.*, 2000, **72**, 2455–2462.
68. Y. Su, Z. Jin and Y. Duan, *Can. J. Anal. Sci. Spectrosc.*, 2003, **48**, 163–170.
69. J. H. Barnes IV, O. A. Grøn and G. M. Hieftje, *J. Anal. At. Spectrom.*, 2002, **17**, 1132–1136.
70. J. J. Giglio, J. Wang and J. A. Caruso, *Appl. Spectrosc.*, 1995, **49**, 314–319.
71. P. A. Fecher and A. Nagengast, *J. Anal. At. Spectrom.*, 1994, **9**, 1021–1027.
72. M. Wu, Y. Duan, Q. Jin and G. M. Hieftje, *Spectrochim. Acta, Part B*, 1994, **49**, 137–148.
73. D. A. Wilson, G. H. Vickers and G. M. Hieftje, *Anal. Chem.*, 1987, **59**, 1664–1670.
74. A. Shinohara, M. Chiba and Y. Inaba, *Anal. Sci.*, 1998, **14**, 713–717.
75. M. Chiba and A. Shinohara, *Fresenius' J. Anal. Chem.*, 1999, **363**, 749–752.
76. Y. Sano, H. Satoh, M. Chiba, A. Shinohara, M. Okamoto, K. Serizawa, H. Nakashima and K. Omae, *J. Occup. Health*, 2005, **47**, 242–248.
77. M. Chiba, A. Shinohara, M. Sekine and S. Hiraishi, *J. Radioanal. Nucl. Chem.*, 2006, **269**, 519–526.
78. A. Shinohara, M. Chiba and Y. Inaba, *Anal. Sci.*, 2001, **17** (suppl), i1539–i1541.
79. A. Shinohara, M. Chiba and Y. Inaba, *J. Anal. Toxicol.*, 1999, **23**, 625–631.
80. H. Minami, W. Cai, T. Kusumoto, K. Nishikawa, Q. Zhang, S. Inoue and I. Atsuya, *Anal. Sci.*, 2003, **19**, 1359–1363.
81. J. Yoshinaga, T. Shirasaki, K. Oishi and M. Morita, *Anal. Chem.*, 1995, **67**, 1568–1574.
82. M. Ohata, T. Ichinose, N. Furuta, A. Shinohara and M. Chiba, *Anal. Chem.*, 1998, **70**, 2726–2730.
83. T. Shirasaki, J. Yoshinaga, M. Morita, T. Okumoto and K. Oishi, *Tohoku J. Exp. Med.*, 1996, **178**, 81–90.
84. T. Kawano, A. Nishide, K. Okutsu, H. Minami, Q. Zhang, S. Inoue and I. Atsuya, *Spectrochim. Acta, Part B*, 2005, **60**, 327–331.
85. C. J. Park, Y. N. Pak and K. W. Lee, *Anal. Sci.*, 1992, **8**, 443–448.
86. T. Shirasaki, H. Sakamoto, Y. Nakaguchi and K. Hiraki, *Bunseki Kagaku*, 2000, **49**, 175–179.

87. P. Read, H. Beere, L. Ebdon, M. Leizers, M. Hetheridge and S. Rowland, *Org. Geochem.*, 1997, **26**, 11–17.

88. A. Chatterjee, Y. Shibata and M. Morita, *J. Anal. At. Spectrom.*, 2000, **15**, 913–919.

89. A. Chatterjee, Y. Shibata, A. Tanaka and M. Morita, *Anal. Chim. Acta*, 2001, **436**, 253–263.

90. A. Chatterjee, Y. Shibata and M. Morita, *Recent Res. Dev. Pure Appl. Chem.*, 2001, **3**, 49–66.

91. A. Chatterjee, Y. Shibata, J. Yoshinaga and M. Morita, *J. Anal. At. Spectrom.*, 1999, **14**, 1853–1859.

92. M. Moini, M. Xia, J. B. Stewart and B. Hofmann, *J. Am. Soc. Mass Spectrom.*, 1998, **9**, 42–49.

93. E. Poussel, J. M. Mermet, D. Deruaz and C. Beaugrand, *Anal. Chem.*, 1988, **60**, 923–927.

94. R. A. Heppner, *Anal. Chem.*, 1983, **55**, 2170–2174.

95. S. P. Markey and F. P. Abramson, *Anal. Chem.*, 1982, **54**, 2375–2376.

96. L. K. Olson, W. C. Story, J. T. Creed, W. L. Shen and J. A. Caruso, *J. Anal. At. Spectrom.*, 1990, **5**, 471–475.

97. N. P. Vela, J. A. Caruso and R. D. Satzger, *Appl. Spectrosc.*, 1997, **51**, 1500–1503.

98. A. M. Zapata and A. Robbat, Jr., *Anal. Chem.*, 2000, **72**, 3102–3108.

99. G. P. Miller and C. B. Winstead, *J. Anal. At. Spectrom.*, 1997, **12**, 907–912.

100. C. Wang, *J. Anal. At. Spectrom.*, 2007, **22**, 1347–1363.

101. C. Wang, S. P. Koirala, S. T. Scherrer, Y. Duan and C. B. Winstead, *Rev. Sci. Instrum.*, 2004, **75**, 1305–1313.

102. C. Wang, S. T. Scherrer, Y. Duan and C. B. Winstead, *J. Anal. At. Spectrom.*, 2005, **20**, 638–644.

103. Y. Duan, C. Wang, S. T. Scherrer and C. B. Winstead, *Anal. Chem.*, 2005, **77**, 4883–4889.

104. Y. Duan, C. Wang and C. B. Winstead, *Anal. Chem.*, 2003, **75**, 2105–2111.

105. T. Stacewicz, E. Bulska and A. Ruszczynska, *Spectrochim. Acta, Part B*, 2010, **65**, 306–310.

106. M. W. Borer and G. M. Hieftje, *Spectrochim. Acta Rev.*, 1991, **14**, 463–486.

107. T. Kantor and G. M. Hieftje, *Spectrochim. Acta, Part B*, 1995, **50**, 961–962.

108. J. E. Freeman and G. M. Hieftje, *Spectrochim. Acta, Part B*, 1985, **40**, 653–664.

109. F. Leis, J. A. C. Broekaert and K. Laqua, *Spectrochim. Acta, Part B*, 1987, **42**, 1169–1176.

110. F. Leis, J. A. C. Broekaert and E. B. M. Steers, *Spectrochim. Acta, Part B*, 1991, **46**, 243–251.

111. K. C. Ng and S. Chen, *Microchem. J.*, 1993, **48**, 383–389.

112. M. W. Borer and G. M. Hieftje, *J. Anal. At. Spectrom.*, 1993, **8**, 339–348.

113. B. W. Pack and G. M. Hieftje, *Appl. Spectrosc.*, 2000, **54**, 80–88.
114. Y. Duan, Y. Li, Z. Du, Q. Jin and J. A. Olivares, *Appl. Spectrosc.*, 1996, **50**, 977–984.
115. Y. Duan, Y. Su and Z. Jin, *J. Anal. At. Spectrom.*, 2000, **15**, 1289–1291.
116. S. J. Ray and G. M. Hieftje, *Anal. Chim. Acta*, 2001, **445**, 35–45.
117. L. Bordera, J. L. Todoli, J. Mora, A. Canals and V. Hernandis, *Anal. Chem.*, 1997, **69**, 3578–3586.
118. L. Ding, F. Liang, Y. Huan, Y. Cao, H. Zhang and Q. Jin, *J. Anal. At. Spectrom.*, 2000, **15**, 293–296.

The Future for Microwave Plasma Spectrometry

There is no doubt that MWP analytical spectrometry will continue to receive the attention of spectroscopists. Moreover in light of recent papers microwave excitation sources may be observed as those just having their renaissance era. MWP-OES/MS has proved to be a versatile analytical technique of wide scope and coverage. Although still approaching the precision and detection limits of ICPs, MWP systems are continually being improved in terms of power handling capabilities, efficiency of coupling and in the analysis of solid, liquid and gas samples. If one compares ICP-OES and MWP-OES, it is clear that in trace analysis of non-metals the MWP method tends to be superior or at least competitive. The present powerful techniques such as ICP-MS lack the capability to determine trace non-metals and some essential elements such as Se, As, K and Ca. Helium MWP sources may find more extensive application. Further improvement of analytical performance by using both nitrogen and helium high-power MWPs as sources for analytical spectrometry should be advocated.

It is difficult to design an instrument that is capable of meeting the needs of all users. However, it seems that analytical spectrometry instruments will become targeted towards increasingly specialized applications. Similarly, it seems that a higher level of integration between sample processing/introduction techniques and spectrometric instruments is likely to become closer. Although solution nebulization is the most common sample introduction technique, it exhibits several shortcomings. The greatest impact on the analytical performance of MWP atomic spectrometry techniques has derived from sample processing and introduction. MWP-based techniques usually require more extensive sample pretreatment. The growing interest in laser ablation and related techniques, as well as modern preconcentration techniques, meets the requirements of MWPs sources. In this respect, MWP technology seems to be highly flexible and ready to bring some interesting solutions in future. As these

RSC Analytical Spectroscopy Monographs No. 12
Microwave Induced Plasma Analytical Spectrometry
By Krzysztof J. Jankowski and Edward Reszke
© The Royal Society of Chemistry 2011
Published by the Royal Society of Chemistry, www.rsc.org

systems continue to become more refined, MWPs will certainly play more prominent roles in commercial atomic emission spectroscopy. Further improvements are definitely possible. The instrumentation and commercialization of MWP-AS techniques should be a hot topic in the near future.

MWP-OES continues to be used primarily as a chromatographic detector. MPD has been accepted and adopted in many laboratories and can be expected to hold its strong position among the established analytical methods within the next years. This is especially true for non-metal detection, because speciation of non-metals such as P, S, Cl, Se and As is extremely important in some environmental and biomedical studies. The use of MWP-OES for particle characterization by elemental analysis and size distribution seems to be a new topic for MWP applications.

The development of MWPs will continue with the use of MW microplasmas based on strip-line technology, being a unique way for realizing electrodeless miniaturized systems. They might become especially useful as specific detectors for gas chromatography and continuous air monitoring for trace components.

Future studies will need to give priority to new MW constructions, including MW-driven ICPs and rotating field assemblies, effect of different power levels (and power conditions, including power modulation as well as spatial distribution of power sources), and last but not least fundamental investigations. It appears that large strides have already been made toward these goals, and systems characterizations are underway. The recent analytical results on the use of these MWPs as multi-element excitation and ionization sources offer considerable promise that will surely encourage their further development. The good detectability to halogens and high sample tolerance give a solid foundation for MWPs with rotating fields to be used as ionization sources for elemental MS. Actually, owing to the low power feature of the MWP, it is also feasible to be used for soft fragmentation and molecular MS. Efforts to obtain atomic and molecular information simultaneously by MWP-MS should be encouraged. The new multi-helix or multi-electrode MWP energizing methods bring the hope for generating the plasma with parameters well suited for MS. Splitting of the energy led to quite a set of new constructions. The most expected among them was a MW-driven ICP. However, at the moment of this publication one has to say that no further research has been performed and the hope is that in the future at least some researchers will find interest in evaluation of those new cavities. Potentially successful may also be the idea of split multi-waveguide energizing of plasma invented by Hammer. However, with the use of three and more phases one can generate rotating electric fields which can heat the plasma in a directionless manner. In the closing remarks one must state that the new possibilities in excitation sources have to be discussed along with the new technical conditions which must accompany the heat sources. Too often though, it is only the detection limits that are discussed in the context of acceptable performance. Researchers should realize that MWPS, and for sure those recognized as classical, routinely call for extensive sample pretreatment or dedicated sample introduction technique for satisfactory low detection limits to be seen.

Appendix

Most used optical emission lines in MIP-OES

Element		Wavelength, nm	DL_{exp} ng mL^{-1}
Ag[a]	I	338.29	9
Al[a]	I	396.15	110
As[a]	I	228.81	90
Au[a]	I	267.59	90
B[a]	I	249.77	110
Ba[a]	II	493.41	110
Be[a]	I	234.86	2
Bi[a]	I	223.06	80
Br[b]	II	470.49	20
C[a]	I	193.09	50
Ca[a]	II	393.37	3
Cd[a]	I	228.80	5
Ce[a]	II	413.77	80
Cl[b]	II	479.54	14
Co[a]	II	238.89	85
Cr[a]	I	357.87	45
Cs[a]	I	455.53	65
Cu[a]	I	324.75	20
Dy[a]	II	364.54	36
Er[a]	II	390.63	44
Eu[a]	II	420.51	12
F[b]	I	685.60	4000
Fe[a]	I	248.32	45
Ga[a]	I	417.21	16
Gd[a]	II	342.25	40
Ge[a]	I	265.12	65

RSC Analytical Spectroscopy Monographs No. 12
Microwave Induced Plasma Analytical Spectrometry
By Krzysztof J. Jankowski and Edward Reszke
© The Royal Society of Chemistry 2011
Published by the Royal Society of Chemistry, www.rsc.org

Appendix (*Continued*).

Element		Wavelength, nm	$DL_{exp}\,ng\,mL^{-1}$
Hf[a]	II	368.22	37
Hg[a]	I	253.65	23
Ho[a]	II	389.10	20
I[b]	I	206.24	60
In[a]	I	451.13	18
Ir[a]	I	380.01	130
K[a]	I	766.49	2
La[a]	II	408.67	60
Li[a]	I	670.78	0.3
Lu[a]	II	261.54	7
Mg[a]	II	279.55	7
Mn[a]	I	403.08	25
Mo[a]	II	202.03	350
Na[a]	I	588.99	0.9
Nb[a]	I	405.89	580
Nd[a]	II	410.95	110
Ni[a]	I	232.00	90
Os[a]	I	201.81	600
P[a]	I	213.61	70
Pb[a]	I	405.78	60
Pd[a]	I	340.46	55
Pr[a]	II	390.84	230
Pt[a]	I	265.95	160
Rb[a]	I	420.18	65
Re[a]	II	227.53	75
Rh[a]	I	343.49	60
Ru[a]	I	372.80	85
S[b]	I	469.41	70
Sb[a]	I	252.85	50
Sc[a]	I	391.18	27
Se[a]	I	203.99	47
Si[a]	I	251.61	75
Sm[a]	II	359.26	170
Sn[a]	I	235.48	190
Sr[a]	II	407.77	7
Ta[a]	II	268.51	750
Tb[a]	II	384.87	340
Te[a]	I	214.28	140
Th[a]	II	401.91	160
Ti[a]	II	334.94	70
Tl[a]	I	535.05	17
Tm[a]	II	384.80	23
U[a]	II	385.96	360

Appendix (*Continued*).

Element		Wavelength, nm	$DL_{exp}\, ng\, mL^{-1}$
V[a]	I	437.92	90
W[a]	II	209.48	500
Y[a]	II	371.03	12
Yb[a]	II	369.42	17
Zn[a]	I	213.86	15
Zr[a]	II	339.20	320

Explanation of Symbols and Abbreviations
[a] – measured with the use of SN-Ar MIP-OES
[b] – measured with the use of CVG-He MIP-OES
The symbols I and II indicate that the spectral lines originate, respectively, from the neutral atom and singly ionized state.
DL_{exp} – experimental detection limit

Subject Index

Page references to *figures* and *tables* are shown in *italics*.